AF764343

Designing and Constructing Prestressed Bridges

ice
Institution of Civil Engineers
Publishing

Designing and Constructing Prestressed Bridges

Jiří Stráský and Radim Nečas
Strasky, Husty and Partners, Ltd and Brno University of Technology

Published by ICE Publishing, One Great George Street, Westminster, London SW1P 3AA.

Full details of ICE Publishing representatives and distributors can be found at:
www.icebookshop.com/contact.aspx

Other titles by ICE Publishing:

Bridge Deck Erection Equipment: A Best Practice Guide
Members of IABSE Working Group 6. ISBN 978-0-7277-6193-4
Bridge Launching, Second edition
Marco Rosignoli. ISBN 978-0-7277-5997-9
Bridge Monitoring: A Practical Guide
Campbell R. Middleton, Paul R.A. Fidler and Paul J. Vardanega. ISBN 978-0-7277-6059-3

www.icebookshop.com

A catalogue record for this book is available from the British Library

ISBN 978-0-7277-6385-3

© Thomas Telford Limited 2021

ICE Publishing is a division of Thomas Telford Ltd, a wholly-owned subsidiary of the Institution of Civil Engineers (ICE).

All rights, including translation, reserved. Except as permitted by the Copyright, Designs and Patents Act 1988, no part of this publication may be reproduced, stored in a retrieval system or transmitted in any form or by any means, electronic, mechanical, photocopying or otherwise, without the prior written permission of the Publisher, ICE Publishing, One Great George Street, Westminster, London SW1P 3AA.

This book is published on the understanding that the author is solely responsible for the statements made and opinions expressed in it and that its publication does not necessarily imply that such statements and/or opinions are or reflect the views or opinions of the publishers. While every effort has been made to ensure that the statements made and the opinions expressed in this publication provide a safe and accurate guide, no liability or responsibility can be accepted in this respect by the author or publishers.

While every reasonable effort has been undertaken by the author and the publisher to acknowledge copyright on material reproduced, if there has been an oversight please contact the publisher and we will endeavour to correct this upon a reprint.

Cover photo: © Jiří Stráský

Commissioning Editor: James Hobbs
Development Editor: Melanie Bell
Production Editor: Madhubanti Bhattacharyya
Marketing Specialist: April Nagy

FSC
MIX
FSC® C007785

Typeset by The Manila Typesetting Company
Index created by Madelon Nanninga-Franssen
Printed and bound in Great Britain by Bell and Bain, Glasgow

Contents

Foreword

Everywhere today an environmentally-conscious public is paying greater attention to the structures that make up their environment, such as bridges, viaducts, railroad overpasses. As people realise these are structures built for over a century of useful life, they express concern that so many of these structures are boring, repetitive, and even ugly. Increasingly, an unhappy public and our clients are tending to hold engineers responsible for the proliferation of unimaginative structures in their environment.

The most common approach taken by clients, especially public agencies, to get more beautiful and aesthetically-pleasing structures has been to turn to architects. It is not unusual for clients to require architectural treatment or architectural mitigation of the structural designs by architects as part of the design process.

It can be argued that this approach usually fails, and does not result in better architectural solutions from architectural treatment, not because architects are professionally incompetent as architects, but because the approach arises from false assumptions. Arguably, the way to improve the architecture of bridges is to produce better structural solutions. In bridge design the responsibility for producing better architectural solutions lies not with architects but with structural designers.

When ICE Publishing asked me to prepare a third edition of my book 'Stress ribbon and cable-supported pedestrian bridges' I came to the realisation that the main portion of my work is not the design of these special structures, but the design of so so-called 'bread-and-butter' bridge structures – typical overpasses, viaducts, river crossings, and where possible, bridges across deep valleys and wide rivers. The intention was always to ensure that these structures were structurally-efficient but also aesthetically-pleasing. That is why I suggested to the publisher that the focus of this book should not be about exceptional structures, but those more typical ones, mirroring core projects in my own work.

As I always have a proactive approach to my life and work, and since prestressing enables us to actively control the structural behaviour, I have always used them in my designs. The layout and level of prestressing can be chosen in the way that dead load is balanced, boundary conditions are changed and supports are created within structures. Prestressing is a radical step away from passive reinforcement to creative thinking and development.

I started my professional career in 1969 in the former socialist Czechoslovakia. I was working as a designer in the state company, designing and constructing highways and bridges. Since there were restrictions in collaborations with other countries and since there was a desire to create progressive structures, the organisation had to develop not only the structures, but also the technology for constructing them, which helped with developing a greater understanding of both prestressing and constructing. After the Velvet Revolution, there was an opportunity to work in the United States where I participated in the design of several outstanding structures designed by the firm T. Y. Lin International, San Francisco. When I returned to my homeland, I started to teach at the Brno University of Technology. and with my colleagues opened our design firm. Since I also own an engineering office in California, I have continued to participate on design of several outstanding bridges built in the USA and Canada. Our firm in Brno is designing bridges not only in the Czech Republic but all over the world (UK, European Union, South America and Asia).

When we started our firm, we wanted to design structurally-efficient bridges that have a high aesthetic standard. We want to design structures whose form is inherent to the site where they are constructed, and structures that express their static functions. To be able to design such structures, the designer must fully understand to the structural behaviour, possess deep knowledge of static and dynamic analyses and be familiar with the technology of their construction. To prepare qualified engineers, my colleagues and I are trying to cover all discussed aspects. Since we hope other students and engineers might be interested in our lectures too, we have summarised their subject in this book.

The book was prepared with the collaboration of Dr Radim Nečas, who worked out the calculations and drawings. Since 2001 Dr Nečas is teaching with me at the University and participating in the structural design and static and dynamic analysis of our bridges. The discussed approach to designing of bridges is illustrated using examples of many structures. Since it is difficult to obtain information about designing and constructing bridges designed by other designers, the structures designed by the author's design office are presented here in many cases. Other structures included here are those which we admire from both architectural and technical points of view.

Jiří Stráský

About the authors

Jiří Stráský has worked in industry since 1969, and is Technical Director at Strasky, Husty and Partners Ltd, Brno, Czech Republic. He also holds a teaching post at the University of Brno. Jiří has won numerous awards for design and technical papers, and has been published numerous times.

Radim Nečas is a full-time academic at Brno University of Technology, Czech Republic. He has worked in research since 2001, and has seen his projects realised in his native Czech Republic, Spain and the USA.

ice Publishing

Stráský and Nečas

ISBN 978-0-7277-6385-3
https://doi.org/10.1680/dcpb.63853.001
ICE Publishing: All rights reserved

Chapter 1
Introduction

The concept of prestressing, an idea born in the 20th century, is the single most significant new direction in structural engineering in any period in history (Billington, 1976). It put into the hands of designers the ability to control structural behaviour, and at the same time enabled – or forced – them to think more deeply about the structure itself. Moreover, the application of prestressing opened up new possibilities regarding the form of structures, and this has had an influence on general culture. The idea of prestressing arose out of bridge design, and the most impressive structural forms involving prestressing, from a purely engineering viewpoint, are evident in bridges.

In reinforced concrete, concrete and steel bars are combined by simply placing them together and letting them act together as they may wish. In prestressed concrete, however, high-strength concrete is combined with high-strength steel in an *active* manner. The prestressing allows the structural engineer to balance the load, change the boundary conditions and create supports within the structure. Prestressing is really revolutionary; it represents a radical step from passive reinforcement to creative thinking and development.

As early as 1925, in the construction of the Tempul Aqueduct at Jerez de la Frontera, Spain, the pioneering engineer Eduardo Torroja substituted two piers by stay cables that transferred the vertical reaction into the pylon (Torroja, 2011). The stay cables, made from twisted high-tensile steel, were anchored in the webs of the channel deck, and were bent on saddles supported by jacks located on the pylon legs. The required tension in the stays was achieved by lifting the saddles. A few weeks later, when the spans were fully overloaded, the space between the saddles and the legs was concreted, and the jacks removed. Finally, the cables were encased in concrete. In this way Torroja not only constructed the first concrete cable-stayed structure, but also intentionally substituted the supports of the continuous structure by the external cable stays, and prestressed the deck (Figure 1.1).

Figure 1.1 Tempul Aqueduct, Jerez de la Frontera, Spain: (a) schematic drawing of the post-tensioning; (b) the completed structure

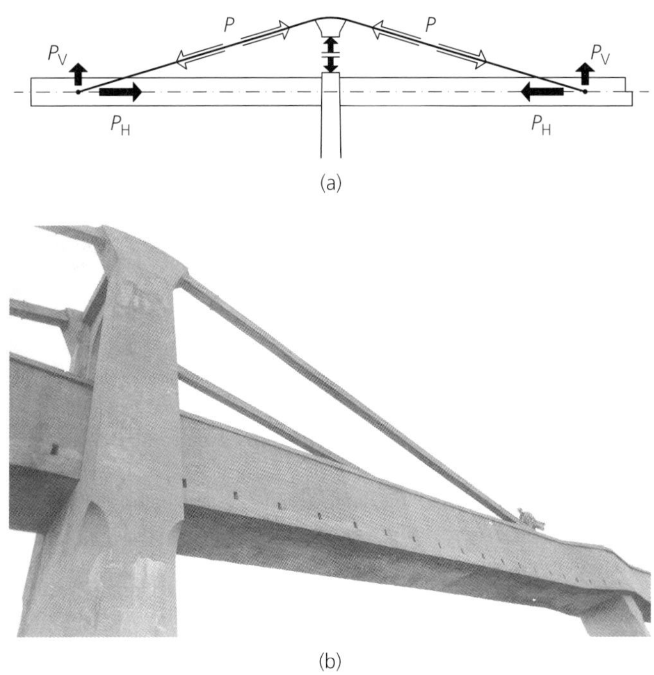

The same principle was applied by Franz Dishinger, both in his 1934 patent for external prestressing (where the prestressing bars or tendons are not encased in the concrete) and in the construction in 1937 of the 303.62 m long Aue Bridge in Saxony, Germany (Aeberhard *et al.*, 1988; Günschel, 1966). The main bridge is formed by a Gerber girder composed of three spans (lengths 25.20 + 69.00 + 23.40 m). The side spans and cantilevers protruding into the main span are prestressed by external cables anchored at the bottom fibres of the deck and bent on saddles situated above the pier diaphragms. The 31.50 m long drop span is prestressed by cables anchored at the support diaphragms and bent at saddles situated under the span diaphragms. Both cables, made from smooth bars, were post-tensioned by the vertical movement of their saddles. In this way both the cantilever and drop spans were prestressed. The forces in the cables could create points of zero deformation at their anchor points or at the saddles. This means that, due to the effects of the self-weight, the structure functions as a nine-span beam (Figure 1.2).

The bridges described above represent the first applications of prestressing, after which the technique was developed further (Billington, 1976; Günschel, 1966). Prestressing represents a proactive approach to the design and construction of bridges. Over time, the prestressing steel has been placed into ducts situated either inside

Figure 1.2 Aue Bridge, Saxony, Germany: (a) schematic drawing of the post-tensioning; (b) the elevation of the main bridge

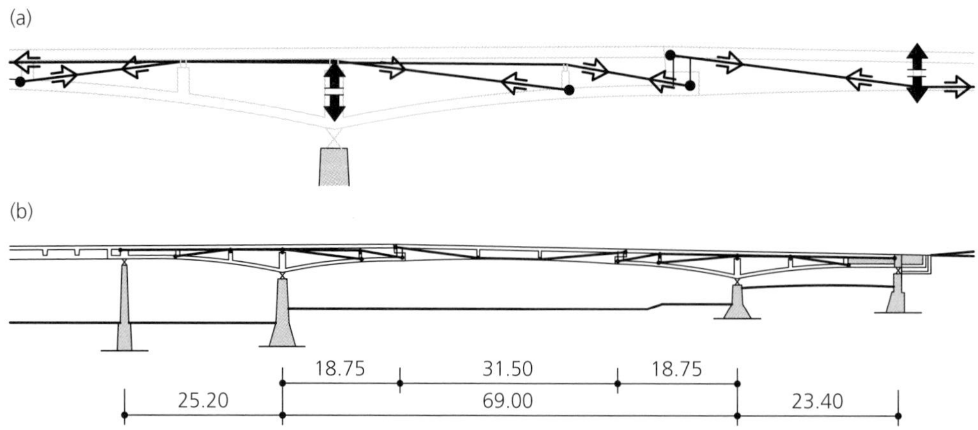

or outside of the bridge deck cross-section, ingenious hardware for anchoring and prestressing concrete structures has been developed, and there has been tremendous progress in the technology used to construct such structures. However, the principles of prestressing remain the same.

It is evident that the external cable can be supported by saddles (deviators) at several points along the lengths of the girders, and the tension force and cable geometry can be determined in such a way that the corresponding radial forces R_i originating at the deviators act against the dead load of the girder (Figure 1.3(a)). Then, with regard to the effects of the dead load, the structure can function as a multi-span beam, and the shear forces and bending moments can be reduced significantly.

If the cable is laid in a duct situated inside the girder, the curve and cable force P of the duct can be determined in such a way that the uniform radial forces r originating along the whole cable length can balance the dead load (Figure 1.3(b)). The bending moments and shear forces due to the dead load and prestress then have the same value, but opposite signs (the resulting bending and shear stresses are zero), and therefore the structure is stressed – *prestressed* – only by the normal force $N = -P_H = -P \cos \alpha$, which creates uniform compression stresses σ_N. If the structure is stressed by an additional load p, the corresponding bending moments create stresses σ_p, which are added to the compression stresses (Figure 1.3(c)).

Of course, the prestressing can balance only a portion of the dead load, or it can create radial forces larger than the dead load. Also, the

Figure 1.3 Girder prestressing: (a) external; (b) internal; (c) stresses due to live load

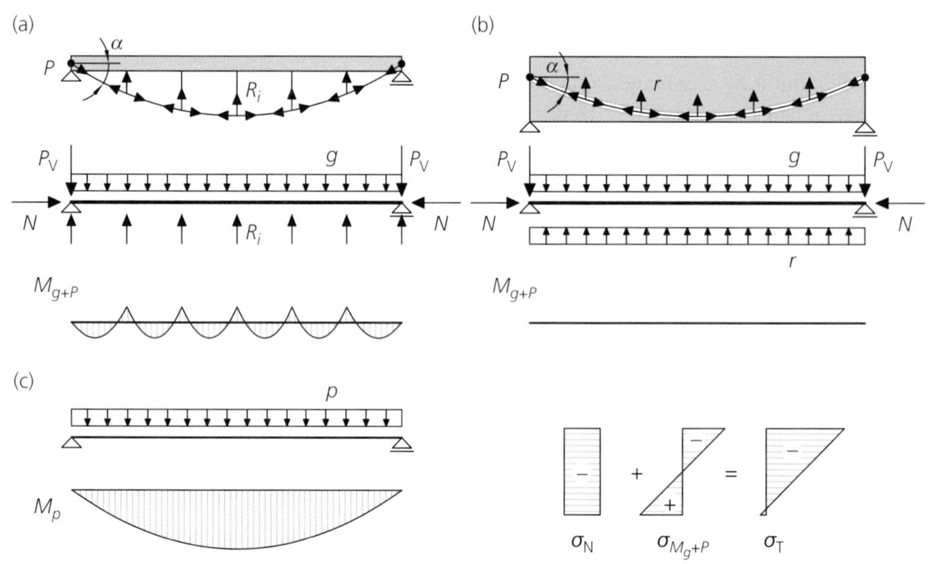

duct's curve need not create exactly uniform radial forces. The forces in the cables are influenced by friction between the prestressing steel and the duct, by elastic deformation of the beam due to progressive tensioning of several cables, by relaxation of the prestressing steel, and by long-term deformation of the concrete due to its shrinkage and creep. These effects are discussed in the other chapters in this book.

The history of prestressed concrete bridges is described in many excellent books (Dyckerhoff and Widman, 1973; Leonhardt, 1984; Troyano, 2003; Wittfoht, 1972). As has been already mentioned, prestressing put into designers' hands the ability to control structural behaviour, and at the same time enabled them – or forced them – to think more deeply about the construction itself. The scope of this book does not allow for a detailed description of the development of concrete structures; therefore, the design, construction and development of concrete structures is illustrated using just several structures. The selection of structures was a personal choice of the authors, and was strongly influenced by their design philosophy. They do not admire large monumental structures, and the mass production of concrete, which on many occasions have damaged our environment. Rather, they prefer structurally efficient bridges that fit their surrounding landscape. Prestressed concrete structures have always responded to the needs of society by through the design and development of the most appropriate solutions.

Bridge decks assembled from precast beams have been designed since the early days of prestressed concrete. The first precast beam bridge was built by Eugene Freyssinet in Algeria in 1936 (Troyano, 2003), and he further demonstrated all the construction possibilities of the new technology in the famous bridges over the River Marne in France. The first of these was the 55 m span Luzancy Bridge, the construction of which was started in 1941 but, due to the Second World War, not completed until 1945. Subsequently, five identical 74 m span bridges were constructed following a similar structural solution (Figure 1.4).

Figure 1.4 Changis-sur-Marne Bridge, France

Bridges assembled from precast beams are still being built all over the word. Precast beams and are used not only for the decks of small bridges and/or long viaducts, but also in the construction of frame overpasses or arch structures. Figure 1.5 shows one of the most beautiful viaducts across the Colombia River in British Columbia, Canada; the bridge was designed by Riccardo Morandi and constructed in 1965 (Bouchet, 1966).

Solving traffic problems in cities requires the building of slender bridges of complex geometry. Prestressed concrete cast on inventory stationary scaffolding enables the construction of structures having smooth, streamlined sections supported by single, one-column piers, requiring minimum space under the bridge. Many beautiful structures have been built in different countries, but the structures built in California are outstanding systems that enable cost-effective construction (Figure 1.6) (California Department of Transportation, 1980).

The technique of span-by-span construction using movable scaffolding has been developed for long motorway viaducts. Originally, the scaffolding was situated under the deck, while later the scaffolding has been also been located above the deck. Since the construction of the first large-scale project of this type, the 1.080 km long Krahnenberg Bridge near Andernach, Germany, in 1964, the technique has been widely used. The beautiful Elz Viaduct, which was constructed using overhead movable scaffolding, completed in 1967, demonstrates that motorway viaducts can be built with minimum effects on the environment (Figure 1.7) (Leonhardt, 1984).

Stationary or movable scaffolding has also been widely used in the construction of structures in which deck is assembled from precast

Figure 1.5 Columbia River Bridge at Kinnaird, British Columbia, Canada

segments, initially with concrete joints and later with epoxy-filled joints. Many interesting structures have been built in the UK, and lately also in countries where UK consulting firms operate. Following the renaissance of external prestressing in France, a new development of the technology began in the USA. An excellent example of this is the construction of the more than 60 km long Dubai Metro viaducts, which were completed in 2009 (Figure 1.8) (Smith *et al.*, 2009).

Bridges assembled from precast segments can also be erected as a progressive cantilever, from one abutment to another. This technology was applied in the construction of the S-shaped 379 m long Linn Cove Viaduct constructed in North Carolina, USA, in 1983. The bridge, which was constructed with minimum impact on the environment, is an excellent example of engineering ingenuity. The viaduct's construction demonstrates that bridges can be built in complete harmony with their natural surroundings (Figure 1.9) (Muller and Barker, 1985).

Figure 1.6 Bridge across I-280, California, USA

Figure 1.7 Elz Viaduct, Germany

Figure 1.8 Dubai Metro viaducts

Figure 1.9 Linn Cove Viaduct, North Carolina, USA

In 1964, a 480 m long bridge deck assembled of precast segments was launched across the River Caroní in Venezuela. Since then, the technology for the incremental launch of structures has been further developed, and it is widely used in construction of bridges that have constant curvature decks (Figure 1.10) (Giovannini, 1974).

Bridges, or their parts can also be cast away from the construction site, and subsequently transported there and lifted into their design position. The 43 m long central part of the beautiful Kylesku Bridge built in 1982 in Scotland was constructed in this way (Figure 1.11) (Ciampoli, 1987). A bridge deck can be also built parallel to the obstacle, and consequently rotated into the design position.

As early as in 1930, a cantilever construction method allowing the progressive casting of the bridge deck independently of the terrain under the bridge was used in Brazil. However, the deck was made from reinforced concrete, which functions when cracks appear in the structural members. This method was developed further after the Second World War, when prestressing, which eliminates cracks, was developed. Since the construction of the Lahn Bridge in Germany, the first prestressed concrete cantilever bridge, cantilever erection has been widely used for bridges with spans longer than 60 m (Figure 1.12) (fib, 2018).

The simplicity of construction of cantilever bridges enables the design of longer spans. However, at a certain length the decks

Figure 1.10 Val Restel Viaduct, Rovereto, Italy

Figure 1.11 Kylesku Bridge, Scotland, UK

Figure 1.12 Valy Bridge, Slovakia

become too deep, and the bridge will create an optical obstacle in the natural environment. Therefore, in several cases the webs of classical box girder bridges have been substituted by trusses, to construct beautiful and 'transparent' bridges (Figure 1.13) (Barlet *et al.*, 2002).

Cantilever construction is also used for the erection of bridges assembled from precast segments. Initially the segments were joined by concrete joints, but later with epoxy-filled joints were used. Special equipment has been developed for the erection of segments, allowing efficient and rapid bridge construction (Figure 1.14).

Cantilever construction is used not only for bridge decks but also for the erection of piers, the inclined legs of frame structures and arches. Recent applications of arch structures in the USA, Portugal and

Figure 1.13 Bras de la Plaine Bridge, Réunion, France

Figure 1.14 Crown Princess Mary's Bridge, Frederikssund, Denmark

Figure 1.15 Colorado River Bridge at the Hoover Dam, Nevada, USA

Spain illustrate the vast possibilities of this technology (Figure 1.15) (Goodyear *et al.*, 2004).

Cantilever construction is widely used in the construction of cable-supported bridges. The first cable-stayed bridges had decks suspended on a few stay cables, formed from locked coil strands or prestressed bars, which resisted the tension caused by the dead load by the prestressing steel. When the deck erection was completed, the steel was encased in concrete, which was also prestressed.

In the construction of the Ganter Bridge in Switzerland in 1980, the prestressed bars were substituted by prestressed walls. The greater stiffness of the walls allowed the height of the pylons to be reduced, allowing for a structurally efficient and very beautiful structure (Figure 1.16) (Billington, 1983; Menn, 1979).

Over the course of time, the few stay cables originally used have been substituted by many cables anchored in the deck at distances of 3–12 m. In this way it has been possible to significantly reduce the bending moment of the deck due to the dead load. Such structures form very economical systems, in which the main structural members are mostly stressed by normal forces: the pylons and decks are compressed and the stays are tensioned.

If the deck is suspended on its outer edges, its depth is determined by its transverse bending. Therefore, very slender bridges can be designed. An example of a bridge of this type is the Pasco-Kennewick Bridge in Washington State, USA, built in 1986 (Figure 1.17) (Leonhardt, 1984). Since then cable-stayed bridges with spans of up to 530 m have been built (Figure 1.18) (Walther *et al.*, 1998).

Cable-supported bridges allow not only the bridging of large obstacles, but also the creation of structures of great beauty. An example of such a structure is the bridge across the Lérez River in

Figure 1.16 Ganter Bridge, Switzerland

Figure 1.17 Pasco-Kennewick Bridge, Washington State, USA

Figure 1.18 Skarnsund Bridge, Norway

Figure 1.19 Lérez River Bridge, Pontevedra, Spain

Pontevedra, Spain, in 1995 (Troyano, 2003). The single inclined pylon is anchored by back stays anchored in anchor blocks arranged perpendicularly to the longitudinal bridge axis. In this way the back stays create a hyperbolic formation, the beauty of which is due to its pure static function (Figure 1.19).

The recent public concern about environmental issues has led to the building of many pedestrian and cyclist trails, in addition to motorways and high speed railways, and these too need to bridge rivers, highways and other man-made obstacles. Such crossings require light and transparent pedestrian structures that are human in scale. Many have prestressed concrete decks that are suspended on steel arches or on stay or suspension cables. The concrete deck gives sufficient stiffness and mass to the structure to reduce the dynamic response to moving people and wind. The deck can be cast on a stationary scaffolding or be assembled from precast segments suspended on stay or suspension cables. Figure 1.20 shows an arch bridge over the Svratka River. The bridge has a span of 58.5 m, and the deck is formed by a spine girder with an overhang that is suspended on a steel arch. The spine girder naturally divides the pedestrian and cyclist lanes (Strásky *et al.*, 2013).

While suspension bridges with a concrete deck are rarely built, constructions of this type are widely used for pedestrian bridges. Such bridges may be earth- or self-anchored structural systems. The Vranov Lake Pedestrian Bridge in Vranov, Czech Republic, was built

Figure 1.20 Svratka River Bridge, Brno, Czech Republic

across the Swiss Bay of the lake in 1993 (Figure 1.21). It is an earth-anchored structure, with a 252 m long deck assembled from precast segments and prestressed by internal tendons. The aerodynamic stability of the bridge is increased by external cables situated at the deck edges (Stráský, 2011).

One of the cleverest structures that forms a self-anchored suspension structural system was completed in 1987 in the small city of Kelheim, Germany. It crosses a new 47 m wide Rhine–Main–Danube Canal in smooth curves, naturally connecting the pedestrian traffic on the two banks (Figure 1.22) (Schlaich and Seidel, 1988). The deck, which has a plan curvature with radius 18.89–37.79 m, is suspended on one suspension cable situated within the plan curvature. Two inclined masts, one situated on each bank, support the suspension cables with hangers. The curved deck that is fixed into the abutments is internally prestressed in the circumferential direction by eccentric internal tendons.

Figure 1.21 Vranov Lake Pedestrian Bridge, Vranov, Czech Republic

Figure 1.22 Kelheim Bridge, Germany

As pedestrian and cyclist trails can follow a variable slope, bridges within such trails can be designed using the stress-ribbon concept. 'Stress-ribbon bridge' is the term that has been coined to describe structures that have a prestressed concrete deck in the shape of a catenary, which is walked or cycled on directly. The concept was first introduced by Ulrich Finsterwalder, who repeatedly proposed such a structure for bridging large spans. The stress-ribbon concept was applied for the first time in 1965 in the construction of the Bircherweid Bridge across freeway N3 near Pfäffikon, Switzerland. Since then, stress-ribbon bridges have been built in many countries. The stress-ribbon deck is usually straight; the first cranked deck alignment is in the bridge built across the River Medway at Maidstone, UK, which opened in 2001 (Figure 1.23). The 101.50 m long bridge has two spans of length 49.5 + 37.5 m, and an angle plan between the spans of 25° (Stráský, 2011).

The brief survey of prestressed concrete bridges given above clearly demonstrates that it is possible to design structurally efficient,

Figure 1.23 Kent Messenger Bridge, Maidstone, UK

cost-effective and beautiful structures that fit into their environment. However, the design of such structures is not straightforward. It requires a deep understanding of the structural forms, structural details and behaviour of prestressed concrete structures post-tensioned by internal tendons and/or external cables.

Modern concrete bridges combine reinforced concrete with different levels of prestressing to obtain the most appropriate behaviour of the structures at both the service and the ultimate load. Post-tensioning of structures enables internal forces to be redistributed within the structure and any stage of stresses to be achieved. However, it is important to be careful. Unlike steel structures, in concrete structures the chosen stage of stresses is redistributed over time due to the creep of the concrete to the natural stage. And, of course, the alignment of prestressing tendons is influenced by the requirements regarding the ultimate capacity and ductility of the structures.

The following chapters discuss several aspects of bridge design that are necessary to obtaining the most appropriate solution. The goal of the book is not to give exact instructions of how to design the structures, but to alert the reader to structural problems that may be encountered.

REFERENCES

Aeberhard HU, Buergi P, Ganz HR *et al.* (1988) *External Post-Tensioning*. VSL International Ltd, Berne, Switzerland.

Barlet E, Causses G and Viallon J-P (2002) A new step forward for composite bridges. The Bras de la Plaine Bridge, Réunion, France. *Concrete Structures in the 21st Century. Proceedings of the First fib Congress, Osaka, Japan*. fib, Lausanne, Switzerland.

Billington DP (1983) *The Tower and the Bridge. The New Art of Structural Engineering*. Princeton University Press, Princeton, NJ, USA.

Billington DP (1976) Historical perspective on prestressed concrete. *PCI Journal* **21(5 special issue)**: 14.

Bouchet A (1966) Le pont en béton précontraint de Kinnaird, sur la Columbia (Canada, Colombie britannique). *La Technique des Travaux* **42(5–6)**: 181–192.

California Department of Transportation (1980) *California Bridges*. University Art Museum, Santa Barabara, CA, USA.

Ciampoli M (1987) The Kylesku bridge in Scotland: a prestressed concrete deck with a partially precast central span. *Industria Italiana del Cemento* **7–8**: 458–471.

Dyckerhoff and Widman (eds) (1973) *Festschrift. Ulrich Finsterwalder 50 Jahre für Dywidag*. G. Braun, Karlsruhe, Germany.

fib (2018) Structural concrete in the Czech Republic 2014–2017. *5th International fib Congress, Melbourne, Australia*. National Report of the Czech Republic.

Giovannini B (1974) Val Restel viaduct for provincial road No. 89 near Rovereto, Trento. Prestressed Concrete Structures in Italy 1970/1974. *Industria Italiana del Cemento* **4**: 60–67.

Goodyear D, Klamerus B and Turton R (2004) The new Colorado River arch bridge at Hoover Dam. An innovative hybrid of concrete and steel. *Concrete Structures: The Challenge of Creativity. Proceedings of the 5th fib Symposium, Avignon, France*, pp. 222–223.

Günschel G (1966) *Große Konstrukteure 1. Freyssinet, Maillart, Dishinger, Finsterwalder. Ullstein Bauwelt Fundamente*. Ullstein, Frankfurt, Germany.

Leonhardt F (1984) *Bridges. Aesthetics and Design*. Deutsche Verlags-Anstalt, Stuttgart, Germany.

Menn C (1979) The Ganter Bridge, Simplon. *IABSE Structures*, **3**: 7–79.

Muller J and Barker JM (1985) Design and construction of Linn Cove Viaduct. *PCI Journal* September–October: 38–53.

Schlaich J and Seidel J (1988) Die Fußgängerbrücke in Kelheim. *Bauingenieur* **63**: 143–149.

Smith DA, Hewson NR and Hendy CR (2009) Design of the Dubai Metro light rail viaducts – superstructure. *Proceedings of the Institution of Civil Engineers – Bridge Engineering* **162(2)**: 55–62.

Stráský J (2011) *Stress Ribbon and Cable Supported Pedestrian Bridges*, 2nd edn. Thomas Telford, London, UK.

Stráský J, Romportl T, Kocourek P and Rayor G (2013) Integral arch bridges. *ARCH 2013*. 7th International Conference on Arch Bridges, Split, Croatia.

Torroja E (2011) *The Structures of Eduardo Torroja. An Autobiography of Engineering Accomplishment*. Literary Licensing, Whitefish, MT, USA.

Troyano LF (2003) *Bridge Engineering. A Global Perspective*. Thomas Telford, London, UK.

Walther R, Houriet B, Walmar I and Moïa P (1998) *Cable Stayed Bridges*. Thomas Telford, London, UK.

Wittfoht H (1972) *Triumph der Spannweiten*. Beton, Düsseldorf, Germany.

ice Publishing

Stráský and Nečas
ISBN 978-0-7277-6385-3
https://doi.org/10.1680/dcpb.63853.011
ICE Publishing: All rights reserved

Chapter 2
Effects of prestressing

The function and technology of prestressing is described in many excellent books (Hampe, 1978; Lin and Burns, 1981; Leonhardt, 1964; Navrátil, 2014; Nilson, 1997). Therefore, only basic information is repeated here. Over time the hardware used for prestressing has been well developed by specialist companies (e.g. VSL, Dywidag and Freyssinet), and these provide comprehensive brochures that both describe their products and give information on the structural arrangement, details, analyses and technology of construction. These brochures are available on the manufacturers' websites.

The design concepts for obtaining durable post-tensioned tendons are presented in *fib Bulletin 33*: *Durability of Post-tensioning Tendons* (Fuzier *et al*., 2006). It is evident that the durability of the prestressing depends not only on the durability of the materials used and on the installation of these materials, but also on the design concept implemented.

2.1. Prestressing – technology

According to the 1996 FIP recommendations (FIP, 1996), prestress is applied through a construction control process (prestressing) that entails stressing tendons (prestressing reinforcement) relative to the concrete member. The prestress is exerted by tendons made of high-strength steel (bars, wires, strands or monostrands) (Figure 2.1). The name 'monostrand' is used for a strand that is protected by a layer of grease and an extruded high-density polyethylene (HDPE) sheath. The monostrands can be encased individually in concrete or as a bundle in grouted ducts. The grease and sheathing provide for low friction against the concrete or cement mortar. The strands can be also protected by tightly extruded sheathing that prevents movement between the strands and the sheath. In this case the grease is substituted by wax. These strands are suitable for ungrouted applications (e.g. stay cables).

The prestressing reinforcement can be pre-tensioned or post-tensioned.

Figure 2.1 Prestressing steel: (a) bar; (b) wire; (c) strand; (d) monostrand

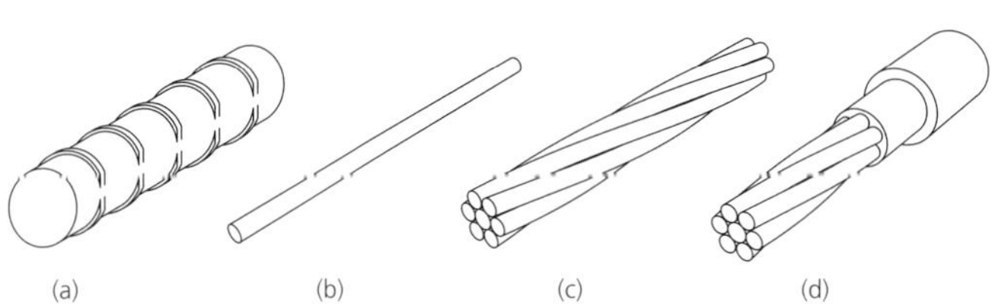

2.1.1 Pre-tensioned structures

In pre-tensioned structures, the prestressing reinforcement is tensioned and temporarily anchored, and the structural element is then cast in a steel form or in a stressing bed. After the concrete has hardened, the temporary anchors are released, which introduces a prestressing force into the concrete. The prestressing reinforcement is anchored through the bond with concrete (Figure 2.2). For efficiency the casting bed is usually designed so that several structural members can be produced in a long line.

The prestressing steel can be straight or draped (Figures 2.3 and 2.4). In the first case it is necessary to de-bond several strands in the area close to the supports in order to control the bending stresses in the structure.

Figure 2.2 Casting bed

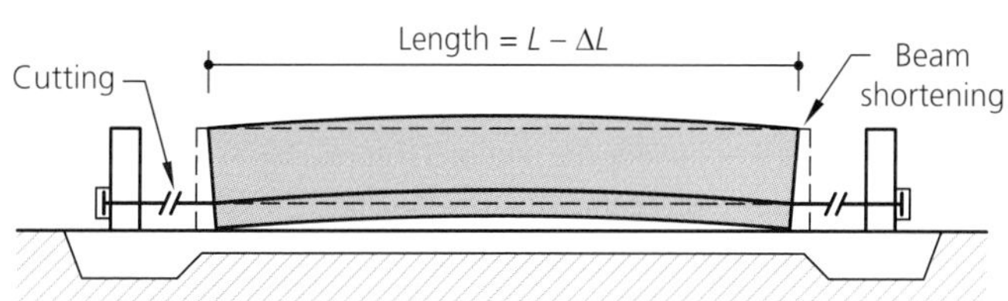

Figure 2.3 De-bond strands

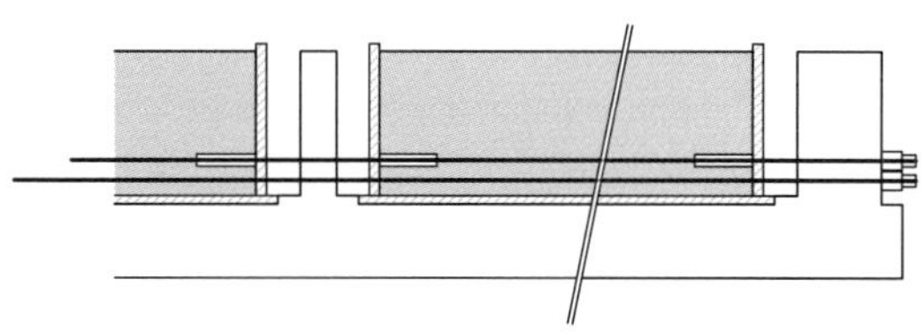

Figure 2.4 Draped strands

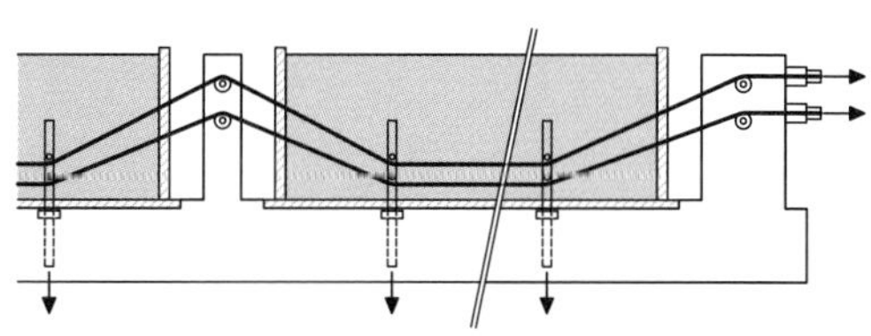

In the bridge industry pre-tensioning is mainly used for the prestressing of precast beams. It has also been used for transverse prestressing of the deck slab of several precast segmental structures and for the floor beams of the beam and slab structures.

2.1.2 Post-tensioned structures

Post-tensioned structures are post-tensioned by tendons situated inside the cross-section (Figure 2.5) or by external cables situated outside the cross-section (Hewson, 2012). The internal tendons can be bonded to the structure by grouting, or left provisionally or permanently unbonded.

In a structure with bonded tendons (Figures 2.5(a) and 2.6) the cement mortar grouted between the prestressing steel and the ducts guarantees that, after post-tensioning, any additional strain is the same for the concrete and the steel in all the sections. Therefore, in the ultimate limit state (ULS), the force in the steel that resists the ultimate load depends on the width of the cracks.

Internal tendons can also be formed of monostrands (greased and sheathed strands). As there is no bond between the steel and the concrete, the strain in the concrete and that in the steel are different. Therefore, in the ULS the force in the steel that resists the ultimate load depends on the total elongation of the tendons between the anchors. That is why the forces in the tendons and the corresponding ultimate bending moments are smaller than is the case with bonded tendons.

External cables can be situated inside the perimeter of the cross-section (see Figure 2.5(a)) or outside the section (Figures 2.5(b) and 2.7). They can also function as a stay cable (Figures 2.8 and 2.9) (Manterola, 2006; Stráský, 2011) or suspension cable (Figures 2.10 and 2.11) (Kasuga,

Figure 2.5 Type of prestressing: (a) internal; (b) external within perimeter; (c) external outside perimeter

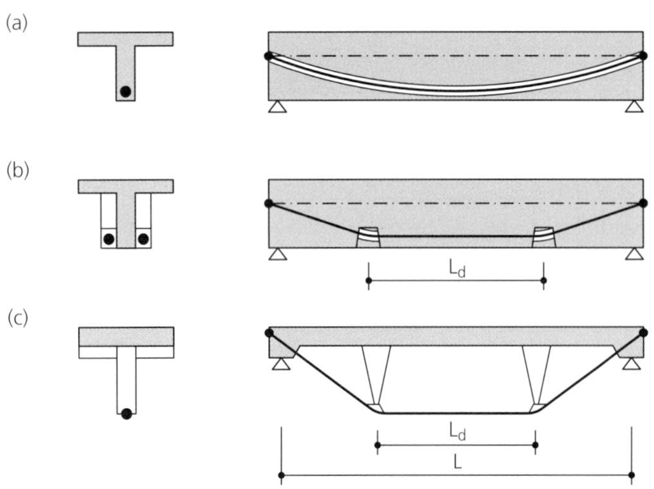

Figure 2.6 Internal bonded tendons

Figure 2.7 External unbonded cables

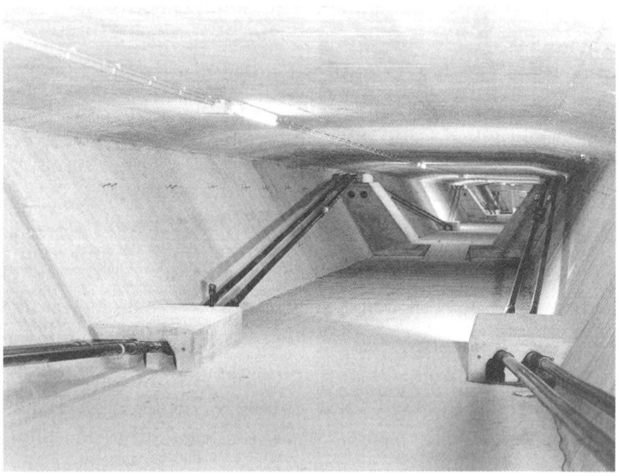

Figure 2.8 External (stay) unbonded cables – Osormort Viaduct, Barcelona, Spain

Figure 2.9 External (stay) unbonded cables – Odra River Bridge, Czech Republic

Figure 2.10 External (suspension) unbonded cables – Seishun Bridge, Japan

Figure 2.11 External (suspension) unbonded cables – Vranov Lake Bridge, Czech Republic

2006; Stráský, 2009). External tendons are usually anchored in so-called 'anchor blocks', and deviated at deviators. The deformations of the structure and the tendons are the same only at the anchor blocks and deviators. The deviator can be rigidly connected to the tendons or it can allow the movement of the prestressing steel in the ducts. At the ULS, the force in the external tendons corresponds to the elongation of the tendons between the points at which tendon slippage is prevented.

Internal tendons and external cables can be combined in individual structures in a way that can exploit their inherent characteristics. Their applications are discussed in Chapters 10 and 11. Both internal tendons and external cables can be designed as electrically isolated structural members.

2.1.2.1 Internal tendons

An internal tendon is formed from a bundle of prestressing reinforcement laid in a duct and post-tensioned between the anchorages (Figure 2.12(a)). The duct can be made of corrugated steel, polymer or polyethylene. While in the early days of prestressing wire was the most commonly used prestressing reinforcement, nowadays the reinforcement is mainly formed by strands. The prestressing steel is protected by a cement mortar grouted into the ducts after post-tensioning.

According to Fuzier *et al.* (2006) the post-tensioning anchorages can be grouped into active or stressing anchorages allowing stressing of tendons, passive or dead-end anchorages not permitting stressing and coupling anchorages to connect tendons. The main methods of anchoring prestressing steel in the anchorages are by wedges, nuts and threads, and bond (Figure 2.13). In the most common solution – anchoring the prestressing steel by wedges – a possible draw-in of wires or strands can occur. This phenomenon, which significantly influences the function of short tendons, has to be considered in the design. The value of draw-in is given by the manufacturer of prestressing system.

Figure 2.12 Typical layout of internal bonded tendons: (a) three-span structure; (b) structure erected span by span

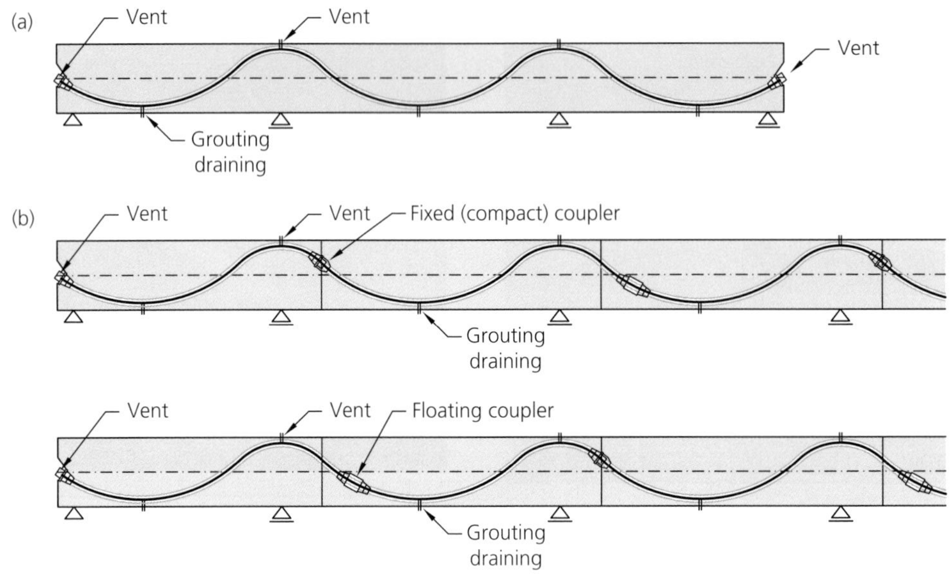

Figure 2.13 Common types of anchorage: (a) wedges; (b) nuts and threads; (c) bond

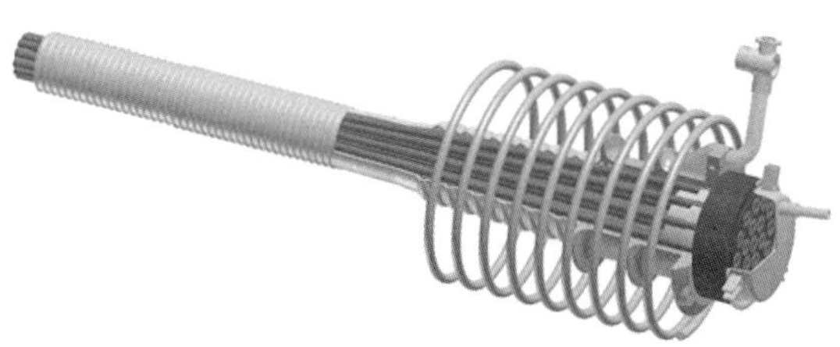

(a)

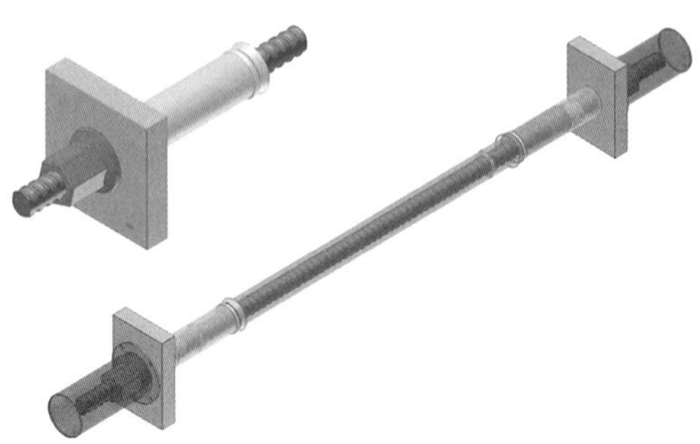

(b)

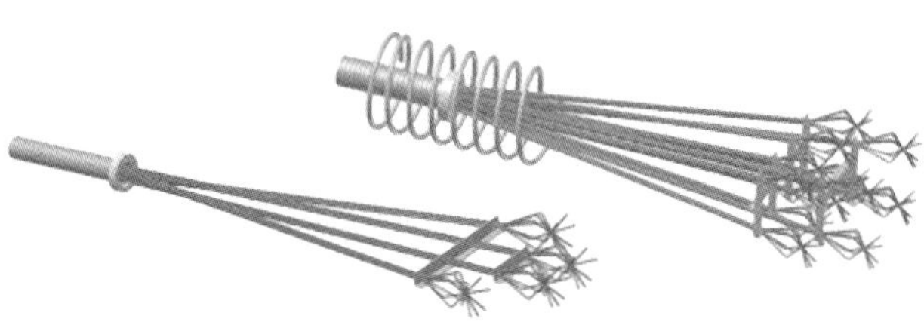

(c)

Tension stresses in the prestressing steel are reduced over time by relaxation of the steel. The relaxation loss, which is given in codes, is usually defined as the percentage ratio of the variation in the prestressing stress over the initial prestressing stress.

While bridge decks of lengths up to 100 m are usually prestressed by tendons anchored at both ends, multi-span structures are usually built segmentally and are progressively prestressed using coupled tendons (see Figure 2.12(b)).

Coupling of tendons is common in certain parts of the world, but is typically avoided in others, such as the USA and France. Fixed (compact) couplers couple tendons onto already installed and stressed tendons. They have two anchor heads connected by a threaded sleeve or an anchor body in which the strands are overlapped. Usually the already stressed tendons are anchored in the centre and the other tendons are anchored along the perimeter (Figure 2.14). Floating couplers connect unstressed tendons (e.g. in bridge decks constructed span by span). The spice chuck consists of

Figure 2.14 Fixed (compact) coupler

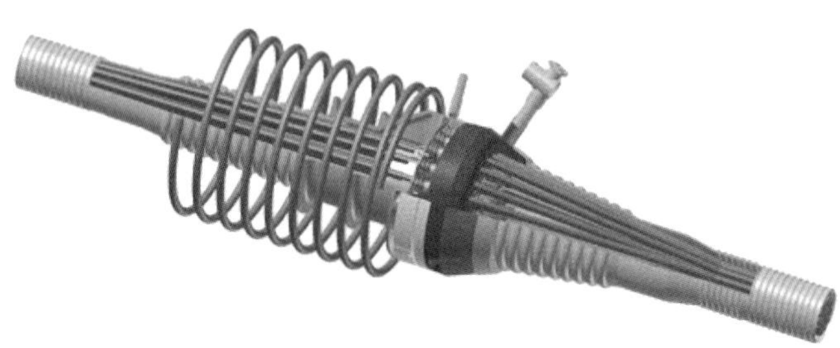

Figure 2.15 Floating coupler

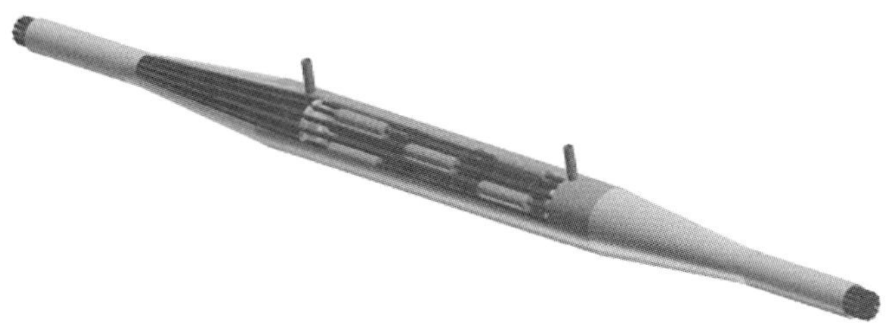

two spring-loaded wedges that connect two strands individually (Figure 2.15). Floating couplers that connect all the tendon strands have been also developed and used.

2.1.2.2 External cables

External cables, which were used in the early days of prestressing, had a renaissance in the 1980s. The cables are assembled of strands or monostrands that are grouted in polymeric or HDPE ducts. They are anchored at anchor blocks and deviated at piers and span deviators. External cables are not usually coupled, but they are anchored at the piers or span diaphragms where they overlap.

External cables can be anchored in each span, or, as HDPE ducts guarantee a small friction coefficient, can be led over two or three spans (Figure 2.16). The cables can be replaceable or not replaceable. In the first case, the HDPE ducts have to be situated in another duct made of concrete or steel in order to allow removal of the cable. If the cable is not replaceable, it is possible to lay parts of the cable within the sections of the deck and thus utilise its larger eccentricity.

2.1.2.3 Stay and suspension cables

Tension members (stay or suspension cables) can made of structural steel or cables developed by the steel or prestressed concrete industries. They can also be composite or prestressed concrete members (Stráský, 2011).

Tension members made of structural steel can be rolled shapes, plates, bars, tubes or pipes. The cables developed by the steel industry are formed of spiral strands, locked coil strands, parallel wires or ropes. They are factory fitted with a combination of socket types to enable load transmission between the structure and the cable. To eliminate the construction stretch, it is important to prestress the cables. Prestressing is conducted using a series of cyclic loadings, typically between 10% and 50% of the strand-breaking force, until an apparent stable modulus of elasticity is achieved. Typical arrangements of the cables and structural details can be found in brochures prepared by the manufacturers (e.g. Bridon and Pfeifer).

The cables developed from the prestressing tendons are formed by parallel prestressing bars, parallel wires or prestressing strands. A typical arrangement for a stay cable formed from prestressing strands is shown in Figure 2.17. Cables were traditionally made from prestressing strands that were grouted in steel or HDPE tubes. More recently they have also been made from epoxy-coated strands or monostrands or HDPE-coated galvanised strands with a wax fill. All types of strands can be grouted in steel or HDPE tubes.

The arrangement of cables varies according to the prestressing system used. Details can be found in the brochures produced by the manufacturers (e.g. Freyssinet, VSL, BBRV and Dywidag).

Figure 2.16 Typical layout of external cables used in construction of a segmental deck erected span by span: (a) elevation; (b) deviator; (c) deviator of the non-replaceable cable; (d) pier segment

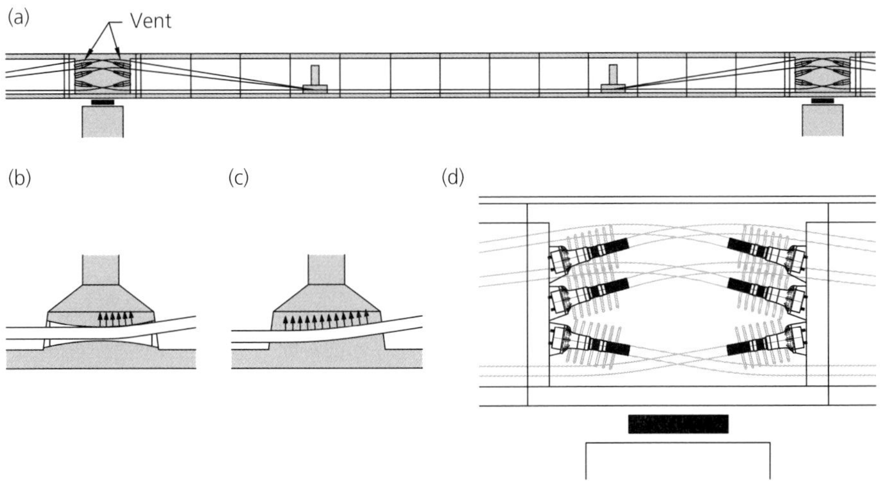

Figure 2.17 Typical arrangement of a stay cable

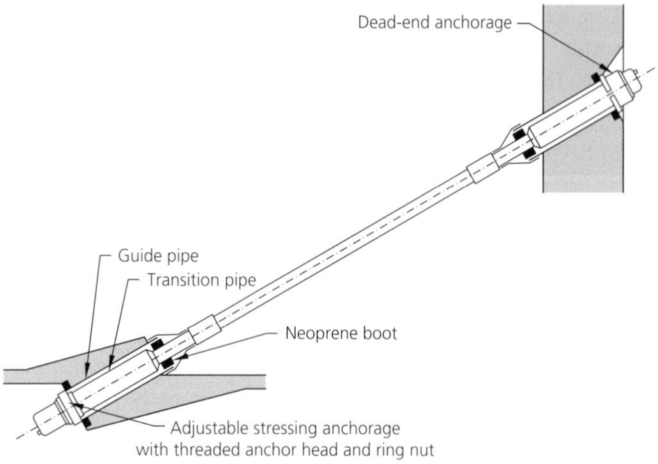

As suspension cables are stressed by relatively low fatigue stresses, they are usually designed as non-replaceable structural members. In contrast, suspenders and stay cables are stressed by large fatigue stresses. Therefore, it is necessary to design their connections to the towers, struts, suspension cables and deck in such a way that the forces in the cables can be adjusted and allow for the replacement of the suspenders and stay cables.

Suspension and stay cables can be bent at saddles situated on the towers or struts (uprights), or they can be anchored there. The anchors can be attached to tension plates or can be made to overlap there. The structural details of the cables at the saddles or at their anchors are influenced by local bending moments that originate there (Stráský, 2011).

Recently, suspension cables assembled from monostrands grouted in steel pipes have been used in the construction of a large suspension pedestrian bridge across the Ume River in Umeå, Sweden. The 180.7 m bridge deck is assembled from precast segments having a depth of only 0.44 m, and is suspended on two three-span (lengths 28.8 + 187.0 + 29.5 m) suspension cables (Figure 2.18). The cables are composed of 55 monostrands grouted in steel pipe of 273/12.5 mm diameter. The monostrands are anchored at typical Freyssinet anchor heads situated at the pylons and anchor blocks. The steel pipes together with monostrands resist the tension force.

Cables made from parallel wires of carbon fibre reinforced polymer (CFRP) or bars of carbon fibre composite have recently been built in Switzerland and France. These cables are seven times lighter than steel cables with the same breaking load. As the cables are resistant to corrosion and chemical attack, no corrosion-inhibiting compound or grout is necessary. However, to protect the wires against wind erosion and ultraviolet radiation, polyethylene sheaths are used for shielding.

Figure 2.18 Steel pipes of the suspension cables on the Ume River Bridge, Umeå, Sweden

2.2. Effects of prestressing

During post-tensioning and during the service of the structure, the effects of prestressing can be expressed by an equivalent load at the anchors by the force $N = -P$ acting in the direction of the tendons and along the length of the duct or at the deviators by radial forces r. These radial forces are the resultant of the normal and tangential forces k and t.

The normal and tangential forces can be determined from the theory of the frictional loss of a cable around a curve. Consider an infinitesimal length ds of prestressing tendon the centroid of which follows an arc of radius $\rho(\alpha)$ (Figure 2.19). The change in angle for a length ds is

$$\mathrm{d}\alpha = \frac{\mathrm{d}s}{\rho(\alpha)} \tag{2.1}$$

$$\mathrm{d}s = \rho(\alpha)\mathrm{d}\alpha \tag{2.2}$$

Along the length ds the force $P(\alpha)$ is changed by d$P(\alpha)$. The forces $P(\alpha)$ and $P(\alpha)$ + d$P(\alpha)$ act on this infinitesimal element.

The action of these forces can be substituted by normal forces $k(\alpha)$ and by tangential forces $t(\alpha)$. Their resultants are

$$\mathrm{d}K(\alpha) = \mathrm{d}s[k(\alpha)] \tag{2.3}$$

$$\mathrm{d}T(\alpha) = \mathrm{d}s[t(\alpha)] \tag{2.4}$$

Figure 2.19 Equivalent forces

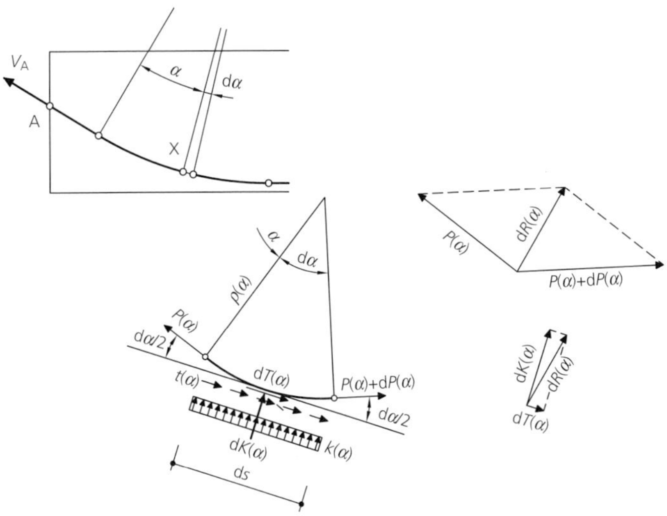

These forces can be determined from the conditions for equilibrium. The equilibrium in the direction of the tangent is

$$P(\alpha)\cos\frac{\mathrm{d}\alpha}{2} - [P(\alpha) + \mathrm{d}P(\alpha)]\cos\frac{\mathrm{d}\alpha}{2} = \mathrm{d}T(\alpha) \tag{2.5}$$

Since

$$\cos\frac{\mathrm{d}\alpha}{2} \cong 1 \tag{2.6}$$

then

$$\mathrm{d}T(\alpha) = -\mathrm{d}P(\alpha) \tag{2.7}$$

Equilibrium in the normal direction is given by

$$P(\alpha)\sin\frac{\mathrm{d}\alpha}{2} + [P(\alpha) + \mathrm{d}P(\alpha)]\sin\frac{\mathrm{d}\alpha}{2} = \mathrm{d}K(\alpha) \tag{2.8}$$

Since

$$\sin\frac{\mathrm{d}\alpha}{2} \cong \frac{\mathrm{d}\alpha}{2} \tag{2.9}$$

$$\mathrm{d}P(\alpha)\frac{\mathrm{d}\alpha}{2} \cong 0 \tag{2.10}$$

then

$$\begin{aligned} P(\alpha)\mathrm{d}\alpha &= \mathrm{d}K(\alpha) \\ \mathrm{d}K(\alpha) &= P(\alpha)\frac{\mathrm{d}s}{\rho(\alpha)} \end{aligned} \tag{2.11}$$

$$\begin{aligned} k(\alpha)\mathrm{d}s &= P(\alpha)\frac{\mathrm{d}s}{\rho(\alpha)} \\ k(\alpha) &= \frac{P(\alpha)}{\rho(\alpha)} \end{aligned} \tag{2.12}$$

The friction $t(\alpha)$ is proportional to the normal compression $k(\alpha)$ and the coefficient of friction μ, thus

$$t(\alpha) = \mu[k(\alpha)] \tag{2.13}$$

The friction force always acts against the direction of movement of the tendon.

The value of the tendon force $P(\alpha)$ at point X is given by the well-known formula

$$P(\alpha) = P_{\mathrm{A}}\,\mathrm{e}^{-\mu(\alpha+kx)} \tag{2.14}$$

where P_{A} is the prestressing force at the anchor, μ is the coefficient of friction, α is the sum of angular displacement along x, k is the unintentional angular displacement or wobble (per unit length) and x is the length of the tendon from the anchor.

With $k(\alpha)$ and $t(\alpha)$ determined with the help of Equations 2.12 and 2.13, the resultant force $r(\alpha)$ can be determined as

$$r(\alpha) = \sqrt{k^2(\alpha) + t^2(\alpha)} \tag{2.15}$$

This force acts along the length of the cable and represents the equivalent load of the tendon. It can be transformed into the global coordinate axes X and Y.

Given that currently most structures are analysed using the finite-element method it is not necessary to determine the function of forces $r(\alpha)$. It is more appropriate to determine the forces at several nodes along the length of the tendons. Figure 2.20(a) shows a typical layout of a tendon formed by two straight parts and one parabolic curve. The tendon is post-tensioned from the left anchorage.

Figure 2.20(b) shows a diagram of the prestressing force, which is influenced by the friction losses and the wedge draw-in at the anchorage. Along the length of the structure the tendon is divided in several elements for which we can substitute a continuous force diagram by a piecewise constant diagram. The length of the elements should be chosen in such a way that sufficient accuracy is ensured.

From the geometry of the tendon and the forces in the elements we can determine the equivalent radial forces R (Figure 2.20(c))

$$\begin{aligned} \tan\alpha_{i-1} &= \frac{y_{i-1} - y_i}{x_i - x_{i-1}} \\ \tan\alpha_i &= \frac{y_i - y_{i+1}}{x_{i+1} - x_i} \end{aligned} \tag{2.16}$$

Figure 2.20 Modelling of prestressing

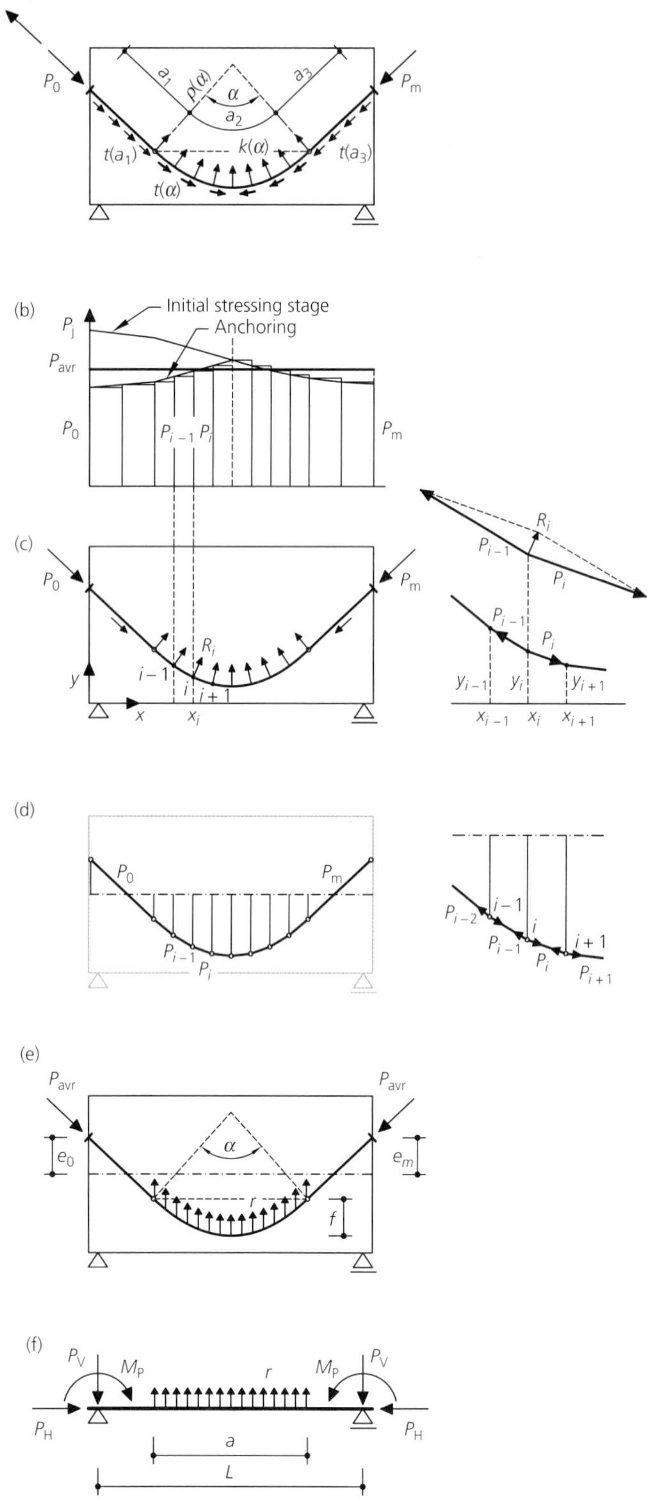

$$R_{xi} = P_{i-1} \cos \alpha_{i-1} - P_i \cos \alpha_i$$
$$R_{yi} = P_{i-1} \sin \alpha_{i-1} - P_i \sin \alpha_i \quad (2.17)$$

$$R_i = \sqrt{R_{xi}^2 + R_{yi}^2} \quad (2.18)$$

Although the above procedure is relatively simple, it requires some effort. It can be easier to model the tendon as a chain of straight, mutually connected elements (Figure 2.20(d)). The elements are linked to the nodes of the analysed structure by rigid links. The tendon elements are pin connected to the rigid links and have zero stiffness ($EA = 0$) during the post-tensioning phase. They are stressed by axial forces determined according to Figure 2.20(b). As the members have zero stiffness, the resultant of forces in each node have the same value and direction as the forces determined according to Equation 2.18.

After the post-tensioning of the structure, it is possible to give the tendon elements their actual stiffness (E_s, A_s) and make the final link to the structure. In this way it is possible to correctly describe the function of the tendon in the structure – during the post-tensioning the tendon is not a part of the structure (it is not included in the stiffness matrix), while after the post-tensioning the tendon contributes to the stiffness of the structure (it is included in the stiffness matrix). Any loads will cause corresponding stresses in the tendons which are proportional to the stiffness, position and connection to the structure.

Modelling the tendons by a chain of straight members also makes it easy to describe the function of the draped tendon in plan curved structures (Figure 2.21). Figure 2.22 shows the calculation model for a structure post-tensioned by internal and external tendons both during the post-tensioning operation (Figure 2.22(a)) and during service (Figure 2.22(b)).

It is necessary to realise that the transverse components of the radial forces of the tendons cause significant torsion that has to be resisted at the supports. For preliminary calculations of simple structures, it can be assumed that the structure is post-tensioned by a tendon for which the horizontal component of the prestressing force P_H is constant and is equal to the horizontal component of the average value of the prestressing force P_{avr} (see Figure 2.20(b)).

$$P_H = P_{avr} \cos \alpha = \text{constant} \quad (2.19)$$

If the curve of the duct is a second-degree parabola (see Figure 2.20(e)) the radial forces r are given by

$$f = \frac{a}{4} \tan \frac{\alpha}{2}$$
$$r = \frac{8f}{a^2} P_H \quad (2.20)$$

Figure 2.21 Curved beam – equivalent load: (a) section A–A; (b) section B–B; (c) elevation; (d) plan

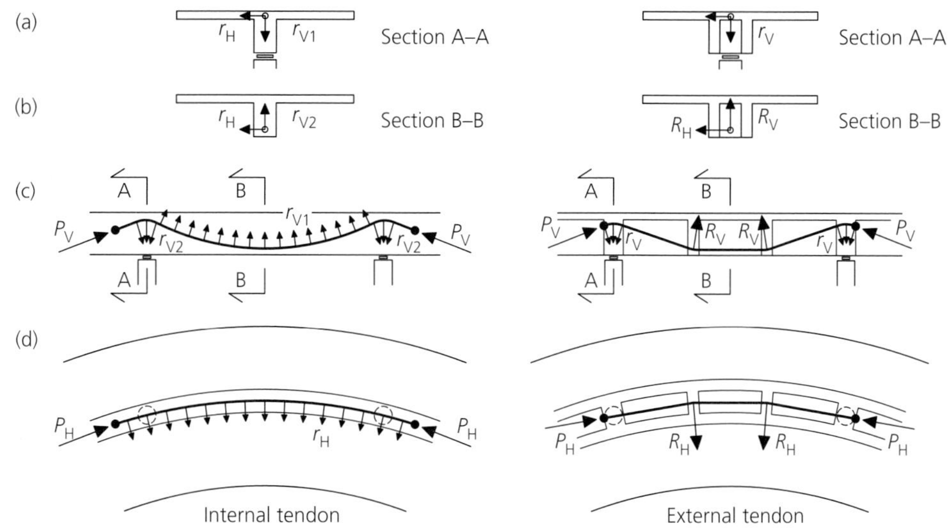

If the structure is analysed as a beam, the anchor forces P must be transferred into its axis and substituted by vertical (P_V) and horizontal (P_H) components and by bending moments ($M_P = P_H e$) acting in nodes 0 and m (see Figure 2.20(f)).

Figure 2.22 Modelling of the curved beam: (a) during post-tensioning; (b) during service

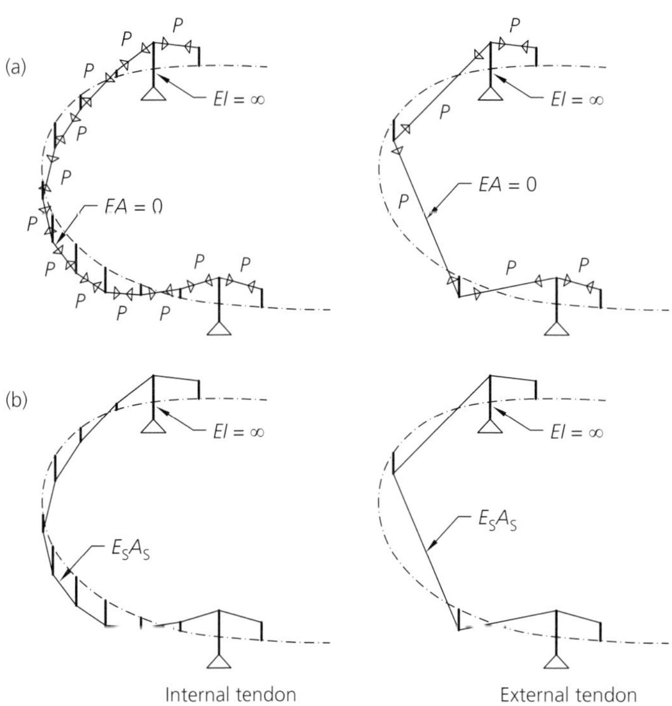

The normal forces N_{Px} and bending moments M_{Px} in the beam can be obtained by analysing the beam for the above load (Figure 2.23). The same bending moment M_{Px} will be obtained if the normal force N_{Px} is multiplied by the eccentricity of a tendon e_x: $M_{Px} = N_{Px}e_x$. However, this is true only for statically determined structures.

Figure 2.23 Equivalent load and internal forces and moments in a statically determined beam

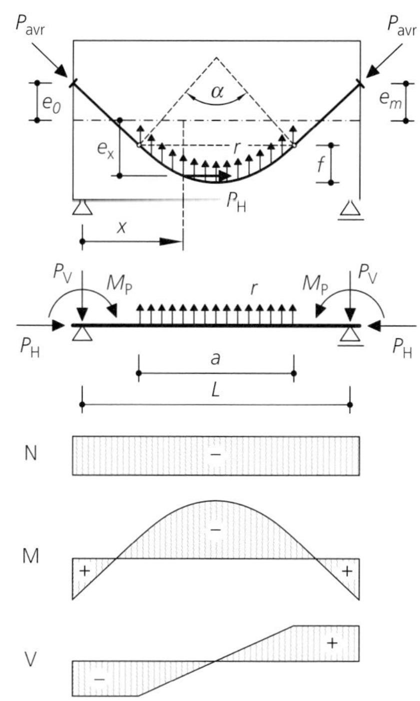

In statically undetermined structures, the prestress bending moment is given by

$$M_{Px} = M_{Px}^0 + M_{Px}' \tag{2.21}$$

where $M_{Px}^0 = N_{Px}^0 e_x$ is the statically determined (primary) bending moment and M_{Px}' is the additional (secondary) bending moment. The meaning of these moments is usually explained using the example of a two-span continuous beam prestressed by straight tendons situated close to the bottom fibres (Figure 2.24).

The prestress effects are analysed in two steps. In the first step the central support is removed (Figure 2.24(a)). Then the structure functions as a simple beam for which the deformation, primary internal forces N_{Px}^0, V_{Px}^0 and moments M_{Px}^0 are determined (Figure 2.24(b)).

In the second step the structure is loaded by a deformation situated at the central support (Figure 2.24(c)). The value of this deformation corresponds to the uplift caused by the prestressing. This deformation creates secondary internal forces V_{Px}' and moments M_{Px}'.

Figure 2.24 Internal forces and moments: (a) primary; (b) secondary; (c) total; (d) internal forces and moments

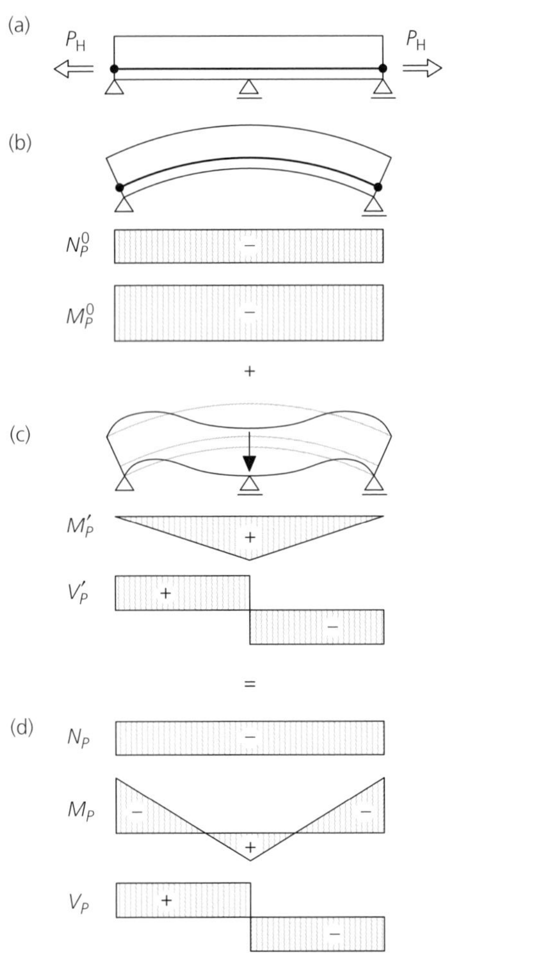

The total prestress effects are the sum of the effects determined in the two steps (Figure 2.24(d)).

$$N_{Px} = N_{Px}^0 \tag{2.22}$$

$$V_{Px} = V_{Px}^0 + V_{Px}' \tag{2.23}$$

$$M_{Px} = M_{Px}^0 + M_{Px}' \tag{2.24}$$

To understand the behaviour of the prestress, it is useful to review several basic examples of post-tensioned structures.

Figure 2.25 shows simply supported beams and beams with restrained ends. One end of the beam is rigidly fixed so it cannot undergo any rotation or displacement, while the other end is prevented from rotating but is free to move longitudinally. These beams of a span length L are either post-tensioned by straight tendons or prestressed by parabolic tendons. It is assumed that the horizontal component of the prestressing force is constant along the length of the cable.

Figures 2.25(a) to 2.25(d) show the equivalent load, normal forces and bending moments in beams that are post-tensioned by straight tendons. In all these cases, the effects of post-tensioning can be expressed by the equivalent load acting on the concrete at the anchorages.

$$N = -P_H \tag{2.25}$$

$$M_a = M_b = Ne_b \tag{2.26}$$

In cases (b) and (d) the tendon has an eccentricity $e_a = e_b$, so the normal force N introduces external bending moments at the beam ends.

In all cases the girders are stressed by a constant normal force N. However, a bending moment

$$M = M_a = M_b \tag{2.27}$$

is seen only in the case of the simply supported beam. In the beam with restrained ends the bending moment is zero, as the equivalent bending moment acting at the beam ends cannot cause any rotation of the beam. In this case the so-called 'primary' and 'secondary' bending moments have the same value but with opposite sign.

$$M_a^0 = -M_a' = M_b^0 = -M_b' = Ne_b \tag{2.28}$$

Figure 2.25 Equivalent load, normal forces and bending moments in a straight beam: (a)–(d) post-tensioned by straight tendons; (e)–(h) prestressed by parabolic tendons

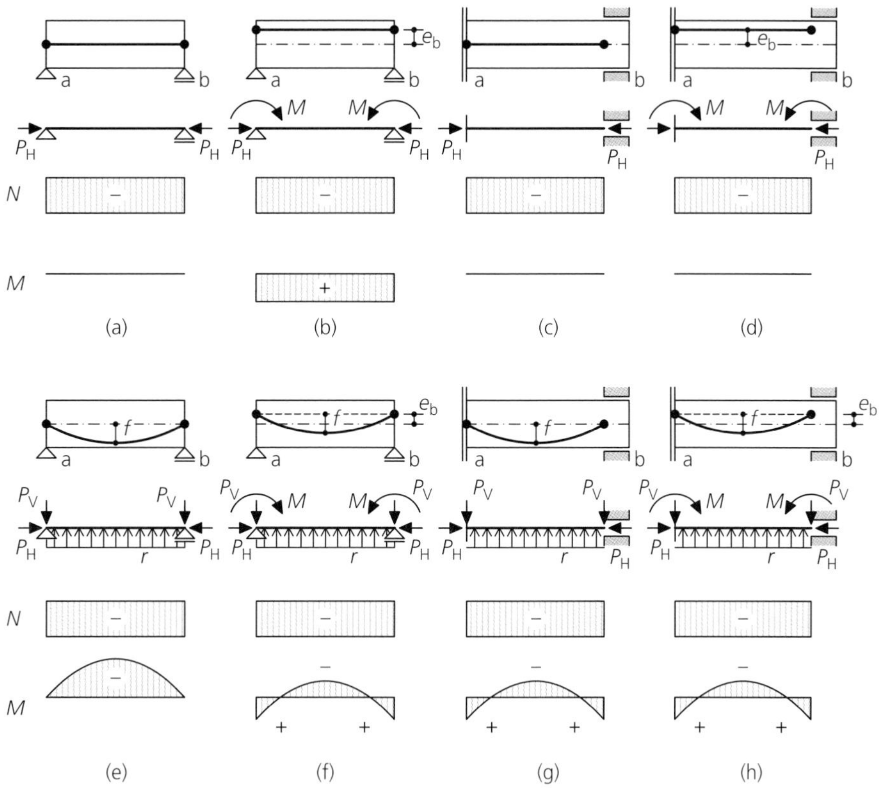

Figures 2.25(e) to 2.25(h) show the equivalent load, normal forces and bending moments in beams that are prestressed by parabolic tendons. In all cases the tendon has the same sag f. Similar to the previous example, the effects of post-tensioning can be described by an equivalent load acting at the anchorages.

$$N = P_H = P_{avr} \cos \alpha \tag{2.29}$$

In cases (f) and (h) the tendon has an eccentricity $e_a = e_b$, so the normal force N loads the beam ends with bending moments

$$M_a = M_b = Ne_b \tag{2.30}$$

Because the tendons are in a parabolic curve, the beams are also loaded by radial forces

$$r = \frac{8f}{a^2} P_H = \frac{8f}{L^2} P_H \tag{2.31}$$

In all cases the girders are stressed by a constant normal force N.

As the radial forces are constant, the bending moment has the shape of a second-degree parabola.

In the case shown in Figure 2.25(e) the bending moment at midspan is

$$M_{L/2} = -\frac{1}{8} rL^2 = -P_H f \tag{2.32}$$

In the case shown in Figure 2.25(f) the bending moment at midspan is

$$M_{L/2} = -\frac{1}{8} rL^2 + M_b = -P_H f + P_H e_b = -P_H (f - e_b) \tag{2.33}$$

It is important to note that the values and shapes of the bending moments are the same for the layout of prestressing steel shown in Figures 2.25(g) and 2.25(h). As also noted above, the equivalent bending moment acting at the beam ends cannot cause any rotation of the beam. Therefore, the bending moments at the ends are

$$M_a = M_b = \frac{1}{12} rL^2 = \frac{2}{3} P_H f \tag{2.34}$$

and the bending moment at midspan is

$$M_{L/2} = -\frac{1}{24} rL^2 = -\frac{1}{3} P_H f \tag{2.35}$$

Figure 2.26 shows curved beams post-tensioned with curved tendons. The tendon is situated either in the beam axis or parallel to the axis. The beams of span L are circular with radius ρ. The beams in Figures 2.26(a) and 2.26(b) are simply supported; the beams in Figures 2.26(c) and 2.26(d) have rigidly fixed ends that cannot undergo any rotation or displacements. It is assumed that the prestressing force P is constant along the length of the cable and that the friction (tangent) forces are zero.

The effects of post-tensioning can be described using the equivalent loads acting at the anchorages and the radial forces. As the tangential forces are zero the radial forces are constant

$$k = \frac{P}{\rho} \tag{2.36}$$

These uniform radial forces create in the beams a constant normal force

$$N = -P = -\rho r \tag{2.37}$$

This is also true for the curved beam in which both ends are fixed. Bending moments originate only in the simply supported beam (see Figure 2.26(b)) that is post-tensioned with an eccentric tendon. The anchor forces create equivalent moments at the ends of the beam

$$M_a = M_b = P_H e_b \tag{2.38}$$

As the reactions due to prestressing are zero, the bending moments are constant along the length of the beam.

It is evident from Figures 2.26(c) and 2.26(d) that the arch structure, which is post-tensioned by a tendon that is parallel to the axis, is stressed only by normal forces. But this is true only for structures in which deformations due to normal forces are not significant.

In slender structures analysed as the geometrically non-linear structures, in which the final state of the stresses is determined on the deformed shape, the bending stresses are significant. From Figure 2.27, which shows the radial forces and deformations of a typical arch (Figure 2.27(a)) and slender arch (Figure 2.27(b)) structure, it is evident that the radial forces change their value and direction. Therefore, any analysis has to include this phenomenon. This can be easily achieved by using a model in which the prestressing tendons are modelled as a chain of members loaded by axial forces (Figure 2.27(c)).

Figure 2.26 Equivalent load, normal forces and bending moments in a curved beam: (a)–(b) simply supported beam; (c)–(d) beam with rigidly fixed ends that cannot undergo any rotation or displacement

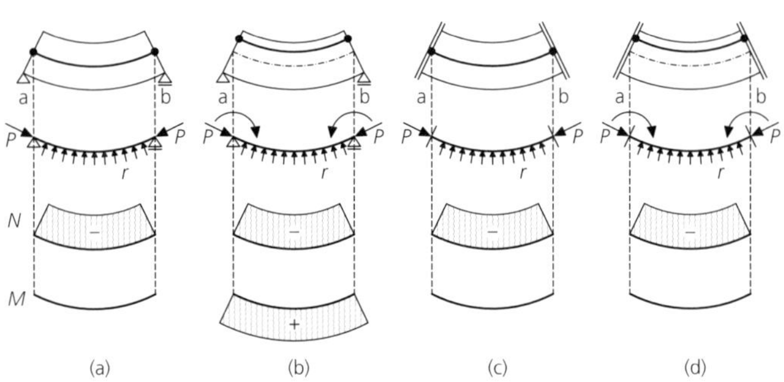

Figure 2.27 Radial forces acting on a stiff and slender arch: (a) typical arch; (b) slender arch; (c) chain of members loaded by axial forces

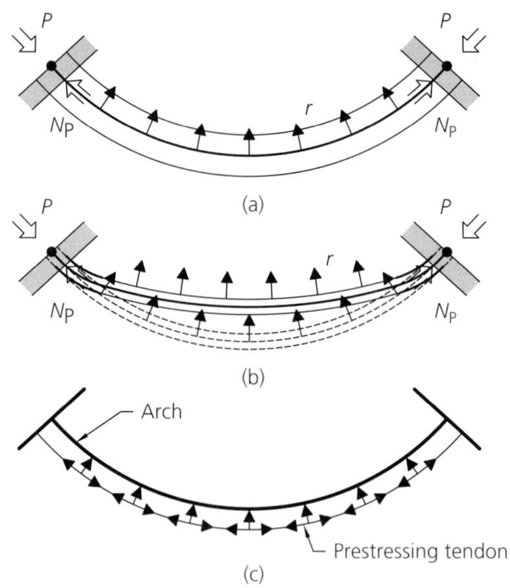

2.3. Prestressing losses

The difference between the jacking force at the anchors of the tendons and the prestressing force in a girder section is the prestressing losses. These losses can be divided into two groups: (*a*) instantaneous losses related to the mechanics of the post-tensioning system and the tendon geometry, and (*b*) time-dependent losses related to the material properties of the concrete and the prestressing steel. These losses are discussed in Chapter 7.

The components of the instantaneous losses are

- duct friction due to curvature
- wobble (unintentional friction)
- wedge set (or anchor set)
- elastic shortening of the concrete.

Friction between the strands and the duct during stressing is related to intended angular changes in the tendon geometry. Figure 2.28(a) shows the trajectory of the tendon within a desired duct profile. As the tendon is stressed, the friction where the tendon contacts the duct wall

Figure 2.28 Internal tendon: (a) tendon position in a curved duct; (b) wobble in a straight tendon

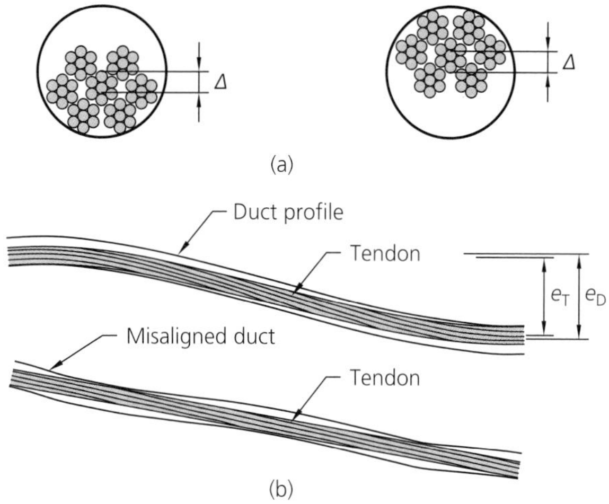

Figure 2.29 Elastic shortening of the tendon: (a) tendons; (b) model for use in calculation

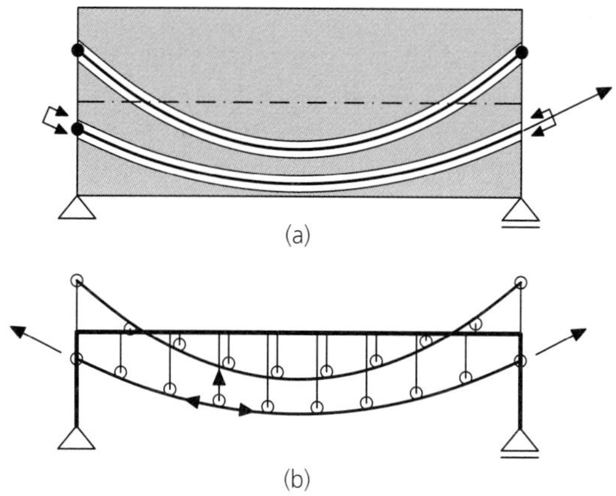

Figure 2.30 Typical beam structure: (a) layout of the tendon; (b) prestressing force *P*; (c) equivalent load; (d) shear forces; (e) bending moments

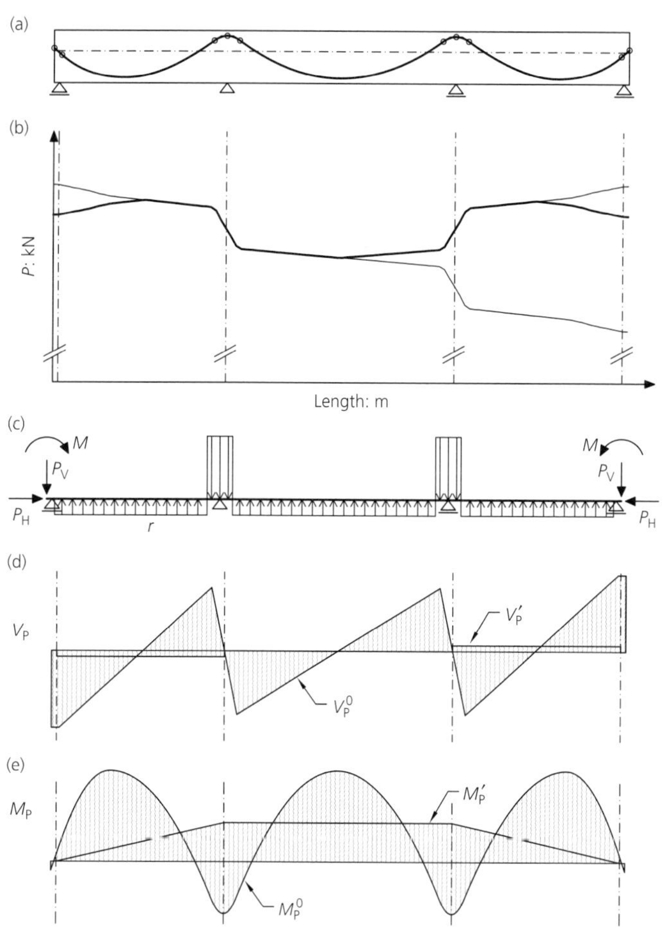

reduces the force in the tendon. The friction coefficient μ is a function of the duct material. Sections A–A and B–B shows the positions of the strands in the duct and remind the reader that distance of the tendon's axis e_T is different from the distance of the duct axis e_D.

Figure 2.28(b) illustrates wobble losses as results of duct misalignment. To simplify the calculation of these losses the same formula is used as for friction losses using the friction coefficient k (Equation 2.14).

The losses caused by the anchorage set are also determined using Equation 2.14, but the distance called the 'length affected by anchorage set' has to be determined first (Navrátil, 2014). The friction and wobble effect act oppositely to the movement of the tendon.

During prestressing, a structural member is elastically deformed. It shortens and deflects. If the member is progressively post-tensioned by several tendons, the stresses in already tensioned and anchored tendons are reduced. This phenomenon can be easily determined for a structure in which the prestressing tendons are modelled as a chain of bar elements connected to the structure by stiff members (Figure 2.29). The tensioned tendons are modelled as elements that have zero stiffness (E_t, $A_t = 0$) during their tensioning. The already tensioned tendons are modelled as a chain of bar elements with actual stiffness (E_s, A_s).

For the purposes of illustration, Figure 2.30 shows the prestressing tendon of a three-span structure, the course of the stresses in the tendon tensioned from the one side, the final course of the stresses after tensioning of the cable from both sides, the equivalent tendon forces P, r and the corresponding primary and secondary shear forces V_P^0 and V'_P, and the bending moments M_P^0 and M'_P. It is evident from the figure that the prestressing creates in the structure not only compression stresses but also shear forces and bending moments that can balance the bending moments originating from the dead load.

While the wedge set (draw-in) usually causes significant prestressing loss in short tendons, in long draped tendons it gives rise to uniformity in the course of the prestressing force (see Figure 2.30). It can be also used to reduce the horizontal forces originating in saddles of cable-supported bridges. The Ebro River bridge (described in Chapter 4, see Figures 4.88 and 4.89) serves as an example. The bridge forms a self-anchored suspension structure that was erected as a cable-stayed structure suspended on steel pipes. By tensioning the suspension cables passing through these pipes the structure was modified into a self-anchored suspension structure. During the tensioning of the strands passing through the pipes, large friction forces originate at the pylon saddles. To reduce these forces, the anchor forces were reduced after the tensioning of the strands (Figure 2.31) (Strásky *et al.*, 2014).

Prestressing by draped tendons can not only balance the dead load but it can also reduce the horizontal forces that load the footings of the frame structures. Figure 2.32 shows the deformations of the frame structure and the forces that load the footings of the frame structure for a deck that is prestressed by draped tendons.

An important aspect in the design of post-tensioned structures is the distribution of the prestressing force from the anchors into the whole of the structural member. The problems are discussed in all basic books describing the design of prestressed concrete structures (Hampe, 1978; Lin and Burns, 1981; Leonhardt, 1964; Navrátil, 2014; Nilson, 1997). Practical examples including an explanation of the division of

Figure 2.31 Prestressing of suspension cables

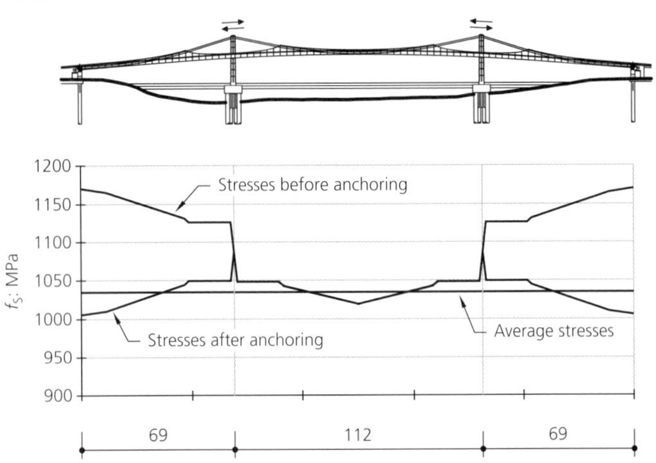

Figure 2.32 Prestressing of a frame structure: (a) deformations of the frame structure; (b) forces loading the footings of the frame structure

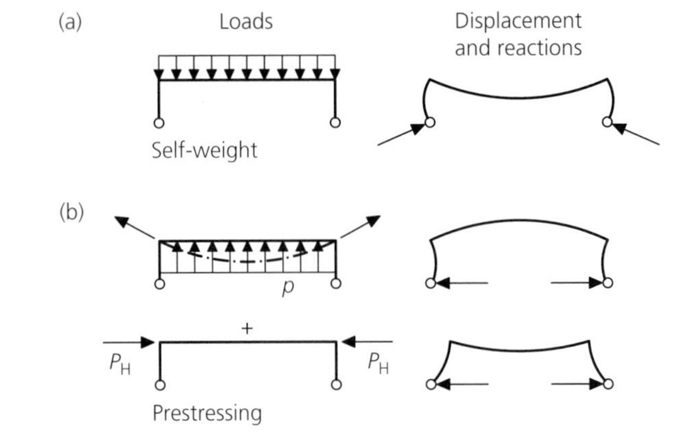

responsibility between the supplier of the prestressing technology and the designer can be found in VSL International (1991).

The supplier of the post-tensioning system is usually responsible for the design of the anchorage device and the local zone immediately surrounding the device. The designer (Engineer of Record) is responsible for the design of the general zone that surrounds the local zone. The way in which the concentrated anchorage zone is spread out over the entire cross-section of the member is usually studied using a strut-and-tie model, which allows for the determination of the concrete compression stresses and the tension forces in steel ties (see also Chapter 8).

REFERENCES

FIP (1996) FIP Recommendations. Practical Design of Structural Concrete. FIP Congress, *Amsterdam, The Netherlands*.

Fuzier JP, Ganz HR and Matt P (ed.) (2006) Durability of post-tensioning tendons. *fib Bulletin* **33**.

Hampe E (1978) *Spannbeton*. VEB Verlag für Bauwesen, Berlin, Germany.

Hewson N (2012) *Prestressed Concrete Bridges. Design and Construction*. Thomas Telford, London, UK.

Kasuga A (2006) Development of composite truss bridge using suspension structure. *7th International Conference on Short and Medium Span Bridges, Montreal, Canada*. Curran, Red Hook, NY, USA.

Leonhardt F (1964) *Prestressed Concrete. Design and Construction*. Ernst & Sons. Berlin, Germany.

Lin TY and Burns NH (1981) *Design of Prestressed Concrete Structures*. John Wiley, New York, NY, USA.

Manterola J (2006) *Puentes, Apuntes para su diseño, cálculo y construcción*, vols I and II. Colección Escuelas, Madrid, Spain.

Navrátil J (2014) *Prestressed Concrete Structures*. VSB – Technical University of Ostrava, Czech Republic.

Nilson AH (1997) *Design of Concrete Structures*. McGraw-Hill, New York, NY, USA.

Stráský J (2009) Cable-stayed bridge across the Odra River, Czech Republic. *26th Annual International Bridge Conference 2009: Meeting Bridge Challenges in Challenging Times*, pp. 352–360.

Stráský J (2011) *Stress Ribbon and Cable Supported Pedestrian Bridges*, 2nd edn. Thomas Telford, London, UK.

Stráský J, Cobo del Arco D, Novotny P and Raventós I (2014) Construction of "Lo Passador": a self-anchored suspension bridge at the Ebro River delta. *Structural Engineering International* **24(1)**: 114–117.

VSL International (1991) *Detailing for Post-Tensioning*. VSL International Ltd, Berne, Switzerland.

ice Publishing

Stráský and Nečas

ISBN 978-0-7277-6385-3
https://doi.org/10.1680/dcpb.63853.027
ICE Publishing: All rights reserved

Chapter 3
Structural types

Apart from overburdened or directly walkable arches or stress-ribbon structures, bridges have always a deck that transfers the load into the supports – piers and abutments (Leonhardt, 1979; Liebenberg, 1992; 1984; Manterola, 2006; Mathivat, 1983; Menn, 1990; Podolny and Muller, 1982; Troyano, 2003). The deck can be supported by bearings situated on the supports, or it can be connected to the supports by a frame or hinge. The deck can be also supported by inclined struts or arches, or can be suspended from arches, stay cables or suspension cables (Figure 3.1).

3.1. Decks of bridges of typical width

A deck of typical width (up to 15 m) can be formed by slabs, T-beams or box girders (see Figure 3.5). Depending on the nature of the crossing and the load, all deck types can be part of more complex structural systems. They can be fixed into the outer supports to form frame structures (Figure 3.1(a)), or they can be supported by vertical or inclined supports (Figures 3.1(b) to 3.1(f)). The deck can be supported by bearings (Figures 3.2(a) and 3.3), or it can be a part of an integral structural system in which the deck is connected to the supports by a frame or hinge (Figures 3.2(b) and 3.4). Most concrete bridges are post-tensioned by prestressing tendons located inside the perimeter of the cross-section. The bridges can be also prestressed by external cables situated outside the perimeter of the cross-section, forming more complex structural systems (Figures 3.1(c) and 3.1(k)).

Even when the deck is supported by bearings, the bridge is stressed by the normal force originating from the friction in the bearings, and until the friction is overcome the structure acts as a frame. It is obvious that the classical division of structures into girder and frame structures is meaningless, as the bridge supports always interact with the subsoil. Therefore, the bridge always forms an integral system. The same applies to slab structures, which, due to prestressing, always act as shell structures stressed by normal and bending stresses.

A solid slab of rectangular or trapezoidal cross-section (Figure 3.5(a)) is suitable not only for small-span bridges but can also be part of arch- or cable-supported systems with spans up to 200 m. The trapezoidal slab is economical for continuous structures up to 21 m. Slab beams (Figure 3.5(b)) or voided slabs (Figure 3.5(i)) are suitable for spans up to 30 m.

For longer spans, beam and slab structures are suitable. T-beam structures, which are assembled from several longitudinal girders, are now only built using prefabricated girders that are mutually connected by a composite deck slab (Figure 3.5(c); see also Chapter 9). Transverse diaphragms are designed only at supports. In cable-supported bridges and railroad bridges the deck is formed by two edge girders that are connected by transverse diaphragms and by a deck slab (Figure 3.5(d)). This bridge type is suitable for crossings where there is limited space available under the bridge. In this case, the transversal diaphragms are sometimes omitted, and the deck slab is transversally prestressed (Figures 3.5(e) and 3.6) (Smith *et al.*, 2009).

For spans up to 45 m very economical decks can be formed using double-T cross-sections (Figure 3.5(f)). Diaphragms are designed only at the ends of the bridges, or where the diaphragms are also used as supports for movable scaffolding. For longer spans or where there is unsustainable foundation soil, concrete beams can be replaced by steel ones (Figure 3.5(g)). The concrete deck distributes the load, guarantees the bridge integrity and contributes to bridge resistance.

For bridges on which the decks are supported by axial supports, structures formed by spine girders with large overhangs are suitable (Figure 3.5(h)). For spans up to 35 m girders of constant depth are used, while for spans up to 45 m girders of variable depth are designed.

Voided slabs (Figure 3.5(i)) are used in some countries for spans up to 30 m. As the voids cannot be inspected, they require perfect waterproofing.

For spans in the range 30–45 m, multi-cell box girders are very efficient (Figure 3.5(j)), and decks assembled of two or more box girders connected by a deck slab have also proven to be very economical (Figure 3.5(k)). By changing the width of the connecting slab, these structures can easily be used to solve the problems associated with the variable deck widths at junctions between bridges with ramps. Nowadays, decks assembled from several box girders are commonly used in structures assembled from precast U-shaped beams and a composite deck slab. Architecturally pleasing solutions can be achieved using structures formed by two- or three-cell box girders with sloping edge cells and without traditional cantilevers (Figure 3.5(l)).

Figure 3.1 Basic structural types: (a) single-span frame structure; (b) continuous structure; (c) continuous structure prestressed by external cables; (d) continuous structure of variable depth; (e) inclined leg frame; (f) cantilever structure; (g) deck arch; (h) through arch; (i) cable-stayed structure; (j) suspension structure

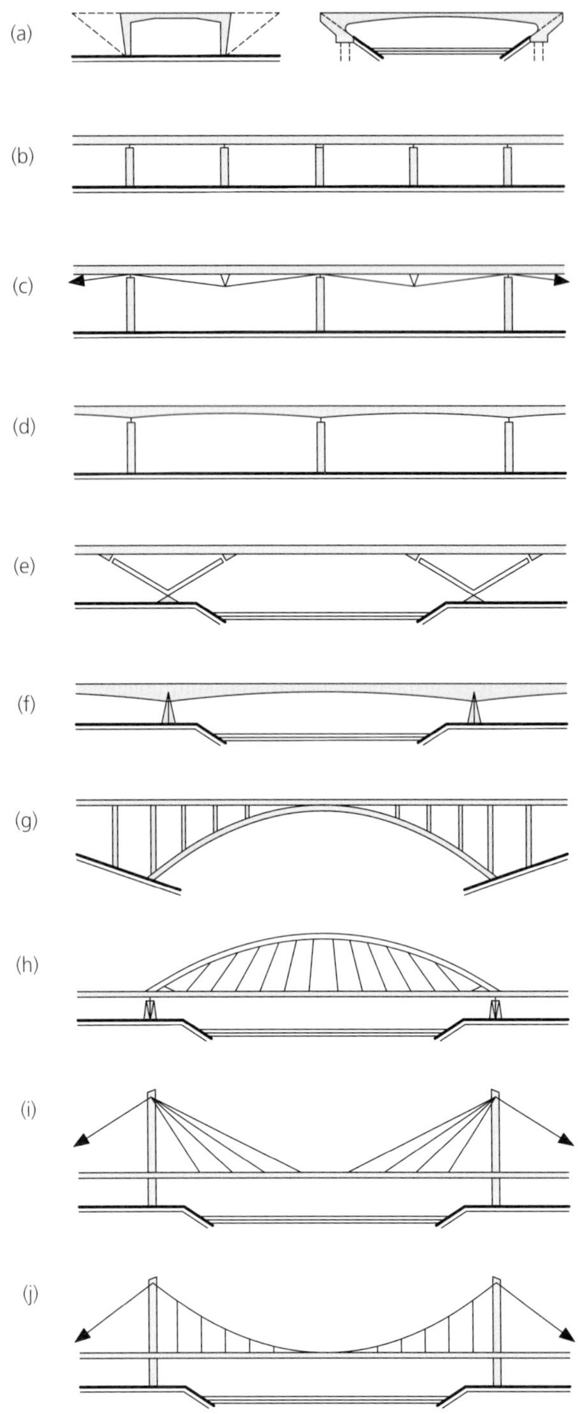

Figure 3.2 Classical (a) and integral (b) structural systems

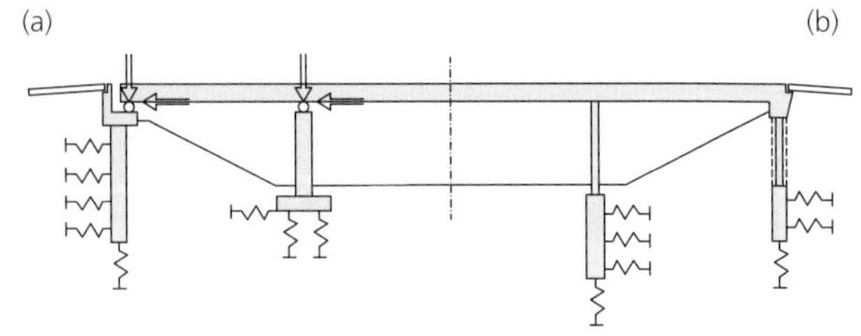

Figure 3.3 Bridge deck supported by bearings

Figure 3.4 Bridge deck integral with the supports

Figure 3.5 Cross-sections of the bridge decks of common width (up to 15 m): (a) slab; (b) slab beam; (c) T-beam; (d) edge beam; (e) trough beam; (f) twin T-beam; (g) steel–concrete composite T-beam; (h) spine girder; (i) voided slab; (j) three-cell box girder; (k) two box girders connected by a deck slab; (l) two-cell box girder with inclined side webs; (m) one-cell box girder; (n) steel–concrete composite one-cell box girder

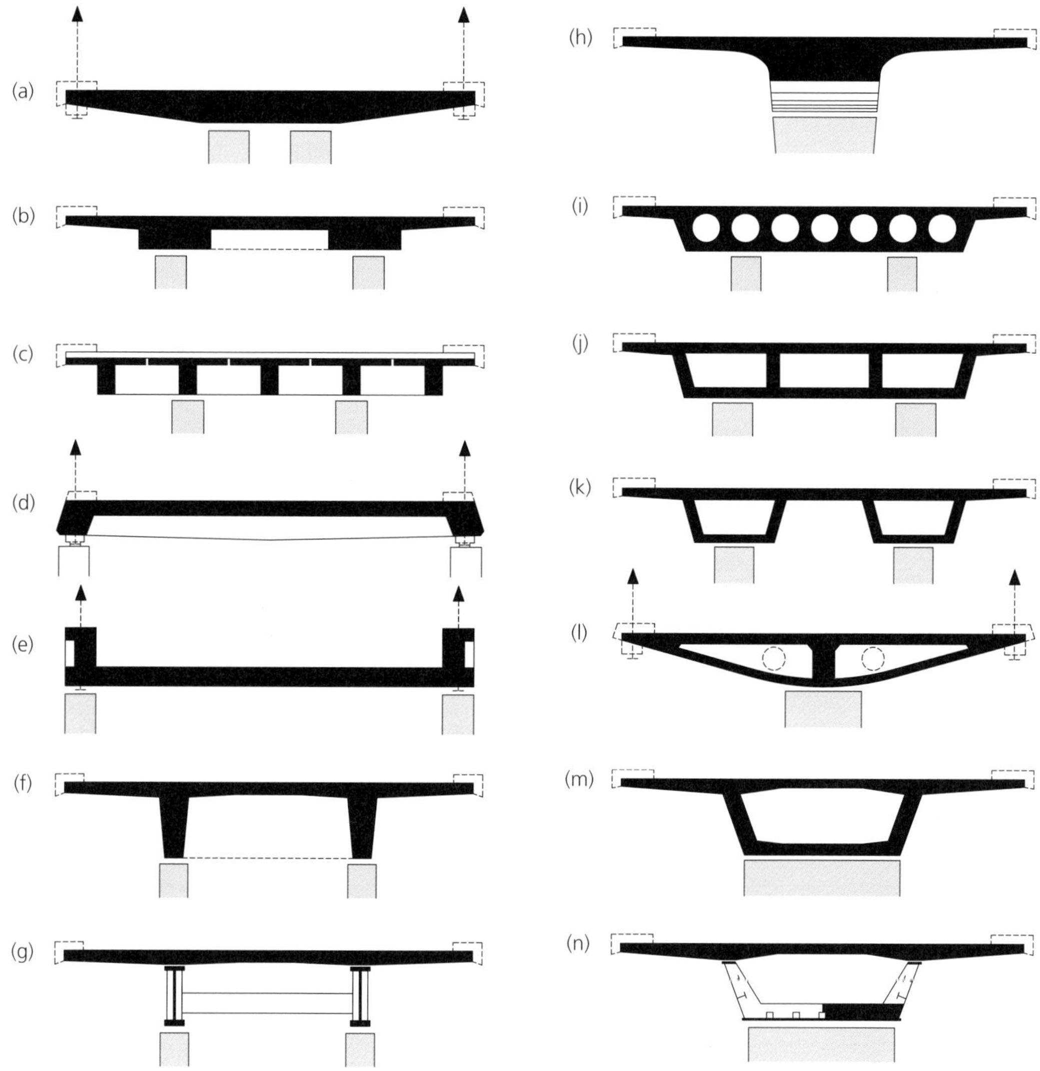

For bridges of spans in the range 45–60 m, a one-cell box girder of constant depth is commonly used, while for longer spans (up to 350 m) a one-cell box girder of variable depth erected with balanced cantilevers represents the most economical solution (Figure 3.5(m)). For these spans composite structures assembled from steel troughs and composite deck slabs are also suitable (Figure 3.5(n)). The concrete deck slab not only distributes the load, but also contributes to the bridge resistance and guarantees the integrity of the structure. Concrete can also be placed on the bottom plates of steel sections close to the supports to reduce the compression stresses in the steel and to increase the stability of the bottom flanges.

Box girder bridges usually have a depth equal to 1/25 of the span length, and T-beam bridges usually have a depth equal to 1/20 of the span length.

Modern composite structures have a hybrid decks assembled from concrete top and bottom slabs, and webs formed from either corrugated plates (Figures 3.7(a) and 3.8) (Maeda *et al.*, 2002), or concrete (Figure 3.9) (Giacomelli and Grennerat, 1989) or steel trusses (Figures 3.7(b) and 3.10) (Barlet *et al.*, 2002). These structures are prestressed by means of external cables. In Japan, several modern structures have webs formed by precast members of a butterfly shape that are not mutually connected (Figures 3.7(c) and 3.11). These panels are made from high-strength fibre-reinforced concrete, do not have traditional reinforcement and are prestressed by means of diagonal prestressing (PT) bars (Kasuga, 2017).

3.2. Deck of wide bridges

Motorway bridges situated well above the terrain or structures suspended on arches or pylons are often, even at widths of about 35 m, made up of a single bridge structure. The decks are usually

Figure 3.6 Dubai Metro viaducts

Figure 3.8 Maetani Bridge, Japan

Figure 3.7 Composite bridges: (a) webs from corrugated plates; (b) webs substituted by truss; (c) webs substituted by precast panels

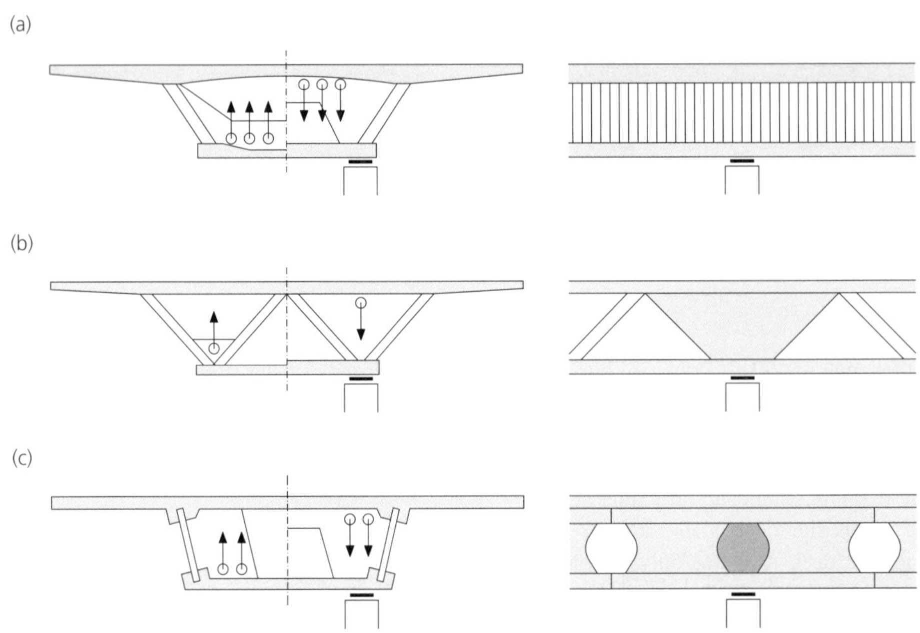

Figure 3.9 Sylans Viaduct, Nantua, France

Figure 3.10 Bras de la Plaine Bridge, Réunion, France

formed by box girders with large overhangs supported by struts (Figure 3.12(a)), or one- or two-cell box girders reinforced by cross-beams, which also support the outer overhangs (Figure 3.12(b)).

The forces from the stay cables or suspenders have to be transferred into the webs. Usually this is done by means of internal ties (Figure 3.13). To avoid deformations of the cross-section, the ties should be formed from prestressed concrete elements or a structural steel. In structures stiffened by cross beams, the forces from the stay cables are transferred by their shear capacity.

Figure 3.11 Shin-Meishin Mukogawa Bridge, Japan

The main cable-supported bridges are often built with long approach viaducts formed by multi-span structures of shorter spans. From an architectural point of view, the main and approach spans should have similar arrangements and have decks of a uniform appearance. This can be easily achieved if the deck of the main span and the approaches is the same (Figures 3.12(a) and 3.12(b)). If the approaches are formed by twin bridges, the main spans can be assembled using the same, mutually connected members. If the approaches are formed by typical one-cell box girders, transversally prestressed truss members can be inserted between them (Figures 3.12(c) and 3.14) (ODOT, 2018). If the approaches are formed by two-cell box girders with inclined edges, the girders can be connected by a transversally prestressed deck slab and precast elements inserted between the box girders (Figure 3.12(d)). If the bridge deck is suspended in the bridge axis, the precast elements are compressed (Figure 3.15) (Stráský, 2009). However, if the bridge deck is suspended on its outer edges, the precast members are tensioned, and therefore it is necessary to prestress them. For wide bridges

Figure 3.12 Cross-sections of wide bridge decks (up to 35 m): (a) one-cell box girder with large overhangs supported by struts; (b) two-cell box girder stiffened by cross beams; (c) two one-cell boxes mutually connected by truss members; (d) two two-cell boxes mutually connected by a top slab and precast struts; (e) edge girders connected by floor beams and a deck slab

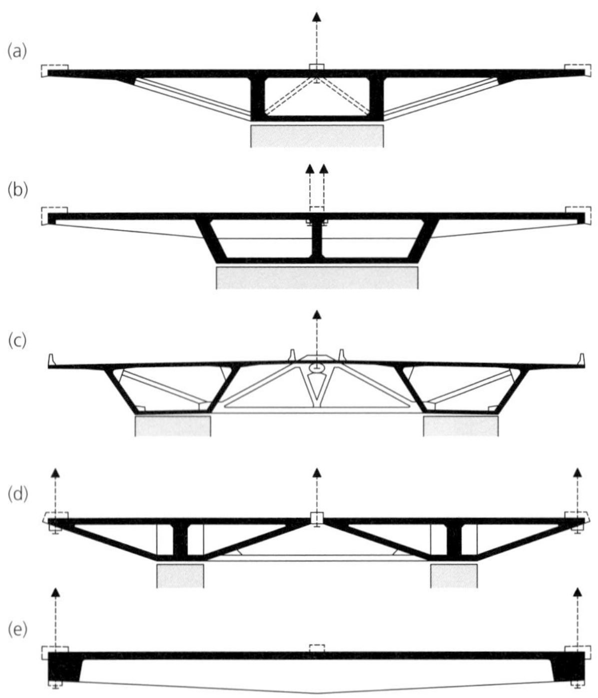

Figure 3.13 Transfer of the cable force into the webs: (a) via a stay cable; (b) via an external cable

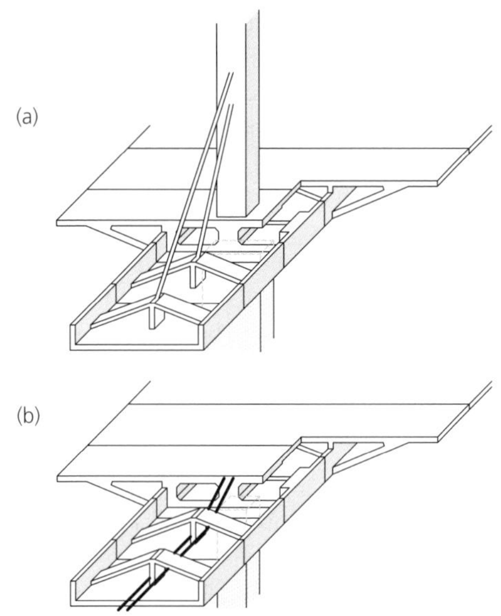

Figure 3.14 Toledo Bridge, Ohio, USA

Figure 3.15 Odra River Bridge, Czech Republic

in which the decks are suspended on their outer edges, decks consisting of edge girders connected by floor beams and a deck slab are very suitable (Figure 3.12(e)).

Structures formed by spine box girders with large overhangs supported by struts are often built progressively. First, the spine girder is built, then the outer struts are erected, and then the deck slab is cast. The supporting struts can be formed from single bars (Figures 3.16(a) and 4.55), a truss (Figures 3.16(b) and 4.59) or slabs (Figures 3.16(c) and 4.60). The slab members can be mutually connected or remain unconnected. The truss and unconnected slab struts, taken together with the top slab, form a pseudo three-cell box girder (Figure 3.17(c)). These struts contribute significantly to the bending and torsional resistance of the structure.

Figure 3.16 Precast struts formed from: (a) single bars; (b) a truss; (c) slabs

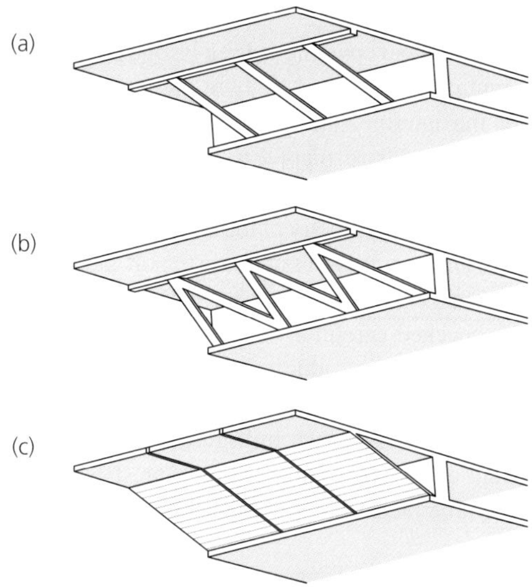

Figure 3.17 Shear flow in: (a) a one-cell box girder; (b) a three-cell box girder; (c) a pseudo three-cell box girder

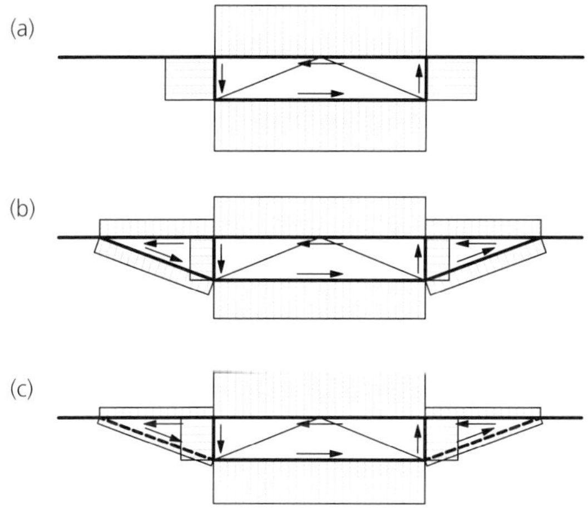

Figure 3.18 Suspension of the deck: (a) in the bridge axis; (b) on both edges; (c) in three planes; (d) on one edge; (e) transverse bending moments for the structural arrangements in (a)–(d)

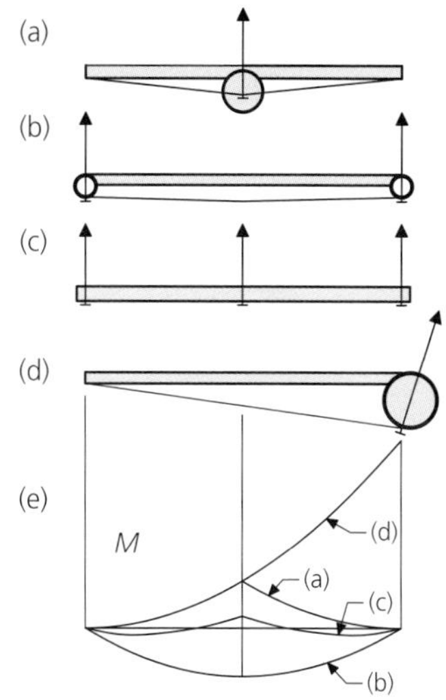

Figure 3.19 Support of a double-T deck

Bridges suspended on arches or pylons may have cables arranged in the bridge axis (Figure 3.18(a)) or at the bridge edges (Figure 3.18(b)). There are also bridge decks suspended in three planes (Figure 3.18(c)) or suspended on only one edge (Figure 3.18(d)). Structures suspended in the bridge axis or on one edge require a torsionally stiff deck. Furthermore, structures suspended on one edge are stressed by the torsion caused by the dead load.

The course of the transverse bending moments of the deck in the various suspended deck arrangements is shown in Figure 3.18(e). It is clear from the figure that the structure suspended in the bridge axis and the structure suspended on the edges are stressed by moments of the same absolute value. It is also evident that the structures suspended on one side are stressed by very large bending moments, and therefore this solution should be used only if necessary.

The bridge deck is usually supported by bearings or concrete hinges situated directly under the beams (Figure 3.19), or supported indirectly via pier diaphragms (Figure 3.20). In both solutions, a pair of bearings resist the torsion originating from the live load located on one side of the deck only.

Figure 3.20 Support of a box girder deck

3.3. Bridges supported in the deck axis

Often, the deck of junction bridges needs to be supported by a single support located in the bridge axis. There are six possible structural arrangements.

(*a*) The bridge deck is supported by single bearings situated on the intermediate supports. The torsion is resisted by a pair of bearings situated at the abutments. In long bridges, the torsion has to be resisted by a pair of supports situated at distance L_T, which is determined by the torsional capacity of the deck (Figure 3.21). These structures lose their order and uniform appearance (see Figure 4.31).

(*b*) The bridge deck is connected to the frame by piers, the footings of which are supported by a pair of bearings situated under the grade (Figure 3.22). Obviously, in this solution it is difficult to inspect the bearings.

(*c*) The bridge deck is supported by one-directional bearings alternately situated at the top and bottom of the piers (Figure 3.23). In this case the torsion of the deck is resisted by a pair of horizontal forces originating at the bearings. This solution requires precise placing of the bearings, which have to resist large horizontal forces, and this is the reason why this solution has only been used in the construction of several bridges. However, the principle of resisting the torsion by pairs of horizontal forces can be used in design of other bridge types.

(*d*) An elegant solution is the arrangement used in the construction of bridges in California, which are assembled from several frames mutually connected by (Gerber) hinges (Figures 3.24, 4.29 and 4.30). Individual frames are composed of integral structures of several spans, to which the deck is frame or hinge connected via piers. It is evident that the main problem associated with these structures is the hinges. The classical solution, which is suitable for sunny California, requires careful maintenance of the expansion joints, and is thus unsuitable for bridges built in less benign climates. However, instead of a classical Gerber hinge, steel members that have been prestressed by PT bars connected to the concrete deck, and that can be replaced in future, can be used (Figure 3.25).

(*e*) It is also possible to construct a long bridge composed of several frames mutually connected by twin piers (Figure 3.26). As the shape of the intermediate and expansion piers can be similar (e.g. the intermediate piers can have vertical recesses), the structure can have a uniform appearance.

(*f*) If the clearance and geometry of the structure allows the piers to be designed with caps, it is possible to support the deck by pairs of closely situated bearings (Figure 3.27). In this case possible uplift has to be checked carefully. In several structures the uplift has been solved by using additional vertical prestressing. In this solution the bending of vertical bars or tendons has to be considered. However, as shown in Figure 4.32, with careful design the pier caps can be small and no vertical prestressing is necessary. In this way, the piers do not influence the junction geometry.

As already mentioned, the bridge deck can be created together with arches or integral cable systems. The arches and/or pylons can be connected to the deck via frames or joints.

3.4. Arch and frame bridges

True arch bridges have their decks supported (Figure 3.28(a)) or suspended (Figure 3.28(c)) on arch structures transferring the arch forces into their foundations. However, the arch structures can also create self-anchored systems, in which the foundations are stressed only by vertical reactions. Figures 3.28(b) and 4.15 illustrate structures in which the arch springs are connected to the end diaphragms of the deck by compression struts that transfer the horizontal component H of the arch force S into the deck. The prestressed concrete deck resists the horizontal force by its tension capacity. The moment

Figure 3.21 Structural arrangement (*a*)

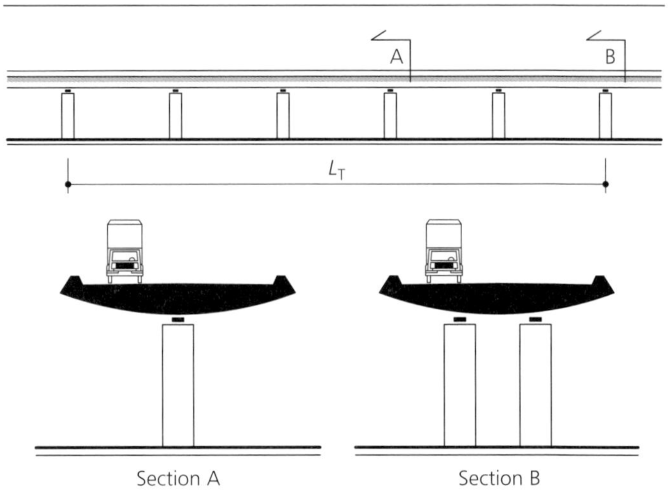

Figure 3.22 Structural arrangement (*b*)

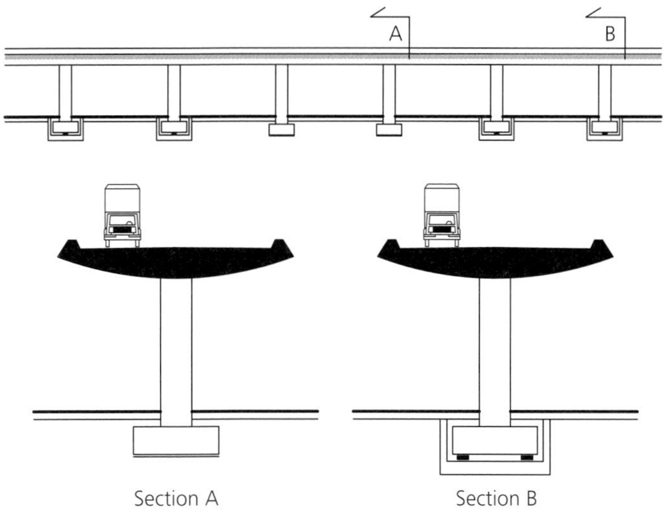

Figure 3.23 Structural arrangement (*c*)

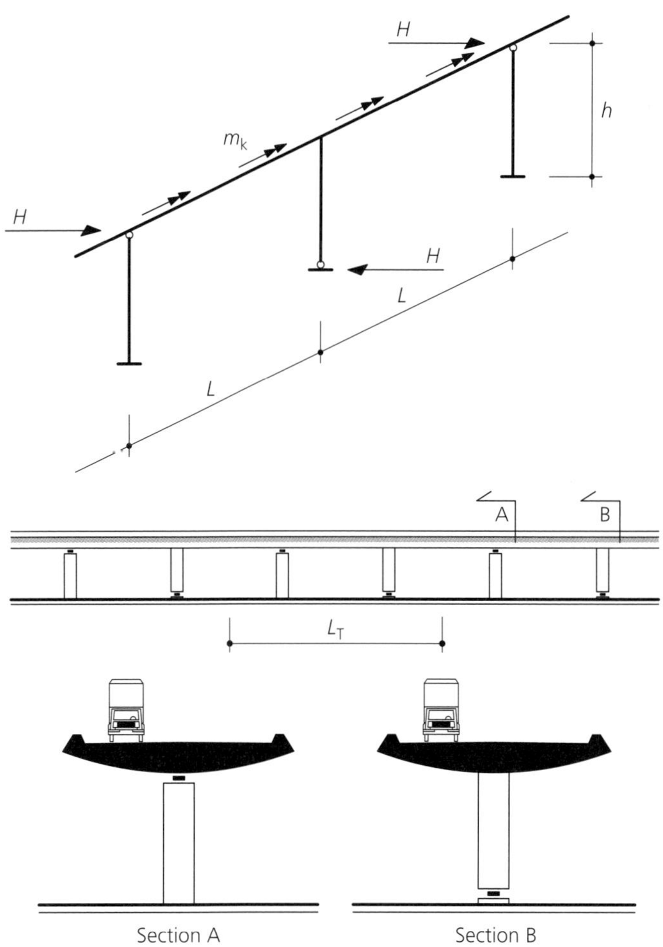

Figure 3.24 Structural arrangement (*d*)

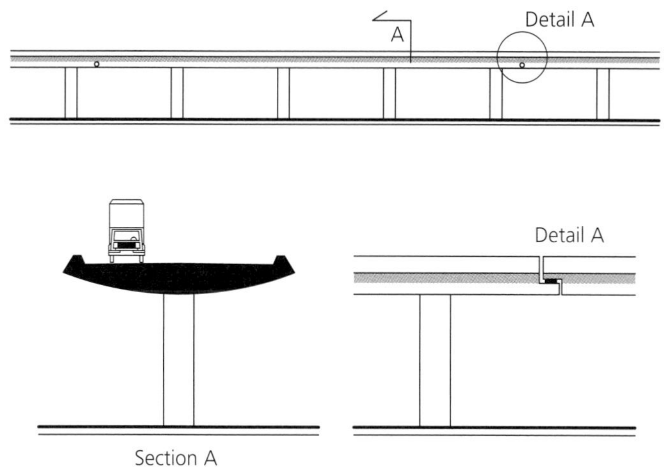

Figure 3.25 Steel hinge: (a) elevation; (b) plan

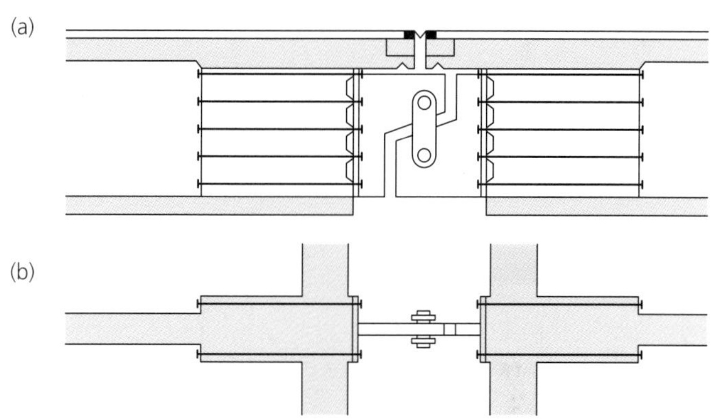

Figure 3.26 Structural arrangement (*e*)

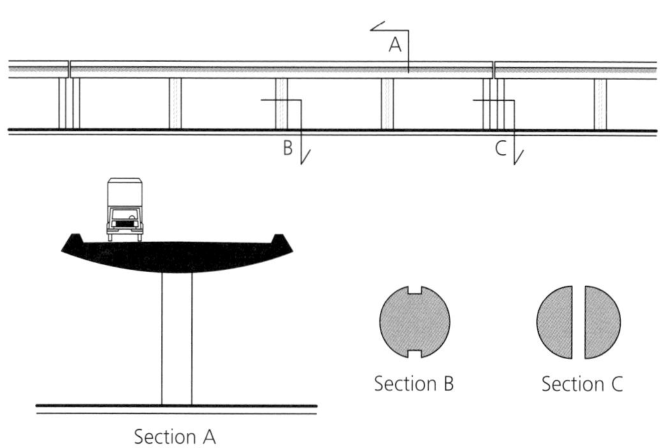

created by a pair of horizontal forces $M_H = Hh$ is resisted by a moment created by a pair of vertical forces $M_V = \Delta Va$. As $M_H = M_V$, the vertical forces are $\Delta V = (Hh)/a$.

Figures 3.28(d) and 4.76 show a tied arch structure called the 'Langer girder'. In this case, the arch is fixed into the prestressed concrete deck.

It is evident that the same arrangement can be used in design of inclined leg (strutted) frames. The bridges can form true frames (Figures 3.29(a) and 3.29(b)) or self-anchored frames (see Figures 3.29(c) and 4.13).

Buried arch structures can be also designed as true or tied arches. The tension ties from the prestressed concrete (stress ribbons) can be situated at the foundation level or above the arches. Then the rear struts connecting the arch footings to the end anchor blocks have to be added (Figure 3.30 and 4.3).

Figure 3.27 Structural arrangement (*f*)

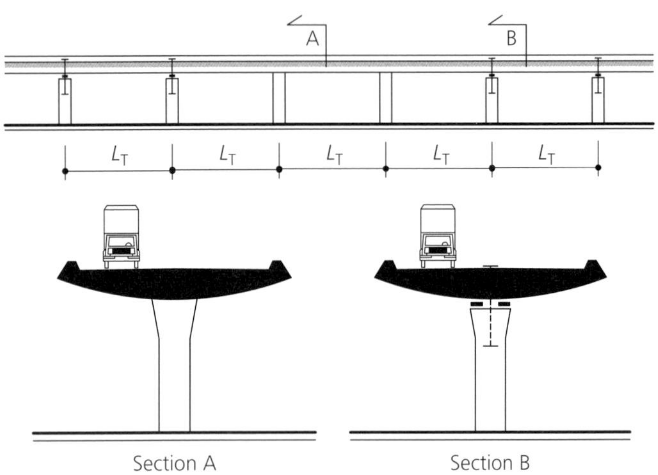

Figure 3.29 Inclined leg frame structures: (a) true frame; (b) true frame with rear legs; (c) tied frame

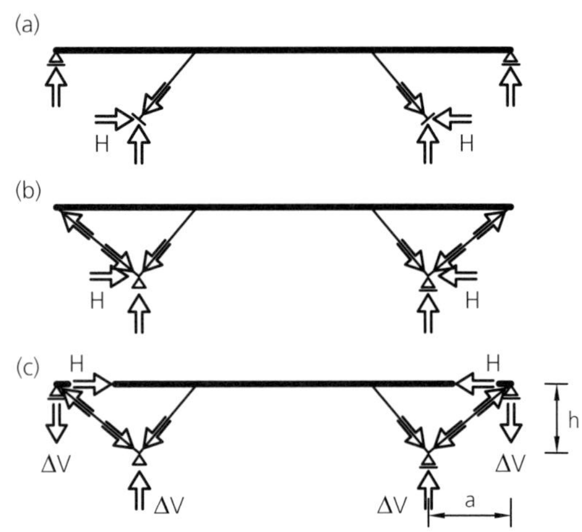

Figure 3.30 Buried arch structures: (a) true arch; (b) tied arch; (c) self-anchored arch

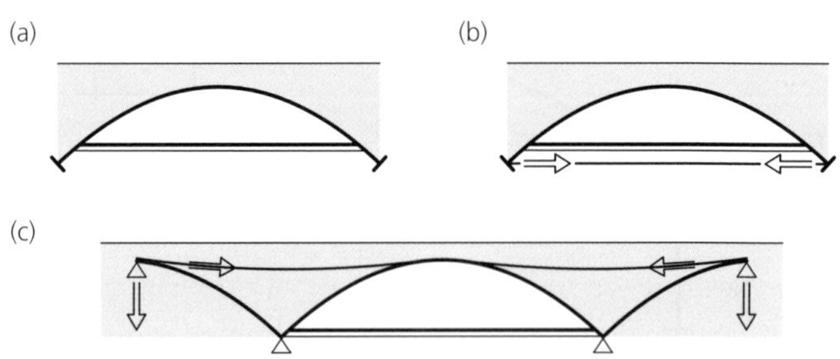

Figure 3.28 Arch structures: (a) true deck arch; (b) self-anchored deck arch; (c) through true arch; (d) through tied (Langer) arch

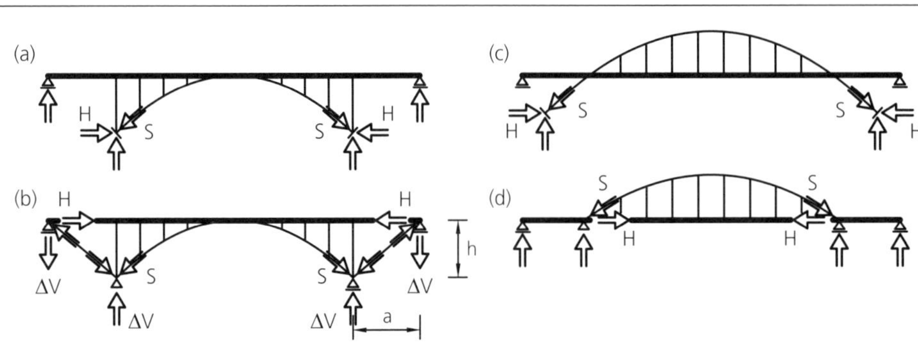

Figure 3.31 Deck arches

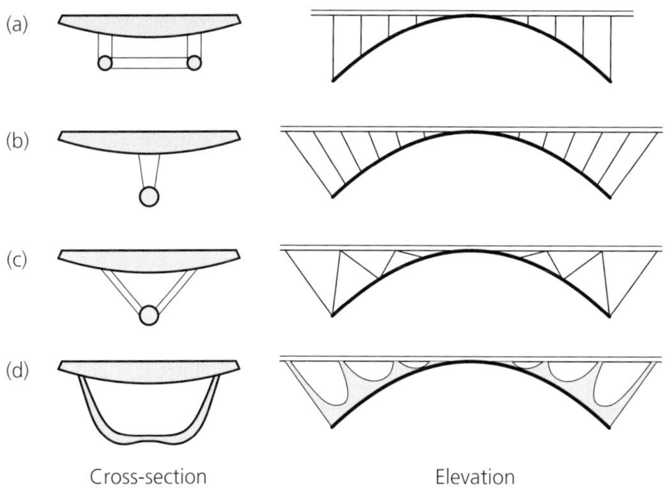

Deck arches (Figures 3.31, 3.28(a) and 3.28(b)) have concrete decks which are supported by struts that can be vertical or inclined, or can form a plane or space truss system (Figures 3.31(a)–3.31(c)). The arches can be formed by one or two concrete arch ribs of a solid or box section. They can be also formed by steel pipes or by shells. In the last option, the arches, together with struts, can form a space shell structure (Figure 3.31(d)).

Structures with concrete decks suspended on concrete or steel arches are structurally very efficient. The arches can be vertical or inclined, and they can be situated in the bridge axis or on the deck edges (Figures 3.32(a)–3.32(e)). While simple structures with vertical arches look boring and industrial, more complex structures with inclined arches have a modern, dynamic appearance. An inward inclination of the arches gives the user a feeling of safety, while an outward inclination gives a feeling of openness. The relatively small width of the decks of pedestrian bridges justifies the design of a diagonal arch arrangement (Figure 3.32(f)), or the suspension of one side of the deck (Figure 3.32(g)). The last solution is suitable for plan curved bridges.

The arches can be substituted by shells, on which the bridge decks are suspended. The suspension can be located on the bridge edges or in the bridge axis (Figure 3.33). Suspenders can be vertical or inclined (Figures 3.34(a)–3.34(c)). Structures in which the decks are suspended on arches via a net of diagonal cables are structurally very efficient (Figure 3.34(d)).

The bending stresses in the arch and in the deck of the arch structures depend mainly on the ratio of their bending stiffnesses. In the past, designs with a stiff arch and a slender deck (conventional arch with moment-resisting ribs; Figure 3.35(a)) or a slender arch with a stiff deck (stiffened deck arch; Figure 3.35(b)) were made to simplify the structural analysis. However, nowadays, with the availability of

Figure 3.32 Through arches

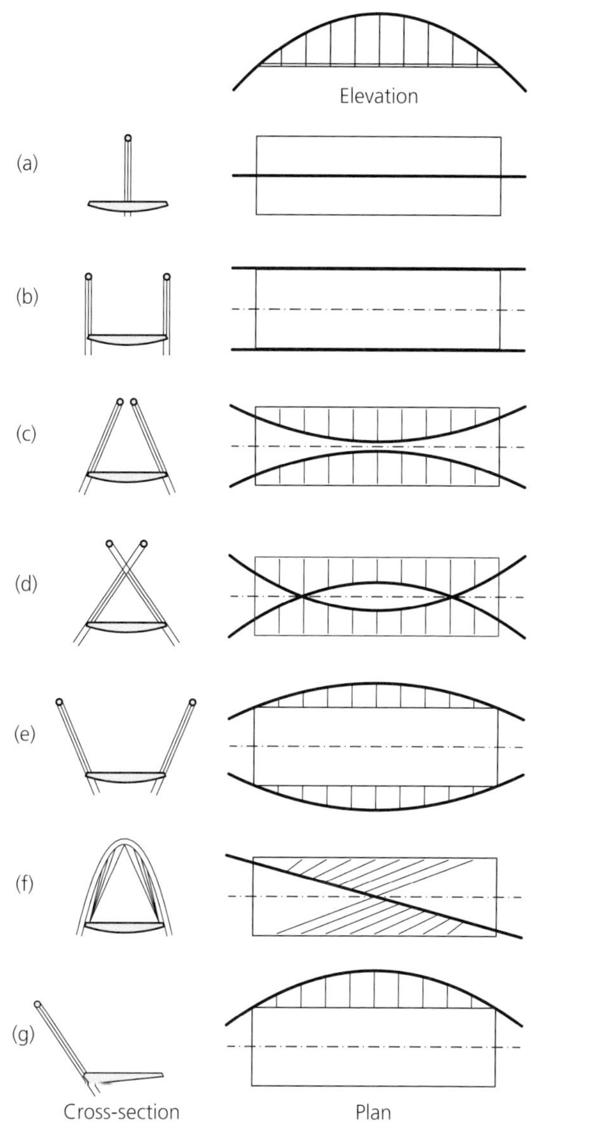

software that can be used to very efficiently solve complex structures, this division does not make sense, and it is possible to design arch structures with optimum stiffness of all the structural members.

3.5. Cable-supported bridges

Cable-supported bridges form a hybrid structural system in which the deck distributes the load and transfers it into a cable system (Figure 3.36). When we think about a cable-stayed structure, we assume that the cables are loaded only at their anchor points (Figures 3.36(a) and 3.36(b)). The stay cables can have different arrangements, which is discussed later. When we imagine a suspension structure, we usually assume that the cable is loaded at several points between the anchor points (Figures 3.36(c) and 3.36(d)).

Figure 3.33 Shell arches

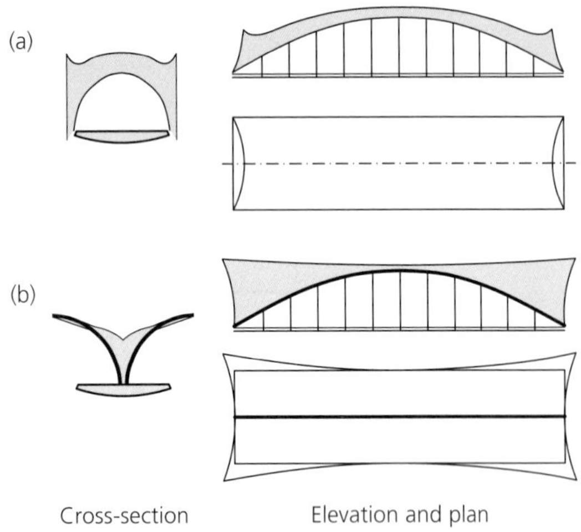

Figure 3.34 Suspender arrangements

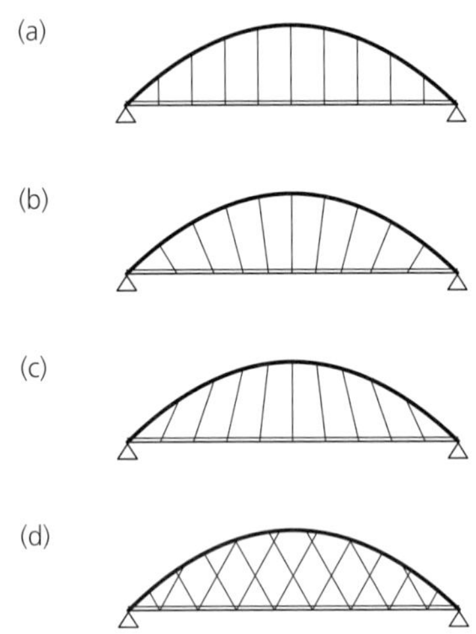

Figure 3.35 (a) Stiff arch; (b) slender arch

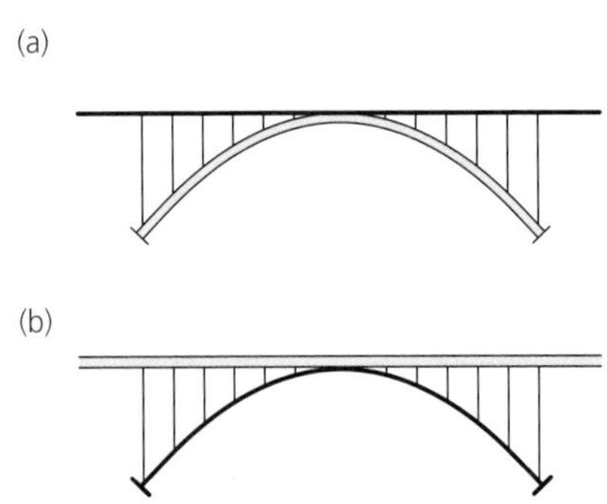

If the suspension cables are situated above the deck, the load is transferred from the deck into the cables by tensioned hangers. In structures with cables situated below the deck, the load is transferred from the deck into the cables by compression struts.

The stay and suspension cables can be anchored outside of the stiffening deck in anchor blocks that transfer the force from the cables into the soil (Figures 3.37(a) and 3.38(a)), or they can be anchored in the deck and create a so-called 'self-anchored system' (Figures 3.37(b) and 3.38(b)). In the latter case, the footings are stressed by vertical reactions only.

Typical arrangements of cable-supported structures with the cables situated above or below the deck are shown in Figures 3.37 and 3.38. It is also possible to combine above- and below-deck suspension and stay cables. Furthermore, systems with suspension or stay cables anchored in anchor blocks can also include cables anchored in the deck. These possibilities are discussed in greater detail in Stráský (2009).

The decision about which type of structure can be used is strongly influenced by the construction possibilities. Figures 3.39(a) and 3.39(b) show a cable-stayed bridge structure suspended on a single pylon. While for the first structure the back-stay cable is anchored in the anchor block, at the second structure the back-stay cable is anchored in the side span. It is evident that the first system is considerably uneconomical, as the footing and deck end diaphragm of the back-stay have to resist large horizontal forces. In the second structure, which forms an economical self-anchored structural system, the horizontal component of the back-stay cables is resisted by the compression capacity of the concrete deck, and therefore the footings are stressed by vertical forces only.

Figure 3.36 Cable-supported structures: (a, b) cable-stayed; (c, d) suspension

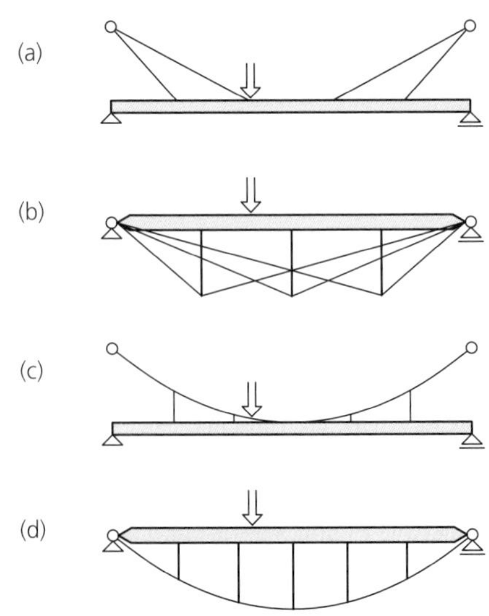

Figure 3.37 Cable-stayed bridges: (a) earth-anchored; (b) self-anchored

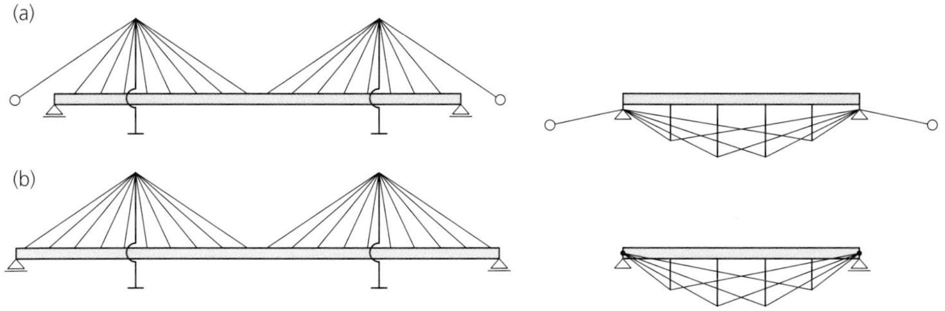

Figure 3.38 Suspension bridges: (a) earth-anchored; (b) self-anchored

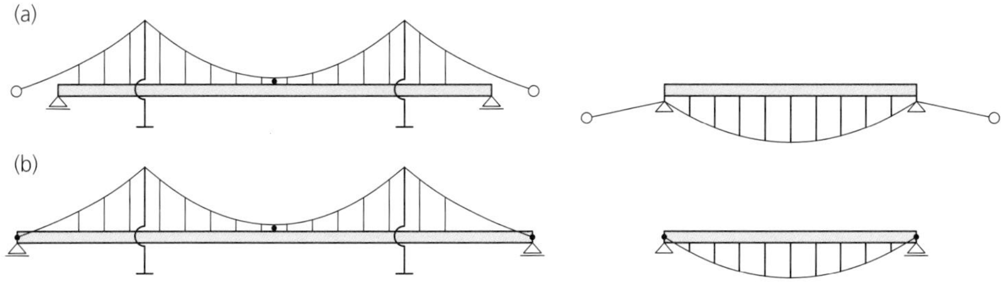

Figure 3.39 Cable-supported structures: (a) earth-anchored cable-stayed; (b) self-anchored cable-stayed; (c) earth-anchored suspension; (d) self-anchored suspension

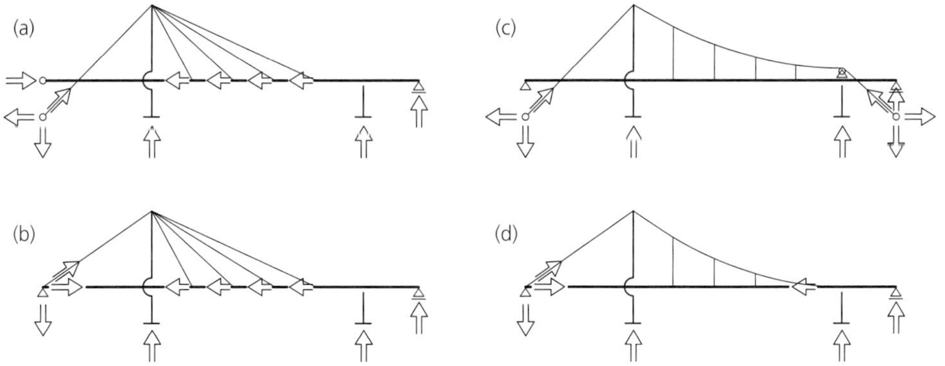

Similarly, Figures 3.39(c) and 3.39(d) show a suspension structure suspended on a single pylon. Although in the case of the earth-anchored suspension structure the cable anchor blocks have to resist large horizontal forces, self-anchored structures are rarely used. The reason for this is that, in this case, the deck has to be erected first. Conversely, the deck of the earth-anchored suspension structure can easily be suspended on the suspension cables, which are erected first.

3.5.1 Cable-stayed bridges

Classical cable-stayed structures have two or three spans suspended on one or two towers (Figure 3.40). The towers can be vertical or inclined. A multi-span cable stayed structure has been also built. In these structures, special attention has to be devoted to the deflection of the deck and to the analysis of the dynamic response.

Figure 3.40 Cable-stayed structure arrangements

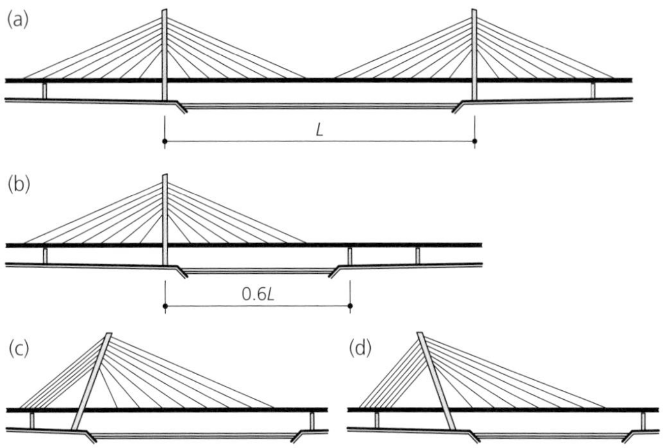

Figure 3.41 Cable-stayed structure – length of the side spans

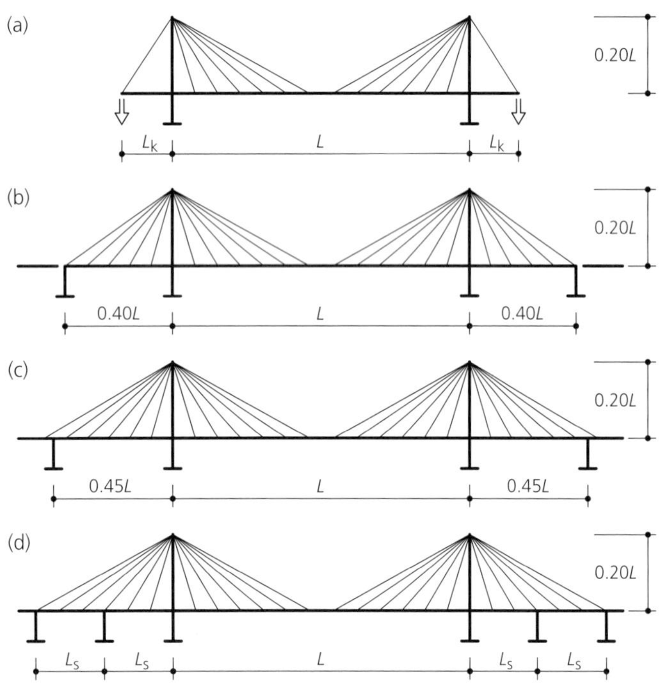

It is possible to suspend only the main span and anchor the back-stays at the short side spans or anchor blocks (Figure 3.41(a)). The optimum span length of the side spans is 0.40–0.45 of the length of the main span. The length depends on the arrangement of the stay cables and the side supports. If the deck has an expansion joint between the side and approach spans, the structure requires shorter side spans and anchoring of back-stays (Figure 3.41(b)).

If the structure is continuous, the side spans can be longer, and the stay cables are symmetrical to the tower and extend into the approach spans (Figure 3.41(c)). The cable-stayed structure can be significantly stiffened by supports situated in the back-spans (Figure 3.41(d)).

The stay cables can have different arrangements, as shown in Figure 3.42. While the statically superior radial arrangement (Figure 3.42(a)) brings structural difficulties associated with the anchoring of all the stays at one point, the statically less advantageous parallel arrangement (Figure 3.42(b)) simplifies structural details. The semi-radial arrangement (see Figure 3.42(c)) represents a reasonable compromise from both a static and structural point of view.

The relatively light load experienced by pedestrian bridges justifies the solution illustrated in Figures 3.42(d) and 3.42(e). The basic arrangements can be combined to create the most suitable solution (Figure 3.43).

To reduce the bending moments in the deck it is necessary to anchor the stay cables at a relatively short distance (3–6 m). If the bending moments due to the dead load are very low, then the deck can be very slender (from 0.30 m).

If the towers have sufficient height ($0.2L$), the stiffness of the cable-stayed structure is given by the stiffness of the system formed by the compressed masts and the deck and by the tension stiffness of the

Figure 3.43 Combination of the basic stay arrangements

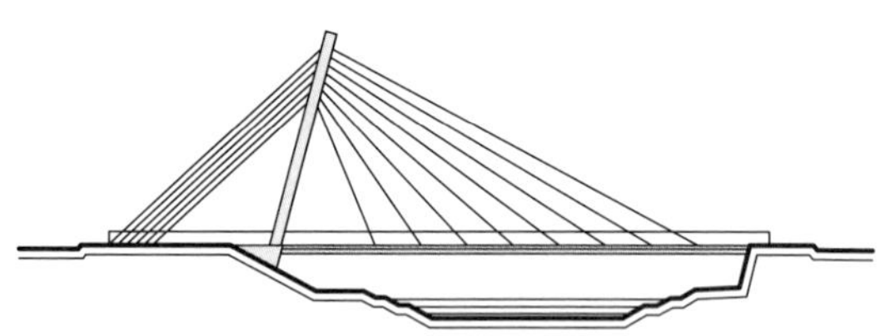

Figure 3.42 Typical stay arrangements

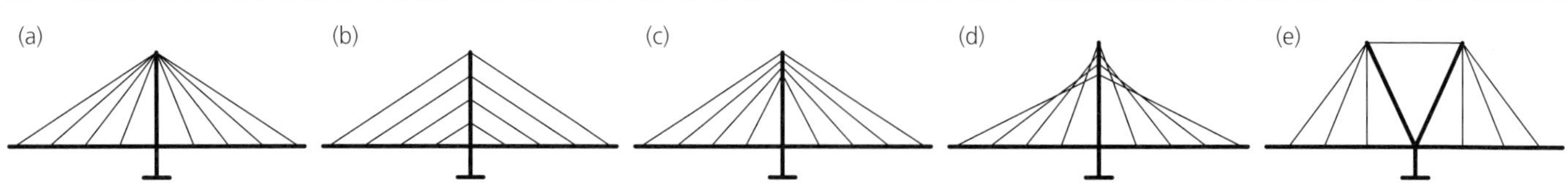

stay cables (Figure 3.44(a)). If low towers are designed, the structure requires sufficient bending stiffness of the deck (Figure 3.44(b)). This system is referred to as 'extradosed'.

The towers can be formed by individual columns, or they can be H, V or A shaped. The possible arrangements of the towers in both the transverse and the longitudinal directions is shown in Figure 3.45.

The deck of highway bridges is usually suspended on two planes of cables – vertical or inclined. Suspension in the bridge axis requires a torsionally stiff girder. In cable-stayed structures of common spans, the concrete deck guarantees the transverse stiffness. Although an outward inclination increases the transverse stiffness, and an inward inclination increases the torsional stiffness of the system, these increases are not significant. However, structures of this type do not vibrate in purely transverse modes. Rather, all the transverse modes are accompanied by a distortion of the deck, which contributes to the stiffness of the system.

Cable-stayed structures do not usually utilise the increase in the stiffness of the structural system due to the anchoring of the deck at the abutments. However, the advantage of restricting the horizontal movement at the abutments was demonstrated by Professor Menn in the design of the Sunniberg Bridge in Klosters-Serneus, Switzerland (Vogel and Schellenberg, 2015). The bridge has five spans with lengths in the range 59–134 m and the radius of the plan curvature is $R = 503$ m. The slender deck, which has a depth of only 1.05 m, runs at 60 m above grade and is suspended on pylons protruding 15.02 m above the deck. The stiffness of the structure comes from the plan curvature of the deck, which is fixed at the abutments (Figures 3.46 and 3.47). While in traditional multi-span cable-supported structures the vertical deflection of the deck has to be controlled by intermediate stiff piers or by the bending stiffness of the deck, in this bridge the vertical deformation of the deck is controlled by the transverse stiffness of the curved deck. Any vertical load causes a horizontal movement of the deck that acts in the horizontal plane as an arch. The transverse movement of the deck creates transverse moments in the piers, forming transverse frames.

The possible arrangements of structures supported by stay cables are shown in Figure 3.48. However, an elegant solution with one upright (see Figure 3.48(a) and 3.49) (Pedrazzini *et al.*, 2005) requires a relatively stiff deck; a slender deck with several uprights (Figure 3.48(b)) is, compared with a self-anchored suspension structure (see Figure 3.36), too complicated.

Figure 3.44 (a) Cable-stayed and (b) extradosed structures

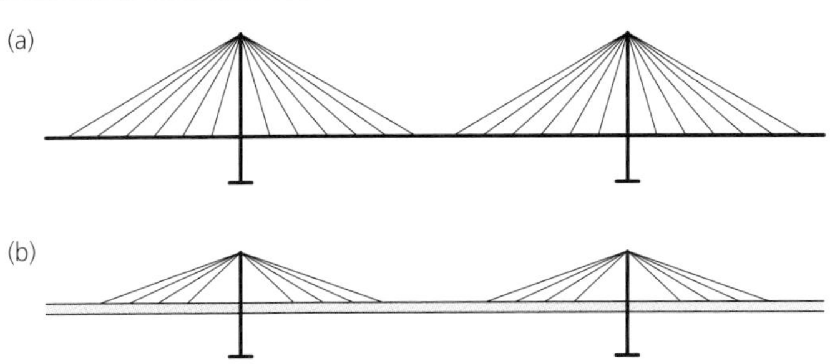

Figure 3.45 Pylon arrangements: (a) partial bridge elevations; (b) cross-sections

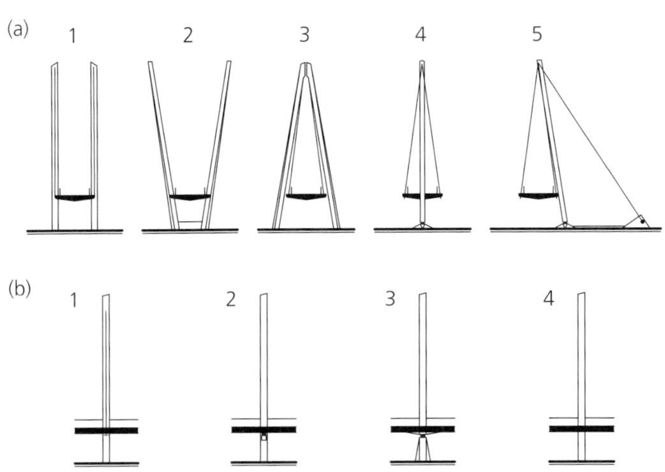

Figure 3.46 Sunniberg Bridge, Klosters-Serneus, Switzerland: (a) bridge cross-section; (b) partial bridge elevation; (c) deck cross-section

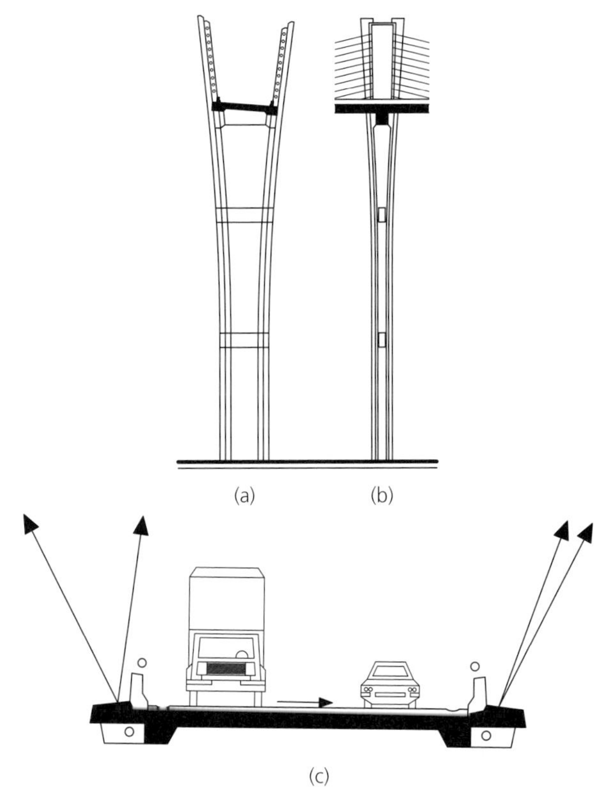

Figure 3.47 Sunniberg Bridge, Klosters-Serneus, Switzerland

Figure 3.48 Cable-supported structures: (a) single upright; (b) several uprights

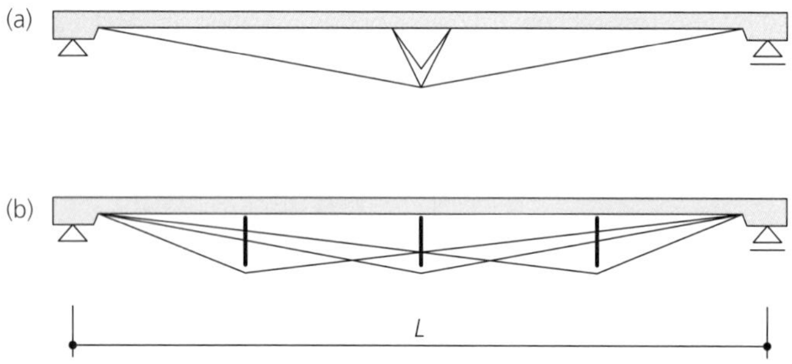

Figure 3.49 Acla Chuoz Bridge, Samedan, Switzerland

3.5.2 Suspension bridges

Suspension bridges usually have steel or steel–concrete composite decks. For medium spans cable-stayed bridges are more economical, so concrete decks have been used only in the construction of several suspension highway bridges. However, many interesting pedestrian bridges with concrete decks have been built. Earth-anchored structures have been constructed for longer spans, while for medium spans both earth- and self-anchored structures that utilise the concrete compression strength have been commonly used.

The suspension cables can be situated above the deck, below deck, or above and below the deck (see Figure 3.38). The suspension cables have a funicular shape, and they balance the effects of the self-weight of the structure and guarantee that the structural members are stressed by normal forces only. For service load, the suspension structure forms a complex system, in which the deck distributes the load and all the structural members contribute to the resistance of the structural system.

Classical suspension structures in which the cables are situated above the deck are usually suspended on one or two towers and have from one to three spans. The side spans can be completely or partially suspended on the suspension cables. The length of the suspended side span is usually less than half the length of the main span (Figure 3.50).

If the deck is suspended in the bridge axis, it has to be torsionally stiff, while if it is suspended on the outer edges it can be very slender. To increase the stiffness of the structural system, the suspension cables should be fixed to the deck at the midspan and the stoppers (shock transition units) should be installed at the end abutments (Figures 3.51(a) and 3.52). Similar effects can be achieved by designing the suspension as a cable net (Figure 3.51(b)).

Similar to cable-stayed structures, the towers of suspension bridges can be formed by individual columns, or they can be H, V or A shaped. The possible arrangements of the towers are shown in Figure 3.45.

Earth suspension bridges are usually assembled from precast members (segments) that are suspended on suspension cables. During erection the precast segments are mutually connected by pins, so the suspension cables automatically have a funicular shape to the given load. When all the segments have been erected, the joints between them are cast and the deck is prestressed.

Self-anchored structures (Figure 3.53) (Stráský, 2011) are usually cast in place on the falsework. The self-weight is transferred into the suspension cables by their post-tensioning, which can be done by jacking at their anchors or by lifting the towers. This operation requires a careful determination of the camber of the deck and the non-tension length of the cables. Use of the system shown in Figure 3.36(d) in suspension structures allows the design of aesthetically pleasing and structurally very efficient bridges (Figure 3.54) (Stráský, 2011).

Figure 3.50 Suspension structures

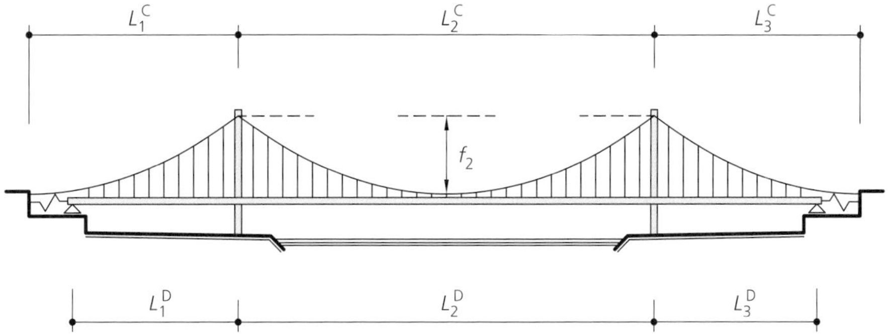

Figure 3.51 Stiffening of suspension structures

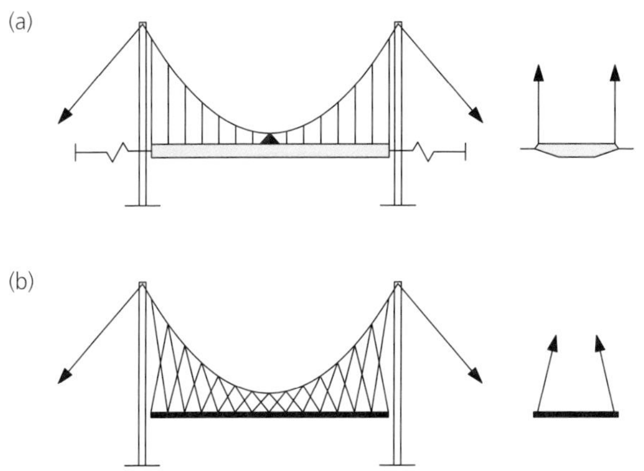

Figure 3.53 Harbor Drive pedestrian bridge, San Diego, CA, USA

Figure 3.52 Vranov Lake pedestrian bridge, Czech Republic

Figure 3.54 Inachus Bridge, Beppu, Japan

3.6. Stress-ribbon bridges

Stress-ribbon bridges (Figure 3.55) are a special type of suspension structure (Stráský, 2011). The term 'stress-ribbon bridge' has been coined to describe structures formed by a directly walked prestressed concrete deck in the shape of a catenary. The bearing structure consists of slightly sagging tensioned cables, bedded in a concrete slab that is very thin compared with the span. The characteristic feature of stress-ribbon structures is a variable slope, which disqualifies the use of this structural type for highway bridges. On the other hand, the variable slope is acceptable for pedestrian bridges.

Stress-ribbon structures can be designed with one or more spans, which can be either cast in situ or formed of precast units. In the case of precast structures, the deck is assembled from precast segments that are suspended on bearing cables (Figure 3.56(a)). Prestressing is applied after casting the joints between the segments to ensure sufficient rigidity of the structures (Figure 3.56(b)).

Due to the limited longitudinal slope, the sag of the structure is small, and therefore the tension forces in the cables is very large, making the anchoring of the cables very expensive. To make stress-ribbon structures more economical, they can be stiffened using suspension cables with a large sag. The cables can be situated either above or below deck (Figure 3.57).

Another possibility is to use arches to stiffen the structure, and thus create a self-anchored structural system (Figures 3.58 and 3.59) (Stráský, 2011).

Figure 3.55 Stress-ribbon bridge in Redding, CA, USA

Figure 3.56 Static function: (a) erection; (b) in service

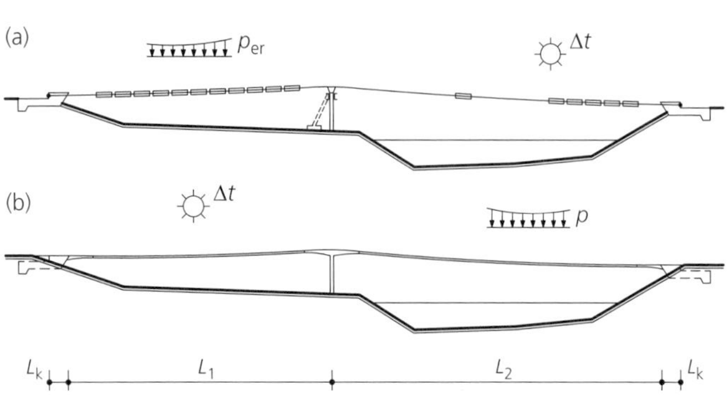

Figure 3.57 Stress-ribbon structure stiffened by cables: (a) cables above the deck; (b) cables below the deck

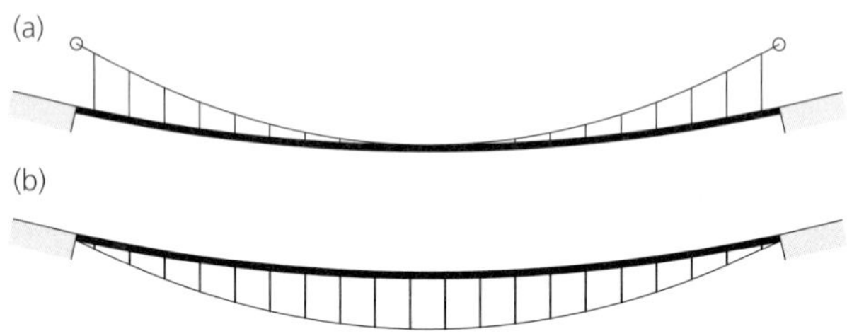

Figure 3.58 Stress-ribbon and arch structures

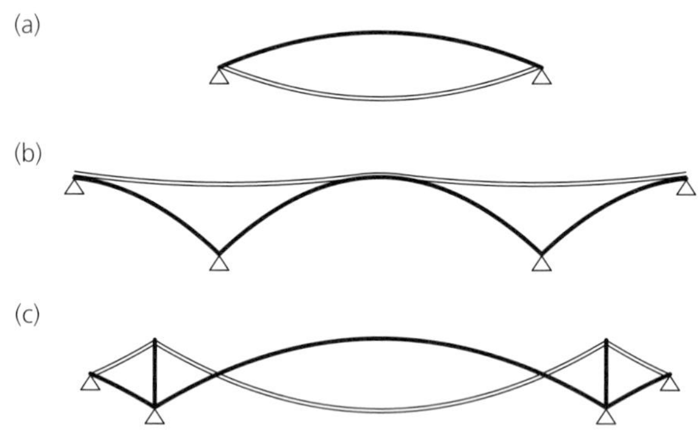

Figure 3.59 Bridge across the D35 motorway near Olomouc, Czech Republic

REFERENCES

Barlet E, Causses G and Viallon J-P (2002) A new step forward for composite bridges. The Bras de la Plaine Bridge, Réunion, France. Concrete Structures in the 21st Century. *Proceedings of the First fib Congress, Osaka, Japan.* fib, Lausanne, Switzerland.

Giacomelli D and Grennerat Y (1989) A40 – Les viaducs de Glacières et de Sylans. *Travaux* **645**: 105–114.

Kasuga A (2017) Butterfly web effects on bridge design. *Structural Concrete* **18(1)**: 128–142.

Leonhardt F (1979) *Vorlesungen über Massivbau. Grundlagen des Massivbrückenbaues.* Springer, Berlin, Germany.

Leonhardt F (1984) *Bridges. Aesthetics and Design.* Deutsche Verlags-Anstalt, Stuttgart, Germany.

Liebenberg A (1992) *Concrete Bridges – Design and Construction.* John Wiley, New York, NY, USA.

Maeda Y, Agawa K, Fujiki Y and Miura Y (2002) Design and construction of the Maetani Bridge with corrugated steel webs and entirely external tendons. Concrete Structures in the 21st Century. *Proceedings of the First fib Congress, Osaka, Japan.* fib, Lausanne, Switzerland.

Manterola J (2006) *Puentes, Apuntes para su diseño, cálculo y construcción*, vols I and II. Colección Escuelas, Madrid, Spain.

Mathivat J (1983) *The Cantilever Construction of Prestressed Concrete Bridges.* John Wiley, New York, NY, USA.

Menn C (1990) *Prestressed Concrete Bridges.* Birkhäuser, Basel, Switzerland.

Pedrazzini A, Pedrazzini E, Baserga N and Mozzetti C (2005) Drei neue Flazbrücken in Samedan. Auβergewöhnliche Aufgabenstellung für einen Wettbewerb. *Beton- und Stahlbetonbau* **100(3)**: 246–252.

Podolny W and Muller J (1982) *Construction and Design of Prestressed Concrete Bridges.* John Wiley, New York, NY, USA.

Smith DA, Hewson NR and Hendy CR (2009) Design of the Dubai Metro light rail viaducts – superstructure. *Proceedings of the Institution of Civil Engineers – Bridge Engineering* **162(2)**: 55–62.

Stráský J (2009) Cable-stayed bridge across the Odra River, Czech Republic. *26th Annual International Bridge Conference 2009: Meeting Bridge Challenges in Challenging Times*, pp. 352–360.

Stráský J (2011) *Stress Ribbon and Cable Supported Pedestrian Bridges*, 2nd edn. Thomas Telford, London, UK.

Troyano LF (2003) *Bridge Engineering. A Global Perspective.* Thomas Telford, London, UK.

Veterans' Glass City Skyway (2018) *Ohio Department of Transportation.* 9 Aug. https://en.wikipedia.org/wiki/Veterans%27_Glass_City_Skyway.

Vogel TH and Schellenberg K (2015) The impact of the Sunniberg Bridge on structural engineering, Switzerland. *Structural Engineering International* **25(4)**: 381–388.

ice Publishing

Stráský and Nečas
ISBN 978-0-7277-6385-3
https://doi.org/10.1680/dcpb.63853.047
ICE Publishing: All rights reserved

Chapter 4
Bridge types

The architectural and structural solution for a bridge should emerge from and is dictated by the basic function that the bridge is to perform. The function of a bridge is to bridge over some particular space in order that some load or traffic may be conveyed over a natural obstacle or another man-made thoroughfare, and the form of the bridge must express this basic function. The best structural solution should be in some way inherent in the site and is the solution that best fulfils the function of bridging the site. The task of the structural designer is to discover a form that can be realised in a way that is both economical and efficient (Leonhardt, 1984; Menn, 1990; Troyano, 2003).

In this chapter the bridge types that in the authors' opinion fulfil the above requirements are discussed. It is obviously not possible to present all the possibilities, and the selection of the structures discussed is that of the authors. As it is difficult to obtain information about the structural arrangement of bridges designed by other designers, in many cases the structures presented are structures designed by the authors' design office.

Prestressed concrete structures are only structurally efficient for spans longer than 15 m. Therefore, short-span structures are not discussed here. The goal of any design should be to develop a structure that has the minimum possible impact on the environment, and therefore the superstructures should be light and transparent, and abutments and wings should have the minimum possible dimensions. As longer spans requires deeper decks, it is necessary to justify the use of longer spans.

Figure 4.1 (a, b) Buried and (c–e) one-span bridges

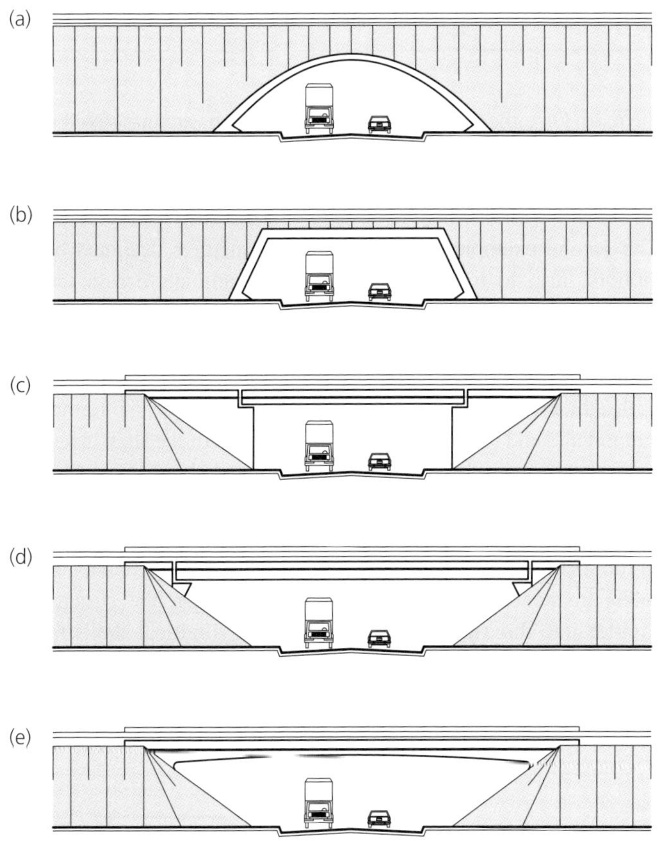

4.1. Buried and one-span bridges

Due to their natural shape, buried arches provide the best fit in the environment, and therefore these should be designed whenever possible (Figure 4.1(a)). The arches can have a simple solid cross-section of a variable depth. As concrete spandrel walls are highly visible, it is reasonable to avoid them, and to design arches of variable width that follow the slope of the highway fill. The problem of resisting arch forces can be solved by using the ties situated under the crossing highway or above the structure. Because it is not possible to check the function of the ties, it is reasonable to design them very conservatively. The author has designed fully prestressed concrete ties with prestressing formed of monostrands grouted in polyethylene ducts.

As examples of the possible solution, two structures are presented here. The first is the Cannstatter Street pedestrian bridge, in Stuttgart, Germany, built in 1977 according to the design by Professor Schlaich (Figure 4.2) (Schlaich and Bergermann, 1992). The bridge is formed by a thin shell arching over a span of 51.20 m, with a thickness of only 400 mm. To increase its stiffness and strength, the longer edges of the shell are folded slightly upward. Thus the structure consists of three curved planes joined along their sides. The width of the central surface is 6 m at the crown, splaying out to 11 m at the abutments. The arch is covered with earth, which forms the walking surface and helps stabilise its shape. The horizontal force of the arch is resisted by prestressed concrete ties buried under

Figure 4.2 Cannstatter Street pedestrian bridge, Stuttgart, Germany: (a) cross-section at the midspan, (b) cross-section at a quarter of the span, (c) elevation

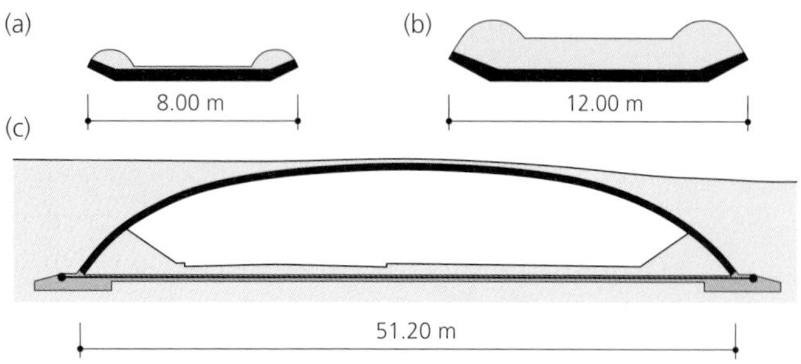

the road. On either side of the walkway, troughs are integrated in the folded-up part of the shell, and plants have been encouraged to grow over the sides.

The second example is the two buried structures designed by the authors' firm to transfer pedestrians and animals across the D1 motorway in the Czech Republic (Figures 4.3–4.5) (fib, 2010). The bridges across the D1 motorway are formed by two-span flat shells with spans of 25.00 m and a rise of 3.30 m. The cylindrical shells, with a thickness of 0.550–0.365 m, are supported by an intermediate support and end footings. The horizontal force of the shell is resisted by prestressed ties – stress ribbons are situated above the shells. The ties are anchored in anchor blocks that are connected to the shell footings by compressed struts. The anchor blocks, the end footings of the shell and the intermediate supports are all supported by drilled piles. In the transverse direction of the bridge, the structure is divided into three parts. The edges of the cylindrical shells follow the slope of the fill, and therefore the side shells have variable width.

Figure 4.4 Bridge across the D1 motorway, Czech Republic

Figure 4.5 Bridge across the D1 motorway, Czech Republic

Figure 4.3 Bridges across the D1 motorway, Czech Republic: (a) cross-section of the tie; (b) prestressing cable and monostrand; (c) elevation

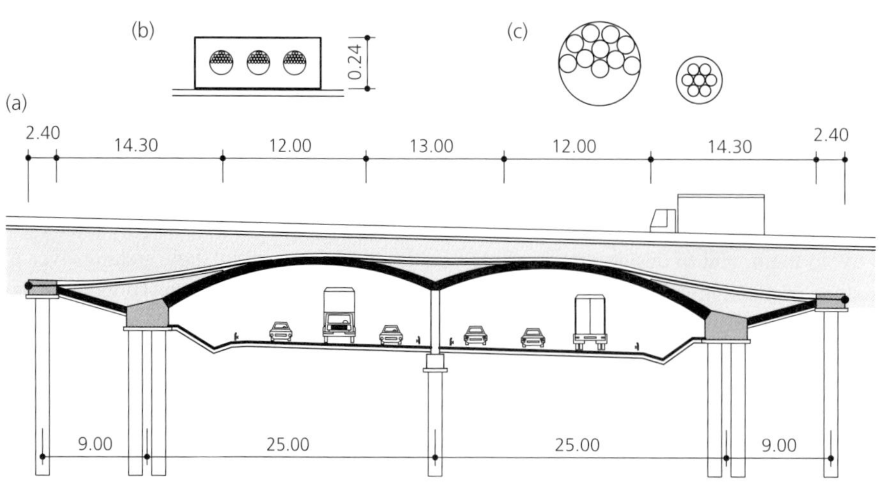

Figure 4.6 A bridge in the city of Bend, Oregon, USA

Figure 4.7 Motorway overpasses

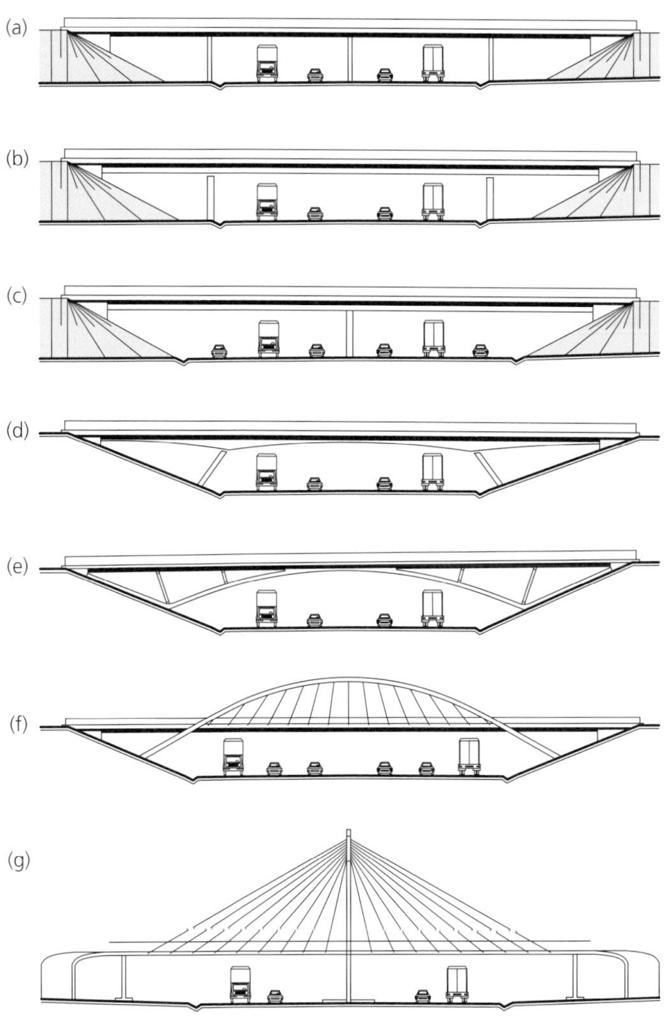

The total width of the bridge is 60.30 m at the crown of the shell, 64.10 m above the central pier and 86.20 m at the footings.

Buried frame structures can be designed similarly (see Figure 4.1(b)). As the decks are stressed by bending, they are usually longitudinally prestressed. Short-span structures usually have decks assembled of precast beams and a composite deck slab. Unfortunately, in many cases the decks are supported by large abutments (see Figure 4.1(c)). However, the authors are convinced that the abutments should be as small as possible (see Figure 4.1(d)) and can be integrated with the deck. In this way they can form elegant frame structures of variable depth (Figure 4.1(e) and 4.6).

4.2. Motorway overpasses

The design of overpasses is based on the nature of the transferred road and the corresponding crossing angle. Four-span slab decks are suitable for local roads that cross the highway at an angle of up to 60° (Figures 4.7(a) and 4.8). The bridge deck is formed either as a trapezoidal slab (Figure 3.5(a)) or as a slab stiffened by wide beams (Figure 3.5(b)). The end diaphragms can either be supported by bearings or can be connected to the frame with a substructure and form, so-called 'integrated structures'.

If it is not possible to construct a pier in the motorway median, three-span structures are usually designed (Figures 4.7(b) and 4.9). As the

Figure 4.8 Bridge across the D1 motorway, Czech Republic

Figure 4.9 Bridge across the D1 motorway, Czech Republic

Figure 4.10 Overpass across a rural road crossed by the D1 motorway, Czech Republic

length of the central span is about 30 m, the bridge deck is formed by beam and slab structures, which are composed of one or more longitudinal beams. The beams have solid or box cross-sections.

A two-span structure with the central pier situated in the motorway median (Figure 4.7(c)) represents a very economical solution. Such structures are suitable for bridges situated at junctions, as the free space under the bridge ensures a safe view into the intersection.

Almost maintenance-free two-span integral structures have been built as overpasses to carry rural roads across the D1 motorway near the city of Olomouc in the Czech Republic (Figure 4.10) (FIP, 1998). The bridge decks of the channel cross-section are connected to the frame by V-shaped central piers, and terminate in end diaphragms supported by drilled piles (Figure 4.11). To reduce the bending stresses originating in the piles due to changes in the concrete volume, the diaphragm piles are weakened at their top portion. The edge girders have the shape of a New Jersey barrier, and the railings are formed by single pipes.

Figure 4.11 Bridge shown in Figure 4.10: (a) cross-section; (b) elevation

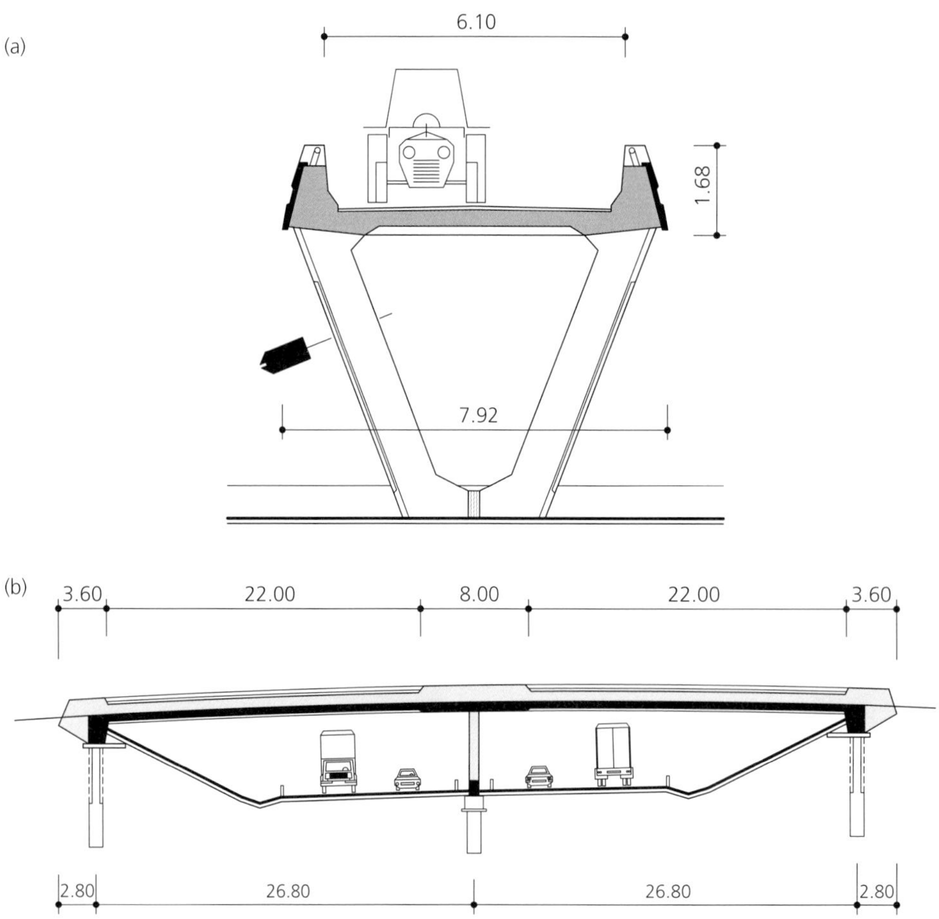

Figure 4.12 Motorway bridge near Olomouc, Czech Republic

An elegant solution is a strutted frame bridge (see Figure 4.7(d)). If a self-anchored structure is used, the footings resist only the vertical forces, and therefore it can be incorporated in designs of structures to be located in less than ideal geotechnical conditions.

Figures 4.12 and 4.13 show an overpass with span of 44.80 m, which was constructed across the D35 motorway near the city of Olomouc in the Czech Republic (fib, 2010). The deck is composed of three spans (lengths 20.10 + 30.00 + 19.50 m) and is formed of a spine girder with large overhangs. The deck has a variable depth in the range 2.16–1.00 m, and has end diaphragms that also serve as abutments. The main span is supported by front inclined struts assembled from two precast members, which are connected to the deck by hinges and fixed into the footings. The end diaphragms are connected to the footings by precast end struts. Both the footings and end abutments are supported by drilled piles of 0.90 m diameter.

Figure 4.13 Bridge in Figure 4.12: (a) cross-section; (b) elevation

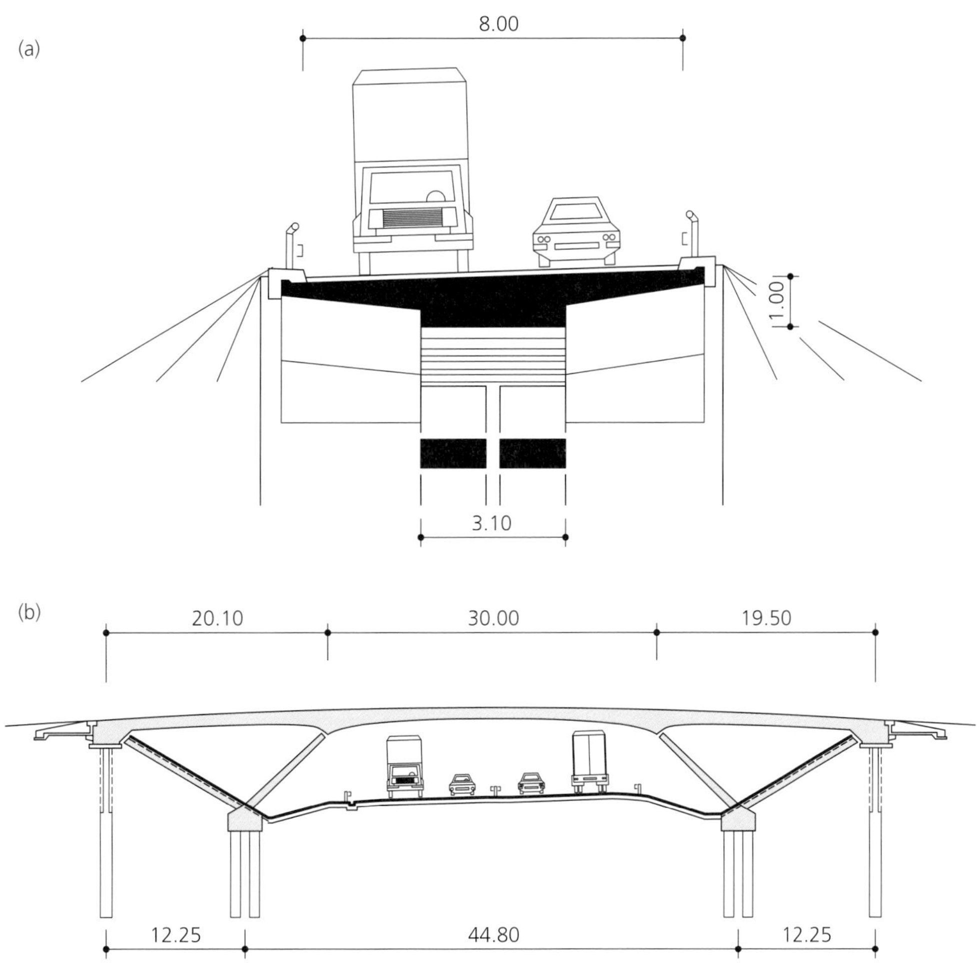

If a bridge crosses a motorway at a greater height than the clearance requirements, an arch structure is suitable (see Figure 4.7(e)). Also, as stated above, the arch structure can form a self-anchored structural system that can be used in less than ideal geotechnical conditions.

Figure 4.14 Motorway bridge near Studénka, Czech Republic

Figures 4.14 and 4.15 show an arch structure built across the D1 motorway near the city of Studénka in the Czech Republic (fib, 2010). The bridge has a deck with a trapezoidal cross-section and a depth of only 0.50 m, which is supported by a flat arch with a span of 53.70 m and a rise of 6.157 m. To obtain a similar span length for the deck, the pendulum piers, transferring the load from the deck into the arch, are inclined. The deck is fixed into the end diaphragms, which also serve as abutments. The arch has a circular shape with a radius of 65.40 m. The 0.60 m thick arch is very narrow, its width being only of 2.50 m. In the middle of the bridge the arch penetrates into the deck and creates an integral connection. Both the footings and the end abutments are supported by drilled piles of 0.90 m diameter, which are arranged in a similar way as described above for other structures.

If an arch bridge crosses a motorway at a skew angle, the arch should be designed to be as narrow as possible. Therefore, an arch formed by a steel pipe filled with concrete represents an optimum solution, and this solution has been applied on the crossing over the D52 motorway near the city of Rajhrad in the Czech Republic (Figure 4.16) (Strásky and Husty, 1999). The bridge is formed of a prestressed concrete deck with a channel cross-section that is supported via steel V-shaped struts by a steel arch (Figure 4.17). The arch is formed by a 900 mm

Figure 4.15 Bridge in Figure 4.14 at (a) cross-section at the midspan; (b) cross-section at a quarter span; (c) elevation

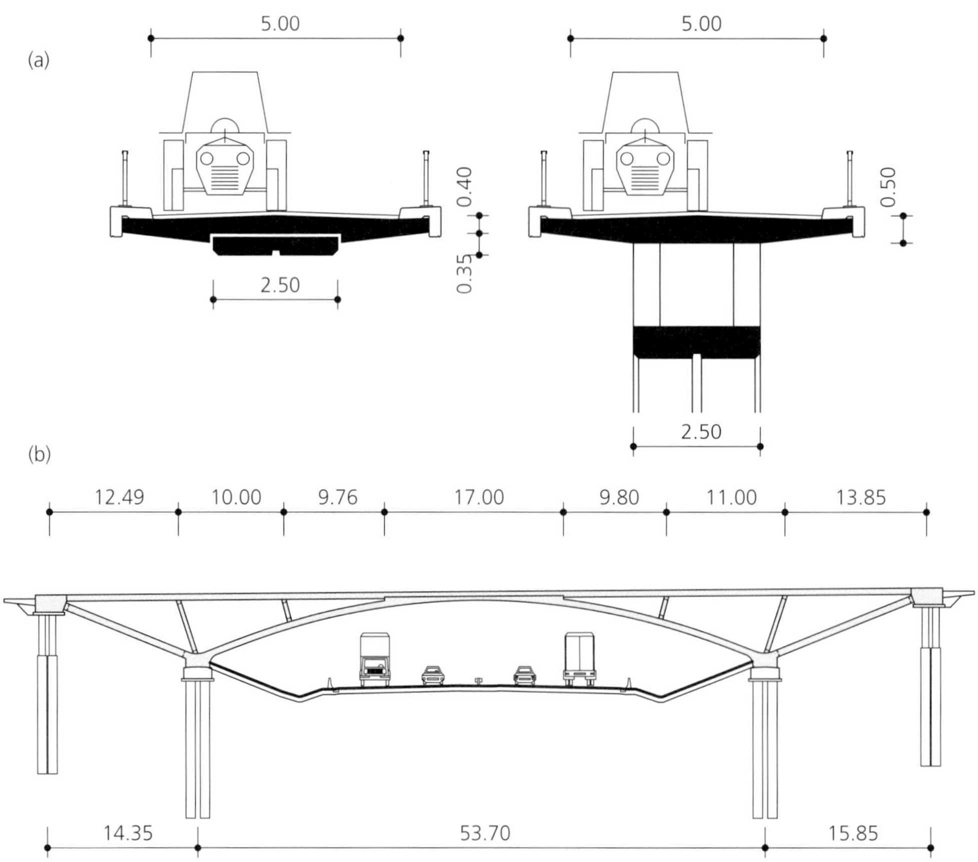

diameter steel pipe infilled with concrete. The arch has a span of 67.50 m and a rise of 8.05 m, and is circular in shape with a radius of 74.75 m. The struts are perpendicular to the arch. The end diaphragms of the deck are supported by concrete struts, which are fixed into the arch footings and transfer the load from the structure into the soil (see Chapter 8, Figures 8.3 and 8.4).

Figure 4.16 Motorway bridge near Rajhrad, Czech Republic

If the bridge crosses not only a motorway but also the junction ramps, a through arch structure crossing all obstacles represents a very suitable solution (see Figure 4.7(f)). It can be also constructed as a self-anchored integral structure. This solution has been used for the crossing over the R1 expressway near the city of Nitra in Slovakia (Figure 4.18) (Stráský *et al.*, 2013). The bridge is composed of

Figure 4.18 Expressway bridge near Nitra, Slovakia

Figure 4.17 Bridge in Figure 4.16: (a) cross-section at the midspan; (b) cross-section at a quarter span; (c) elevation

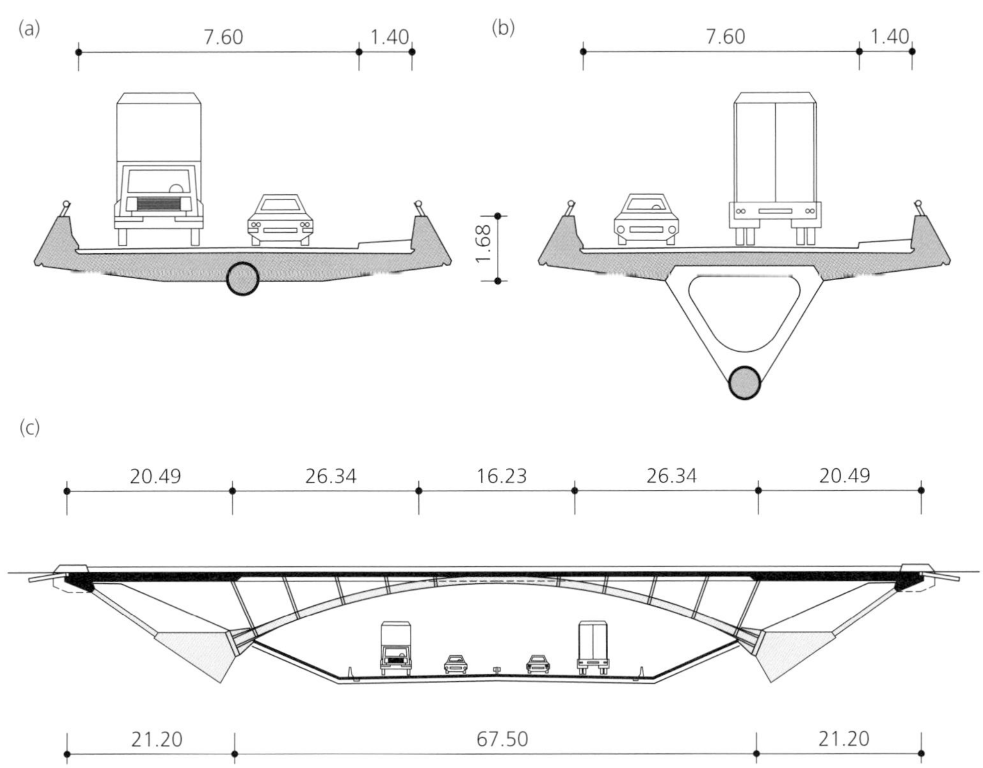

a steel–concrete composite deck that is suspended on arches formed by steel boxes infilled with concrete (Figure 4.19). The arch span is 70.57 m and its rise is 12.04 m. The suspenders are formed of Macalloy bars and are arranged as a fan with the intersection situated above the arch centre. Therefore, the funicular shape of the arch is close to the shape of a fourth-degree parabola. The deck is ended by diaphragms connected to the arch footings by concrete struts. Both the arch footings and the end abutments are supported by 1.20 m diameter drilled piles arranged in a similar way as described above for other structures.

Sometimes overpasses are situated close to important sites, and can thus form a signature point at the entry into an area. The importance of the site can justify increased costs and the overpass can be designed as an interesting arch or cable-supported structure (see Figure 4.7(g)). An example is the pedestrian bridge constructed near the city of Bohumín in the Czech Republic. The bridge crosses the D1 motorway, a local road and a creek (Figure 4.20) (fib, 2014).

The total length of the bridge 115.260 m, and it is composed of two spans (lengths 54.94 + 58.29 m) that are suspended on a single V-shaped pylon located in the area between the motorway and the local highway. Due to heavy bicycle traffic in the city, it was necessary to separate the pedestrian and bicycle pathways. Therefore, the deck is formed of a central spine girder with non-symmetrical cantilevers carrying the pedestrians and bicycles. The pylon and stays are situated in the bridge axis (Figure 4.21). To balance the dead load, the shorter cantilever carrying the 2.25 m wide pedestrian pathway is solid, while the longer one carrying the 3.00 m wide bicycle pathway is formed of a slab that is unweighted by waffles.

Figure 4.20 Pedestrian bridge near Bohumín, Czech Republic

Figure 4.19 Bridge in Figure 4.18: (a) cross-section; (b) elevation

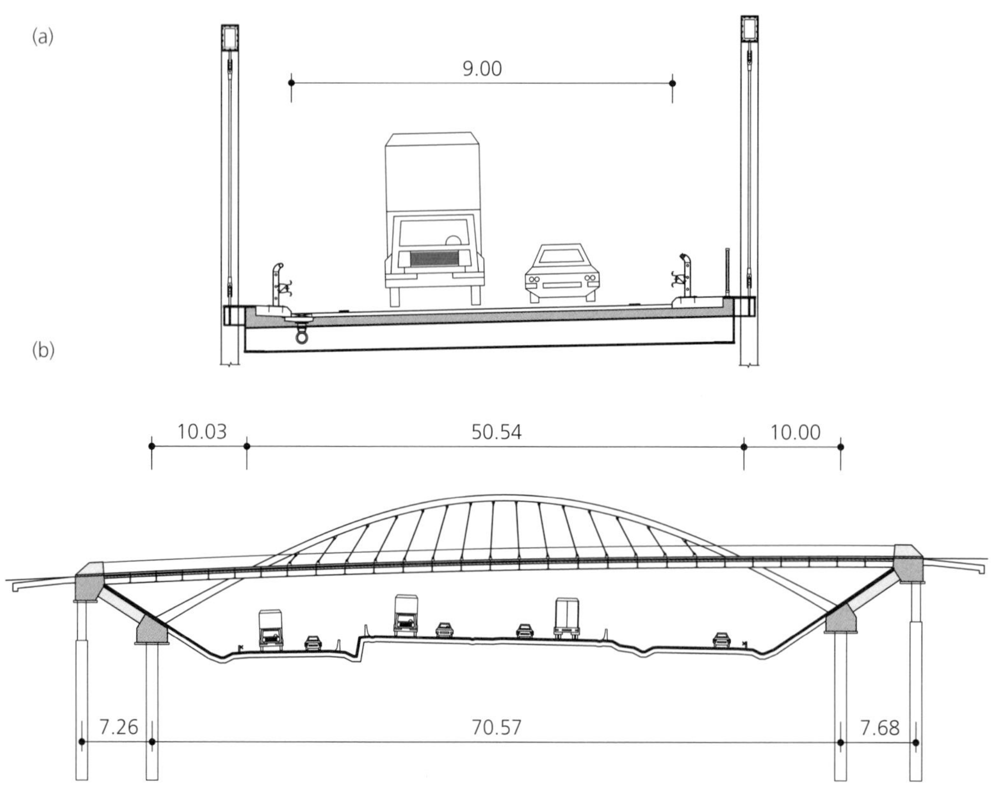

Figure 4.21 Bridge in Figure 4.20: (a) deck cross-section; (b) bridge cross-section; (c) elevation

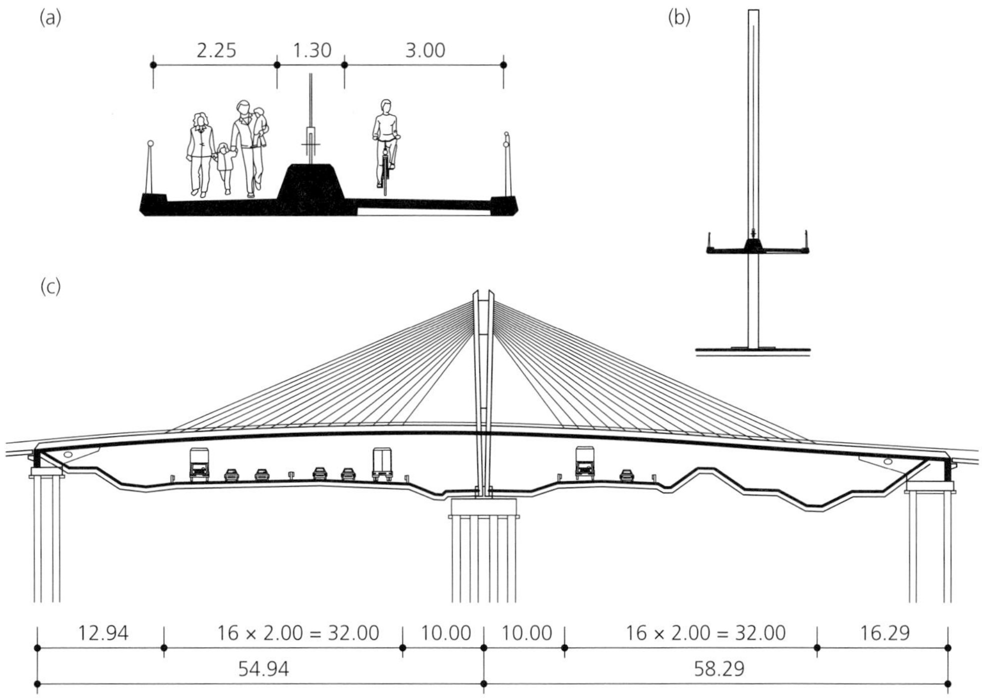

The bridge forms an integral structural system. The deck is fixed into the end abutments and the 25.48 m tall pylon is connected via the frame to the deck and fixed into the footing. Both the end abutments and pylon footing are supported by drilled shafts of 0.90 m diameter. The pylon is formed by two inclined columns of one-cell box sections (800 × 600 mm) that are tied by top and bottom steel plates. The boxes are infilled with concrete that was pressed from the footings to the tops of the columns.

4.3. Skew crossings

Where there is a small crossing angle or when a bridge is situated at a junction, it should be considered that the bridge will be seen at an oblique angle. If the bridge deck is supported by a row of columns at each support, a forest of supports will be seen in the oblique view. Not only will this make it difficult to navigate to such a crossing, but this solution will also damage the aesthetic impression of the crossing. Therefore, it is advisable to support the structure only by columns situated in the longitudinal bridge axis, and to capture the torsion only at the abutments (Figure 4.22).

For lengths of individual spans L of up to 30 m and a distance to the abutments L_T of up to 120 m, a spine girder or a solid cross-section can be designed (Figures 4.23 and 4.24). For longer spans and distances to the abutments of up to 160 m the use of box girders is suitable. From an aesthetic point of view, two- or three-cell box girder structures with inclined or curved outer webs have proved to be very efficient (Figure 4.25). The side cells significantly increase the torsional stiffness. Therefore, these structures can be designed with single-column support for lengths L_T up to 200 m.

For a very small angle of crossing the deck of the overpass is usually supported by transverse frames situated perpendicular to the motorway (Figure 4.26). As designs of this type require careful detailing, not all solutions will be aesthetically pleasing. Therefore, for the highway crossing over the prepared D52 motorway situated in a beautiful recreational area in south Moravia, Czech Republic, the first author designed the overpass to be formed of a slender concrete deck suspended on an A-shaped pylon bridging the motorway (Figure 4.27). The overpass has five spans (lengths 15 + 17 + 108 + 17 + 15 m), the

Figure 4.22 Skew crossing: (a) cross-sections; (b) plan

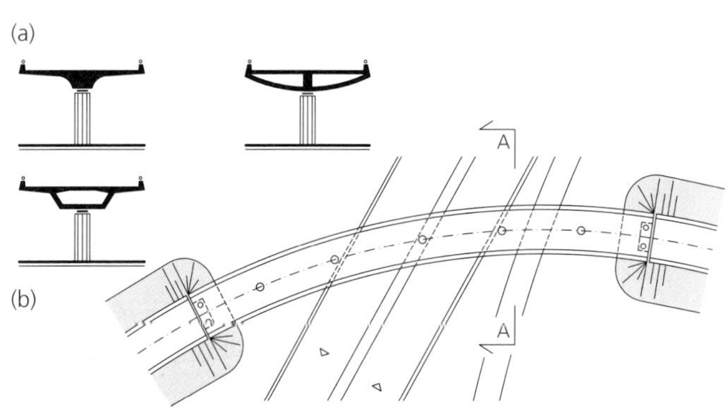

Figure 4.23 Bridge with a spine girder, near Olomouc, Czech Republic

Figure 4.24 Bridge with a solid cross-section, near Olomouc, Czech Republic

Figure 4.25 Bridge across the expressway in a city of Brno

Figure 4.26 Bridge in the San Francisco Bay area, California, USA

Figure 4.27 Bridge near Pasohlavky, Czech Republic

depth of the deck is 0.90 m and the pylon height is 30 m. The 17 m wide bridge deck has a square plan; in the main span it is suspended on its edges, and in the side spans it is supported by single supports situated in the bridge axis. While the suspended span is lightened by waffles, the side spans have a solid cross-section.

4.4. Urban viaducts

Urban viaducts need special attention. Due to the aesthetic requirements and limited space, they should be as slender as possible. Beam and slab structures are therefore not suitable. For smaller spans, spine girders with large overhangs or slabs with curved soffits are suitable, while for longer spans box girder structures are appropriate (Figure 4.28). The box girder structures not only have a higher bending capacity but also a considerably greater torsional stiffness. This allows a design with single column supports in order to minimise the space under the bridge. From an aesthetic

Figure 4.28 Urban viaducts: (a) cross-sections; (b) plan

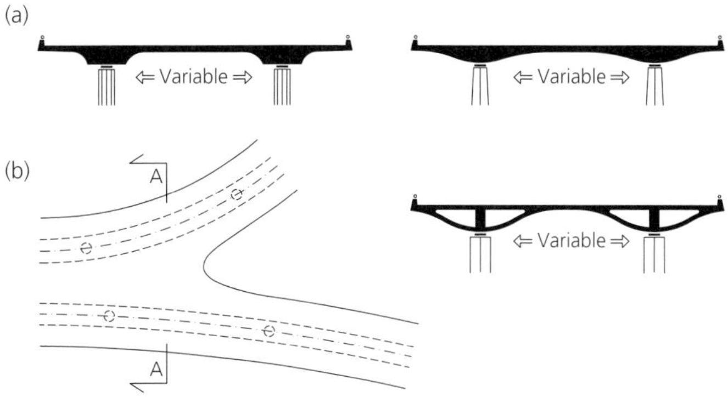

Figure 4.29 Route 92/101 interchange, San Mateo, California, USA

Figure 4.30 Route 92/280 interchange, San Mateo, California, USA

point of view, multi-cell decks with inclined (Figure 4.29) or curved outer webs (Figure 4.30) have proved to be very successful (California Department of Transportation, 1980). As urban bridges are led in different elevations and they mutually cross, they require single supports situated in the bridge axes. Figures 4.29 and 4.30 show the beautiful bridges at the Route 92/101 and 92/280 interchanges situated south of San Francisco, California, USA. The arrangement of axially supported bridges is discussed in Chapter 3.

Bridge decks are usually supported by single bearings situated on several intermediate supports, and the torsion is resisted by pairs of bearings situated at the abutments. In long bridges the torsion is resisted by bearings situated on two column piers (Figure 3.21). These structures lose visual order and are non-uniform in appearance (Figure 4.31). Therefore, structures supported by pairs of bearings situated on single piers with caps represent a better solution (Figure 3.27). The transverse distance of the bearings and the size of the caps is determined by the requirements necessary to prevent uplift of the bearings. The bridge at the Kostiviarska junction on the R1 expressway at Banská Bystrica, Slovakia, shows that, with careful design, the size of the column caps can be small (Figure 4.32) (fib, 2014).

Urban viaducts often form part of intersections, and therefore have complex geometry. They are often of variable width and split into two or more independent structures. As the bridges at a junction should have a uniform appearance, the concept should be developed early on in the design stage, according to the junction geometry (fib, 2006). Figures 4.33 and 4.34 show the bridges at the D1 motorway junctions at Rudná and Místecká, near Ostrava, Czech Republic. The bridges at the Rudná junction are formed of spine girders of variable widths, which are gradually divided into two independent

Figure 4.31 Motorway junction at Sverepec, Slovakia

Figure 4.32 Motorway junction at Kostiviarska, Slovakia

Figure 4.33 Motorway junction at Rudná, near Ostrava, Czech Republic

Figure 4.34 Motorway junction at Místecká, near Ostrava, Czech Republic

Figure 4.35 Bridge across Jan-Wellem-Platz, Düsseldorf, Germany

structures. The bridges at the Místecká junction are formed by two girders mutually connected by a deck slab of variable width (see Figure 4.28).

Many urban bridges have to be built with noise barriers. If these barriers are added to deep girder structures, the bridges look too heavy. The logical answer to this problem is to situate the main load-bearing member on the deck edges, and in this way combine the two functions. Unfortunately, not many bridges have been built in this way. A rare example is the bridge across Jan-Wellem-Platz in Düsseldorf, Germany (Figures 4.35 and 4.36) (Dywidag, n.d.). The twin bridges have a total length of 440 m, with nine spans of lengths in the range 50–60 m. The clear width of each bridge is 13.50 m, and the height of the edge girders above the deck slab is 3.50 m.

On the other hand, channel cross-sections with edge girders, which function as both a load-bearing member and a noise barrier, have been used in many light rail viaducts built all over the world. An excellent example is the Dubai Metro viaducts (Figure 4.37) (Smith *et al.*, 2009). The project includes approximately 46.5 km of viaducts on the Red Line and 14.6 km of viaducts on the Green Line. Most of the viaducts are simply supported bridges with spans up to 44 m.

Where longer spans are required, such as to cross highway interchanges or creeks, two-span or three-span continuous bridges are used. Most of the viaducts utilise U-shaped trough girders with a width of 10 m and a total depth of 2.04 m (Figure 4.38(a)). The three-span bridges have span lengths of 44 + 72 + 44 m. Their decks consist of U-shaped trough girders only near end span, and a combination of U-shaped trough and box girders near the midspan piers (Figure 4.38(b)). This solution enables aesthetic continuity with adjacent spans. The profiled edge beams result in both an economic and an aesthetically pleasing structure.

Figure 4.36 Bridge in Figure 4.35: cross-section

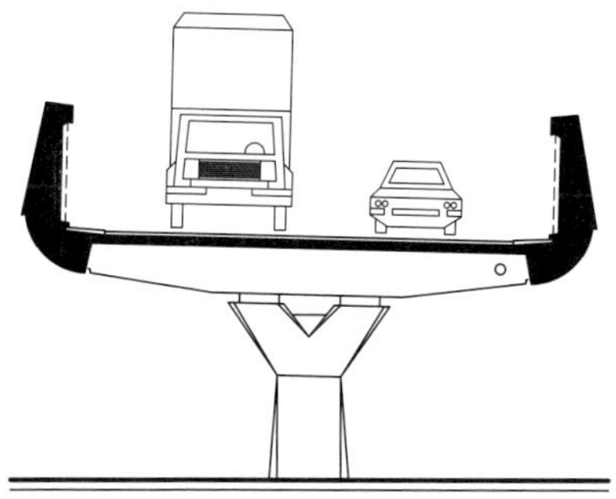

Figure 4.37 Metro viaducts, Dubai, UAE

Figure 4.38 Viaducts in Figure 4.37: (a) cross-section of typical spans; (b) cross-section of long spans at interchanges

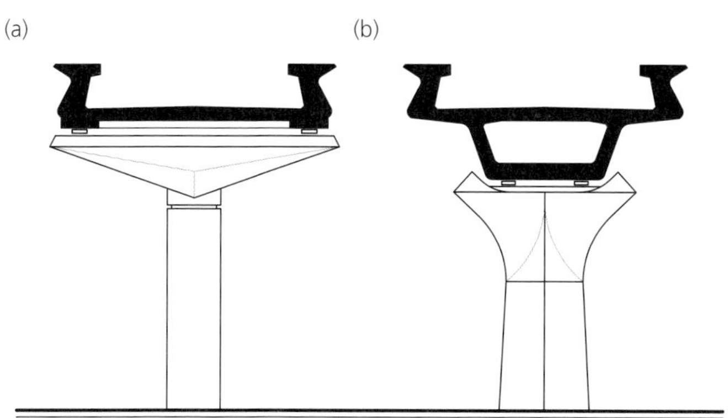

4.5. Motorway and highway viaducts

The design of motorway viaducts depends on the required span, the height above the terrain and the nature of the surroundings. It is not advisable to design unnecessarily large spans requiring deep decks.

For small-span bridges, slab beams or precast beam structures are suitable. Double-T beam decks directly supported by slender columns are very economical for bridges with spans of up to 45 m (Figure 4.39(a)). In the case of poor foundation soil or longer spans, it is advisable to design a steel–concrete composite structure in which the concrete beams are substituted by steel plate girders. This arrangement has been used in many structures. The 840 m long motorway viaduct built in 2009 near the city of Hladke Životice in the Czech Republic can serve as an example (Figure 4.40) (fib, 2010). The twin bridge comprises 21 spans (lengths 31.40 + (19 × 40.80) + 31.40 m) and crosses the railway and several local highways. Each bridge deck is formed by a continuous beam of double-T cross-section formed by two girders and a deck slab (Figure 4.41). Due to the problem of clearance above the railway, the depth of the girders varies from 2.60 m at the supports to 1.70 m at the midspans. The haunches have the shape of a second-degree parabola. Transverse diaphragms are located only at the two ends of the bridge deck.

When a structure is visible from different points of view, a forest of columns may be seen in an oblique view and the bridge loses visual

Figure 4.39 Motorway viaducts

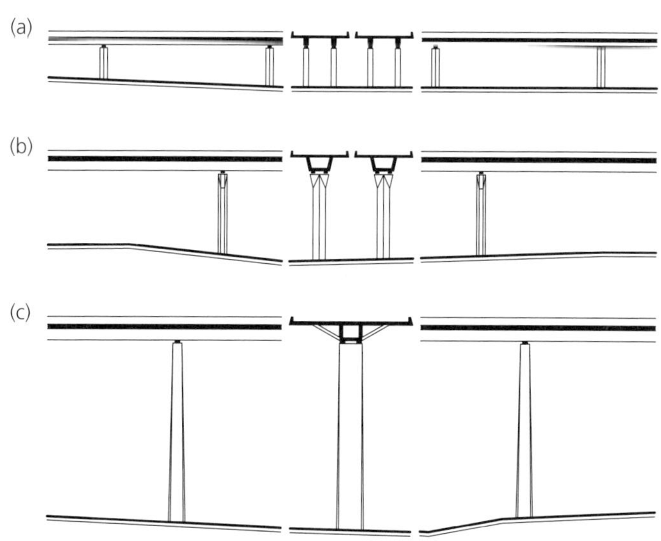

order. Therefore, it is appropriate to design a structure that is supported by narrow piers situated in the bridge axis. It is possible to design an indirectly supported double-T beam deck, or spine girder structures with overhangs.

Many indirectly supported beam structures have been built, for example over the R1 expressway near the city of Zlaté Moravce in Slovakia (fib, 2014). The bridge crosses a local highway and a creek (Figure 4.42). The twin bridge has a total length of 273 m and forms a prestressed concrete continuous structure of seven spans (lengths of 33.0 + (5 × 41.0) + 33.0 m). The deck has a variable depth, from 1.90 m at the midspans to 2.60 m at the intermediate supports (Figure 4.43).

In structures with spine girders, the girders can have a solid or box cross-section. It is possible to design solid beams of constant depth with spans of up to about 35 m, or of variable depth with spans of up to about 45 m. This arrangement has been used in many structures, for example, the viaduct over the D8 motorway at Knínice, in the Czech Republic, close to the border with Germany (Figure 4.44) (fib, 2006). This twin viaduct consists of two parallel bridges with lengths of 1027 m and 1077 m. The bridges form continuous structures with a typical span length of 42 m (Figure 4.45). The decks of both bridges are formed by spine girders with large overhangs, and have a total width of 14.80 m. The spine girder has a solid cross-section of variable depth, from 2.6 m at the supports to 1.4 m at the midspan. The haunch has the shape of a fourth-degree parabola. The deck, which was designed as a partially prestressed structure, is post-tensioned in both the longitudinal and transverse directions of the bridge.

For longer spans, one-cell box girder structures are commonly used. Bridges with spans up to 60 m usually have a constant depth (see Figure 4.39(b)), while longer span bridges have variable depth. The box girders are usually supported by two bearings situated on simple H- or box-section piers. An example of the typical arrangement is

Figure 4.40 Motorway viaduct at Hladke Životice, Czech Republic

Figure 4.42 Expressway viaduct near Zlaté Moravce, Slovenia

Figure 4.41 Viaduct in Figure 4.40: cross-section and partial elevation

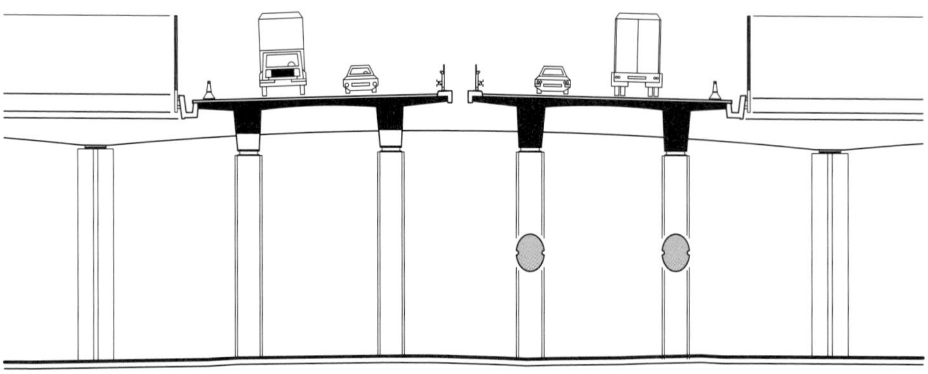

Figure 4.43 Bridge in Figure 4.42: cross-section and partial elevation

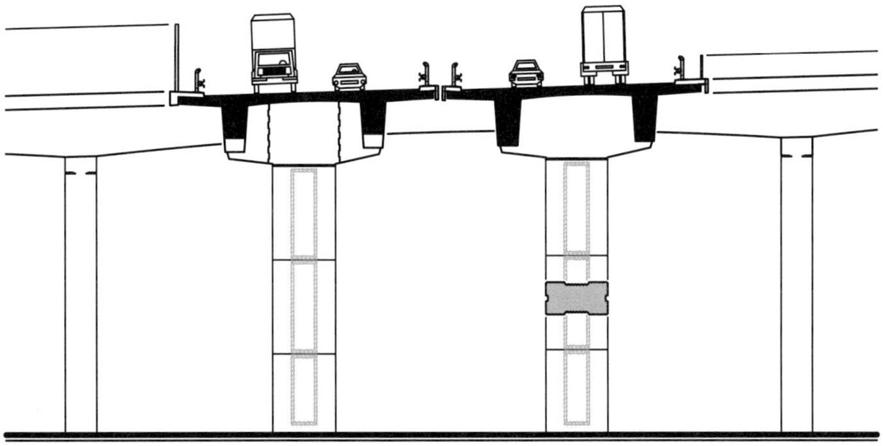

Figure 4.44 Motorway viaduct at Knínice, Czech Republic

Figure 4.45 Viaduct in Figure 4.44: cross-section and partial elevation

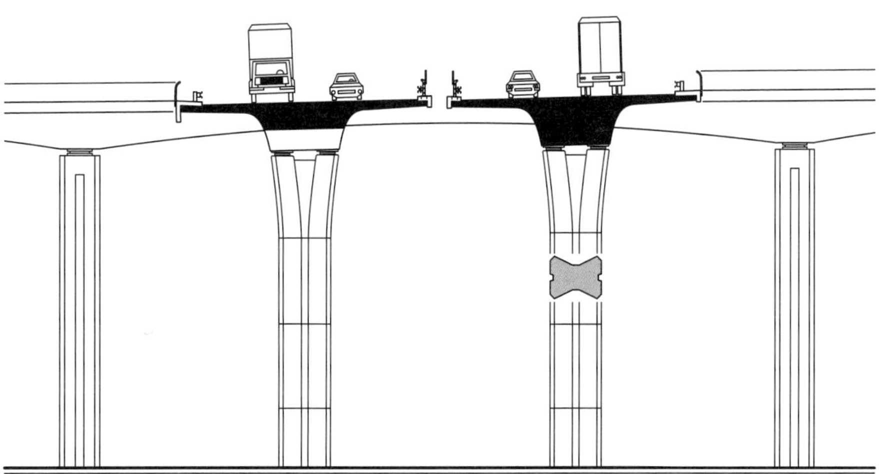

the viaduct over the D1 motorway at Bitarová, Slovakia (fib, 2018). The 625 m long viaduct is formed by a twin bridge of 11 spans (lengths of 41.75 + (9 × 58.50) + 41.75 m) (Figure 4.46). The 13.55 m wide and 3.00 m deep decks are formed of one-cell box girders; the piers of the H-section are formed of two columns connected by shear walls (Figure 4.47).

The torsional stiffness of the box girders allows for the design of viaducts that have complex geometry, such as the viaduct at Linn Cove over the Blue Ridge Parkway in North Carolina, USA (Figure 4.48) (Muller and Barker, 1985). The S-shaped 379 m long precast segmental bridge makes both vertical and horizontal curves with radii as small as 76 m. The deck is formed of a continuous one-cell box girder comprising eight spans (lengths 30.02 + 49.68 + (4 × 54.86) + 49.68 + 30.02 m), and it is supported by cast-in-place end abutments and precast segmental piers (Figure 4.49).

Multi-cell box girders for spans up to 80 m are commonly used on bridges constructed on the West Coast of the USA. Figures 4.50 and 4.51 show the viaduct which transfers Freeway I-5 north of the city of Redding in California, USA over a deep valley. The space between the bridges is reserved for a future structure that will be built if necessary.

Figure 4.46 Motorway viaduct at Bitarová, Slovakia

For bridges that run high above the terrain, it is advisable to carry both directions of the motorway on a single structure formed from a box girder with large overhangs supported by inclined struts (see Figures 4.39(c) and 3.16). Then the box girder can be supported by single piers situated in the motorway axis. These structures have also proved to be successful in the case of long bridges located on mountain slopes.

Figure 4.48 Motorway viaduct at Linn Cove, North Carolina, USA

Figure 4.47 Viaduct in Figure 4.46: cross-section and partial elevation

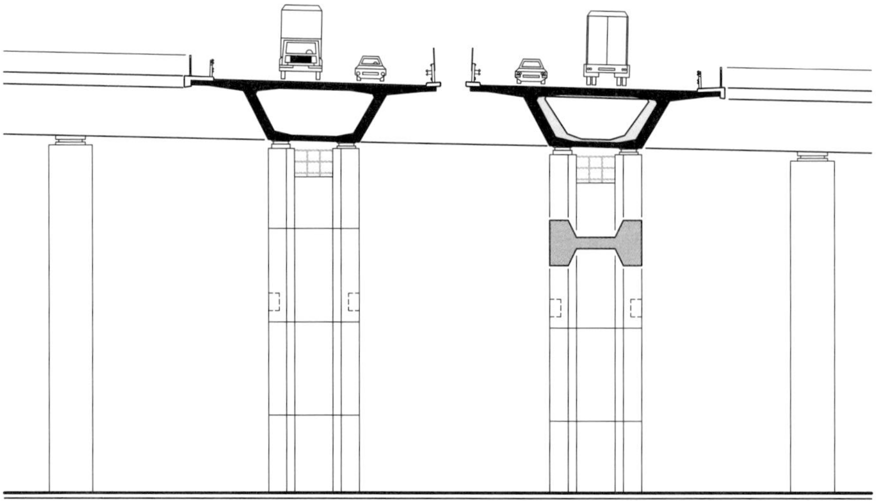

Figure 4.49 Viaduct in Figure 4.48: cross-section and partial elevation

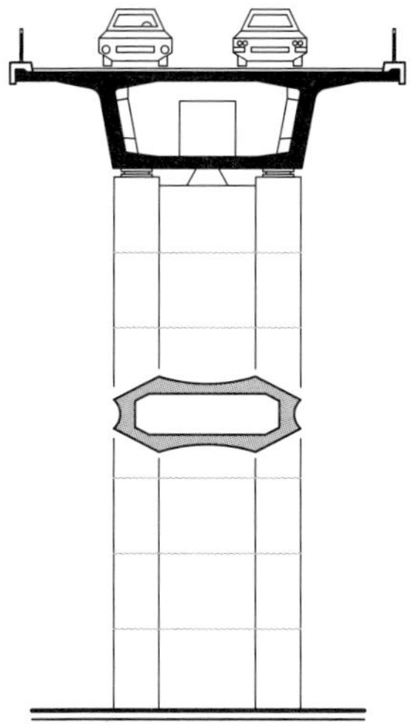

Figure 4.50 Freeway viaduct near Redding, California, USA

Figure 4.51 Viaduct in Figure 4.50: cross-section and partial elevation

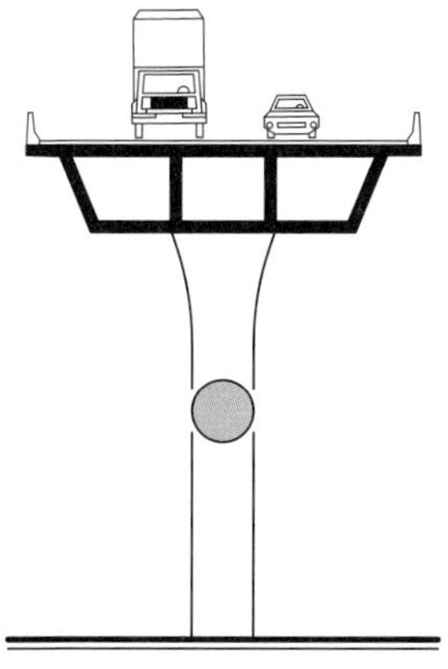

The advantage of the single support is evident from Figures 4.52 and 4.53, which show motorway viaducts that run about 40 m above the terrain. Part (a) in these figures shows the perpendicular view of the bridges (like what can be seen in technical drawings), and part (b) shows the skew view, which is what pedestrians can usually see. Figure 4.52 shows a twin bridge with a typical span of 60 m; Figure 4.52 shows a single bridge with a typical span of 45 m. Although from the perpendicular view, the bridge with the 60 m spans looks better proportioned, the skew view confirms that the single bridge with the shorter 45 m spans looks even better than the twin bridge of the longer spans, and has a clean and uniform order (see Figure 4.53(b)).

It is well known that the cost of a bridge deck increases with its span length. Also, comparative studies have confirmed that for spans longer than 45 m and pier heights of 40 m, single wider piers supporting a single structure is more economical than two piers supporting twin structures. Therefore, in many cases single bridges are more economical.

The advantage of twin bridges is that during the repair of one structure it is possible to transfer both traffic directions to the other structure. For a single bridge, however, there is no guarantee that both traffic flows can be transferred to one side of the bridge while the other side is repaired (Figure 4.54). Therefore, this situation has to be carefully checked in the design.

Single bridges in which decks are formed of box girders with large overhangs supported by struts were developed in Germany, where the outstanding 1.128 km long Kocher Viaduct, near Schwäbisch Hall, was built as early as 1979. The 30.5 m wide bridge deck has nine spans of lengths ranging from 81 to 138 m, and it is supported by piers of up to 178 m high (Leonhardt, 1984).

As discussed in Chapter 3, the deck of these bridges is usually formed from a one-cell box girder with large overhangs supported by struts. The struts can be formed by single bars, a truss or slabs.

For example, single bars were used in the construction of the D8 motorway bridge across Rybny Creek close to the border between the Czech Republic and Germany (fib, 2006). The 356 m long and 31.10 m wide bridge is formed by a continuous box girder of seven spans ranging in length from 34 to 58 m. In a bid project design, the crossing was formed by a traditional twin bridge; however, an alternative design, in which the deck is formed by an incrementally launched one-cell box girder with large overhangs, was accepted for construction (Figure 4.55). The deck is formed by a relatively narrow box girder with large transversally prestressed overhangs (Figure 4.56). At distances of 4.00 m, the overhangs are supported by single struts with dimensions 0.40 × 0.50 m. The deck sections above the piers are stiffened by additional cast diaphragms.

Truss struts were used in the construction of the extradosed viaduct on the D1 motorway at Považká Bystrica in Slovakia (Figure 4.57)

Figure 4.52 Twin bridge: (a) perpendicular view; (b) skew view

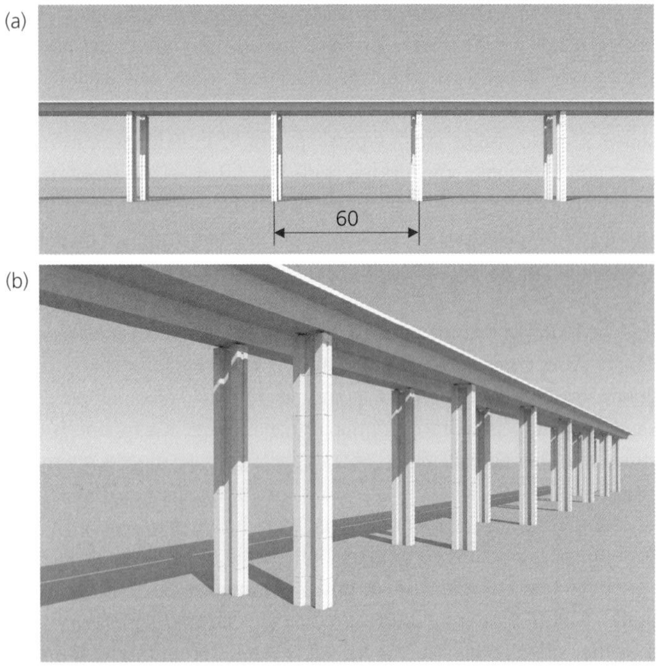

Figure 4.53 Single bridge: (a) perpendicular view; (b) skew view

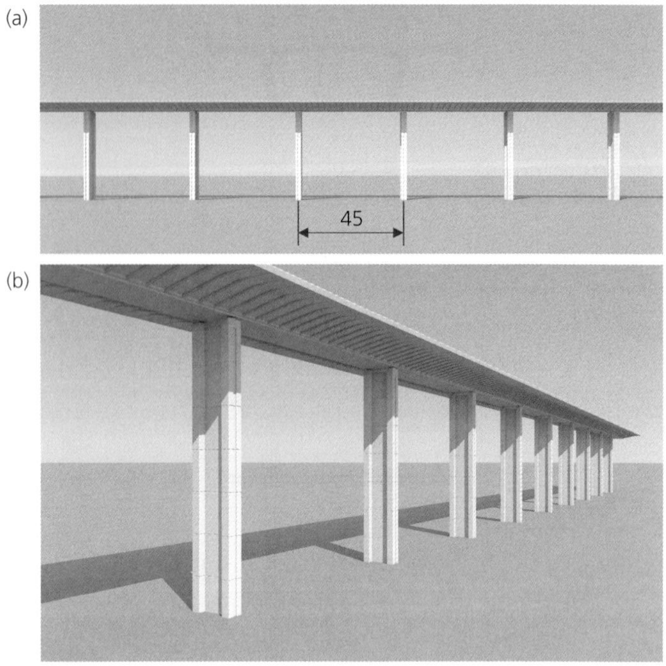

Figure 4.54 Bridge loading: (a) service; (b) repair

(Strásky, 2010). The viaduct, which has a total length of 968 m, is composed of a continuous girder of ten spans (lengths 34.2 + 48.8 + 70.8 + (6 × 122) + 68 m). The deck is formed by a one-cell box girder with large overhangs supported by precast struts. In the bridge axis the deck is suspended on seven low (14 m high) pylons. The stays have a semi-radial arrangement. The girder is supported by four pot bearings situated in two rows on H-shaped piers formed by two mutually connected inclined columns (Figures 4.58 and 4.59).

Figure 4.55 Motorway bridge across Rybny Creek, Czech Republic

Figure 4.57 Motorway viaduct at Považká Bystrica, Slovakia

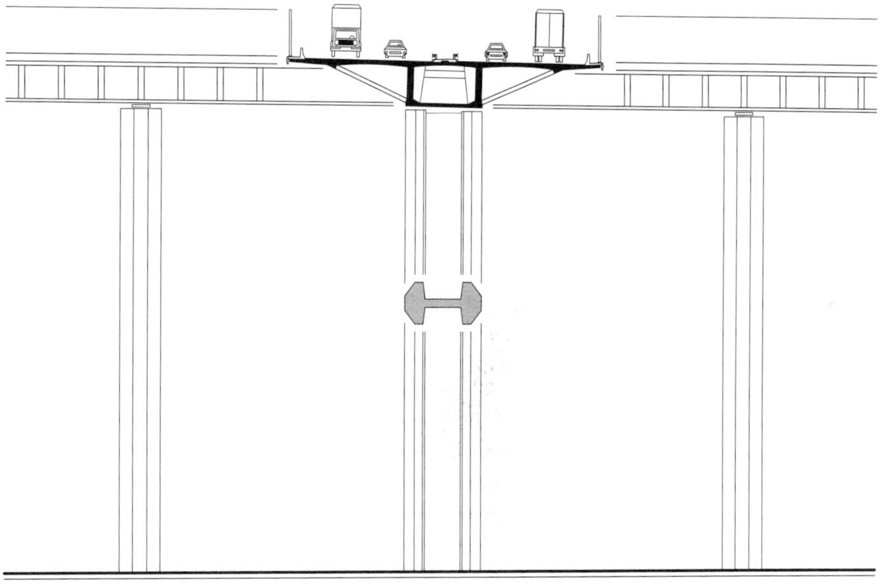

Figure 4.56 Bridge in Figure 4.55: cross-section and partial elevation

The suspended spans were progressively cast, segment by segment, in seven symmetrical cantilevers.

The 30.40 m wide box girder has a variable depth, from 6.00 m at the supports to 4.70 m at the midspan. The relatively narrow box girder has a variable thickness of the webs and bottom slab. The large overhangs are supported by V-shaped precast struts. At the supports of the suspended span the deck is stiffened by A-shaped diaphragms that transfer the load from the pylons into two rows of bearings. The bearings on the two central piers are fixed; the remaining bearings are movable in the longitudinal direction of the bridge.

Slab struts were used in the design of the viaduct across the Hostovsky Creek Valley on the R1 expressway near Nitra in Slovakia (Strásky, 2015). The bridge, which has a total length of 960 m, forms a continuous girder of 17 spans of lengths ranging from 33.0

Figure 4.58 Bridge in Figure 4.57: (a) cross-section; (b) partial elevation

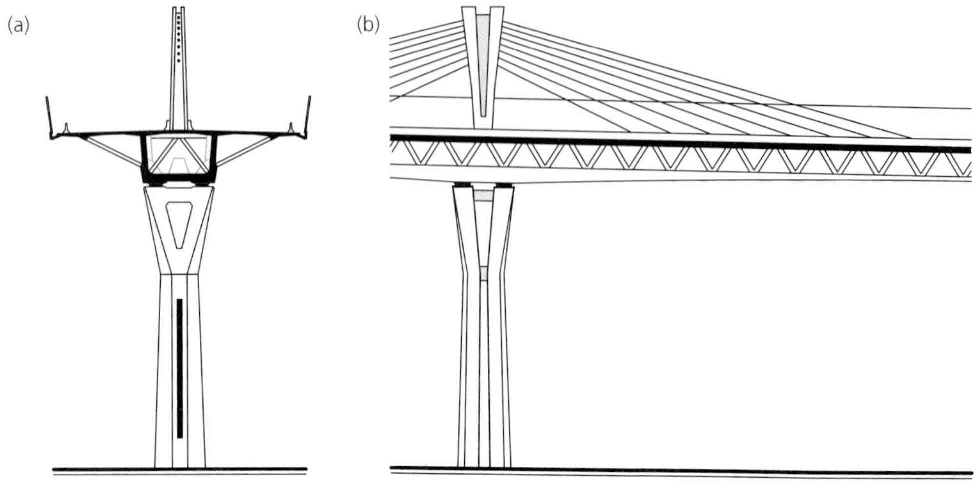

Figure 4.59 Bridge in Figure 4.57: deck and piers

Figure 4.60 Motorway viaduct across Hostovsky Creek Valley, near Nitra, Slovakia

to 69.0 m (Figures 4.60 and 4.61). The 25.66 m wide bridge deck is formed of a spine, one-cell box girder with large overhangs supported by precast struts. The spine girder has a variable depth, from 4.0 m at the supports to 2.6 m at the midspan. The deck was progressively erected in both the longitudinal and transverse directions (see Chapter 5).

In 1970, a 481.5 m long bridge was built over the deep Nusle Valley in Prague to carry the north–south expressway and city metro (Figure 4.62) (Michalek and Klimes, 1970). The bridge, designed by PUDIS Prague, has five spans (lengths 68.25 + (3 × 115.00) + 68.25 m). It is formed by a 21.70 m wide one-cell box girder, which was widened to 26.70 m by casting the sidewalks supported by transverse ribs (Figure 4.63). While the top slab of the girder carries the vehicular traffic, the metro is situated inside the box, and is carried by bottom slab.

The 6.55 m deep box girder is connected via the frame to twin piers, each formed by two transversally inclined columns that directly

Figure 4.61 Bridge in Figure 4.60: cross-section and partial elevation

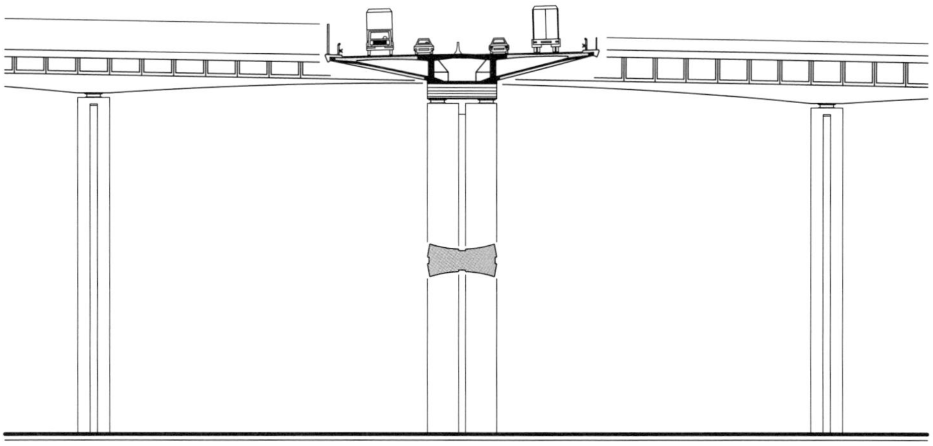

Figure 4.62 Motorway and metro bridge over Nusle Valley, Prague, Czech Republic

support the girder webs. Due to the metro, the girder does not have diaphragms. The box girder is only stiffened locally by the columns, which are situated partially outside the webs (Figure 4.64). As the girder is supported by bearings only at the abutments, the bridge forms a semi-integral structural system. The bridge was constructed using the cantilever method.

4.6. Bridges over rivers

For single spans, frame bridges of variable depth are suitable. For shorter spans, slab beam decks are suitable, while, for longer spans, box girders are useful; the bottom slab can be omitted at the midspan area.

A simple double-T beam bridge was used for the D1 motorway bridge that crosses the River Olse (which forms a border of the Czech Republic) and interchange ramps (Figure 4.65) (fib, 2014). The bridge has a total length of 181 m and is a continuous structure comprising four spans of lengths 37–45 m. Because of the limited clearance at the site, the deck has a variable depth, ranging from 3.4 m at the supports to 1.8 m at the midspan (Figure 4.66). The bridge deck was cast on stationary scaffolding.

If a bridge is required to cross not only a river but also the adjacent inundation area, the main bridge is usually designed as a box girder structure of variable cross-section, and this is attached to either box girder or double-T beam viaducts (Figure 4.67(a)). For spans greater than 60 m, the main spans are usually formed by balanced cantilever structures.

Figure 4.63 Intermediate pier of the bridge in Figure 4.62: (a) partial elevation; (b) cross-section; (c) axonometric view

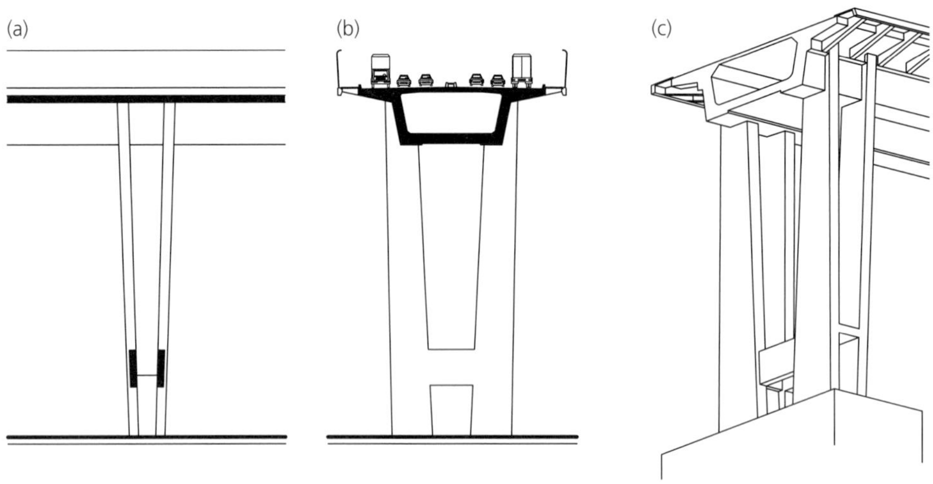

Figure 4.64 Bridge in Figure 4.62: (a) cross-section; (b) elevation

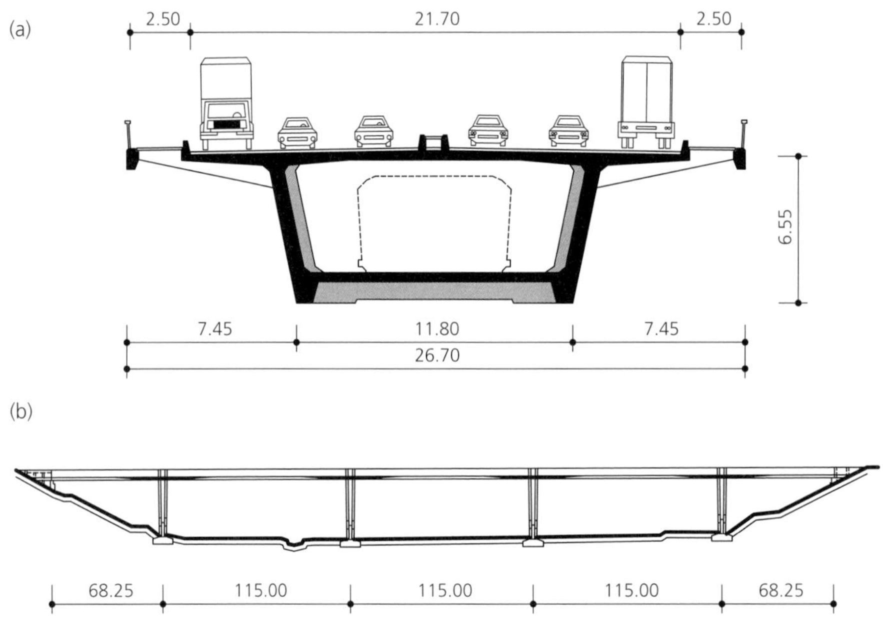

The approach spans should be connected to the main spans, and together they should form one continuous structure. This can be easily done if the approach spans are formed by the same box girder as the main span. However, this is also possible if the main span is composed of a box section and the approach viaducts are composed of a double-T section, as shown by the viaduct across the Strazov Pond and the Hricov reservoir for the River Vah, on the D3 motorway near Žilina in Slovakia (Figure 4.68) (fib, 2018).

The viaduct is formed by twin bridges. The eastbound bridge has a total length of 1493 m and is formed by a continuous structure of 31 spans (lengths 30.51–110.00 m); the westbound bridge has a total

Figure 4.65 Motorway bridge across the River Olse, Czech Republic

Figure 4.68 Viaduct across the Strazov Pond and the Hricov reservoir, near Žilina, Slovakia

Figure 4.66 Bridge in Figure 4.65: cross-section and partial elevation

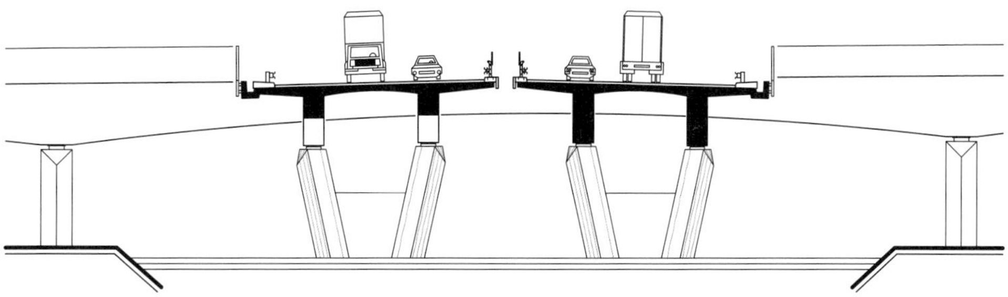

Figure 4.67 Bridges over rivers

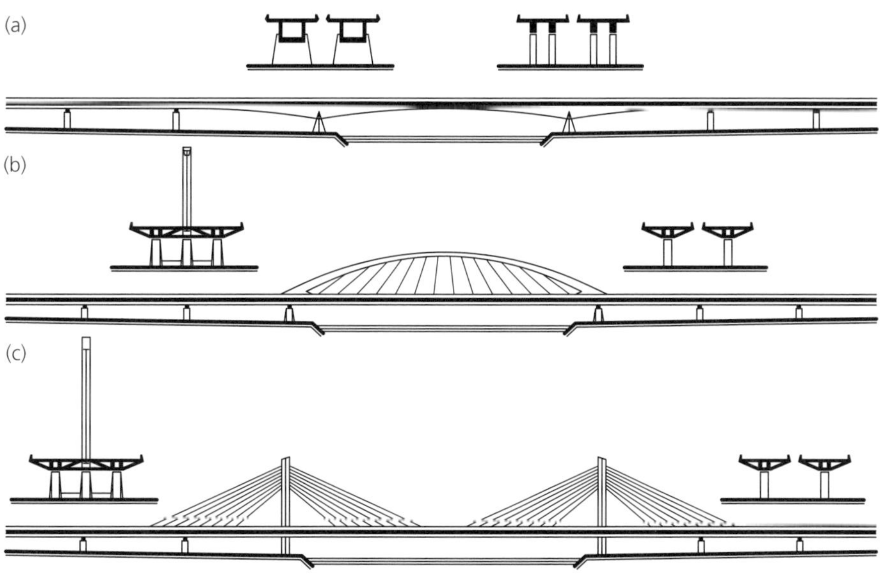

length of 1437 m and is formed by a continuous structure of 30 spans (lengths 27.00–110.00 m). It is the longest span bridge over the River Vah waterway.

Each bridge is formed by a continuous girder that is assembled from two types of structure: (*a*) the main four-span structures bridge the waterway, and are formed from a single-cell box girder of variable depth, from 3.0 m at the midspans to 6.0 m at the supports; and (*b*) all the other spans are formed from a double-T girder with a constant depth of 3.0 m (Figure 4.69). The connection of the two types of structure is achieved by overlapping of the prestressing tendons. As at the connections at the outer perimeters of both the structure types are the same, the bridges have a uniform appearance, expressing a unity of the architecture and the structural solution. While the main spans are supported by twin piers formed by two transversally inclined columns that are connected to the girders by the frame, the double-T girders are indirectly supported by pairs of pot bearings situated on slender piers of elliptical cross-section.

The main spans were segmentally cast in symmetrical cantilevers starting from pier tables, and the remaining spans were progressively cast, span by span, in formworks supported by stationary or movable scaffolding. The main portion of the approach viaducts, which have typical span lengths of 42 m, were erected using two types of overhead movable scaffolding that required advance construction of the pier tables (see Figure 10.32).

A similar problem was solved in the design of the R1 expressway viaduct across the Nitra River in Slovakia (Figure 4.70). The viaduct consists of a 358.8 m long main bridge composed of five spans, which were progressively cast in symmetrical cantilevers, and a 813 m long approach bridge of 20 spans, the deck of which was incrementally launched. The span lengths in the main bridge are in the range 50.0–85.0 m, and those in the approach are 28.0–45.0 m.

Due to the limited space available for the construction of the piers, the main bridge is formed by a single structure supported by narrow piers, while the approach bridge is formed by a twin structure. The main bridge is composed of a three-cell box girder with a width of 26.50 m; its depth above the piers is 4.14 m and its depth at the midspan is 2.44 m. The approach bridges are formed by 13.05 m wide and 2.44 m deep one-cell box girders (Figure 4.71). At the

Figure 4.69 Viaduct in Figure 4.68: (a) cross-section of main span; (b) cross-section of approach span

(a)

(b)

Figure 4.70 Expressway viaduct across the Nitra River, Slovakia: (a) main span; (b) approach span

(a)

(b)

Figure 4.71 Bridge in Figure 4.70: (a) cross-section of main span; (b) cross-section of approach spans

(a)

(b)

connection, the two structures match perfectly, and therefore create a viaduct of uniform appearance.

Multi-span cantilever structures have been used many times to bridge wide rivers with navigation channels. An example of such a structure is the Benicia–Martinez Bridge across the Sacramento River, California, which was built in 2007 (Figure 4.72) (Murugesh and Cormier, 2007). The bridge has a total length of 2.489 km, and is composed of the 1971.2 km main bridge and the north and south approaches. The main bridge is formed of three frames (lengths 515.8 + 644.0 + 811.4 m) that are multiple span structures (span lengths 161.0–200.8 m). The frames are hinge connected (see Figure 11.4(b)).

The deck is formed by a one-cell box girder frame connected to piers (Figure 4.73). The girder is 25.07 m wide and of variable depth. At the midspan the depth is 4.5 m, the depth above the piers is 11.40 m for the 200.8 m span and 9.00 m for all other spans. The top slab is stiffened by the transverse ribs, which also support the outer overhangs. Above the piers the box section is stiffenned by two vertical diaphragms.

Figure 4.72 Benicia–Martinez Bridge across the Sacramento River, California, USA

For simplicity of construction, balanced cantilever structures are being designed for ever longer spans. Therefore, the decks have large depths, and the structures are becoming very heavy and robust. As these bridges do not fit within the goal of achieving fine dimensions, in the authors' opinion they should not be used for spans longer than 160 m. Also aesthetically problematic is the solution of using expansion piers to support structures of different structural depths and widths, which some designers are applying for purely economic reasons.

In such cases, if the flood water level allows, deck arches still represent a very suitable solution. An example of the recently constructed structure of this type is the bridge across the Willamette River, at Eugene, Oregon, USA, which was opened in 2013 (Figure 4.74) (Stráský, 2016). The 604.95 m long bridge, designed by consulting engineers OBEC and the authors of the book, comprises a main bridge with a two-span arch structure and the approaches. The lengths of the arch structure spans are 118.88 and 126.79 m, and those of the approach spans are 25.52–65.53 m; the bridge width is 2 × 20.47 m. The deck of the main bridge is formed of two girders and a deck slab stiffened by floor beams, while the decks of the approaches are formed by five-cell box girders (Figure 4.75). At the connections the decks have the same depth and perimeter, and therefore the bridge has an overall uniform appearance. The bridge was cast on a stationary scaffolding.

Figure 4.73 Bridge in Figure 4.72: (a) cross-section; (b) partial elevation

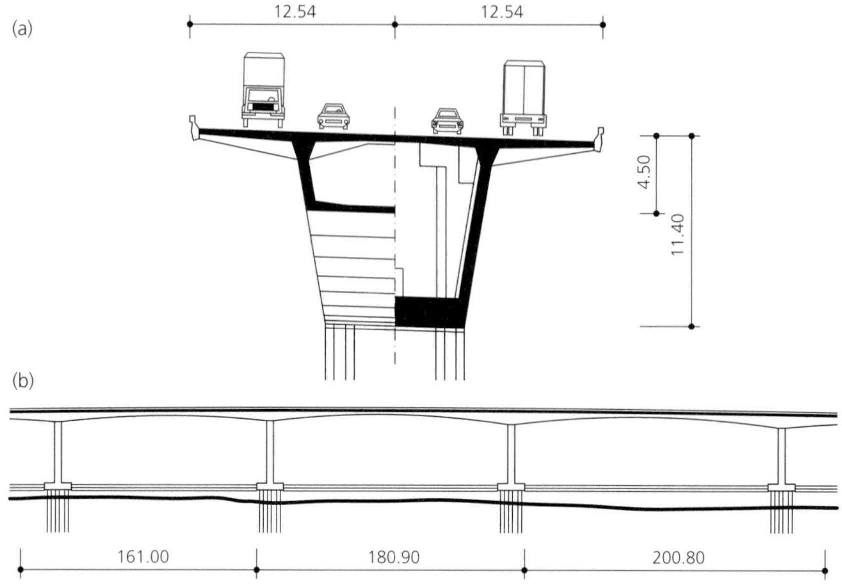

Figure 4.74 Bridge across the Willamette River, at Eugene, Oregon, USA

Figure 4.75 Bridge in Figure 4.74: (a) cross-section of main span; (b) cross-section of approach span

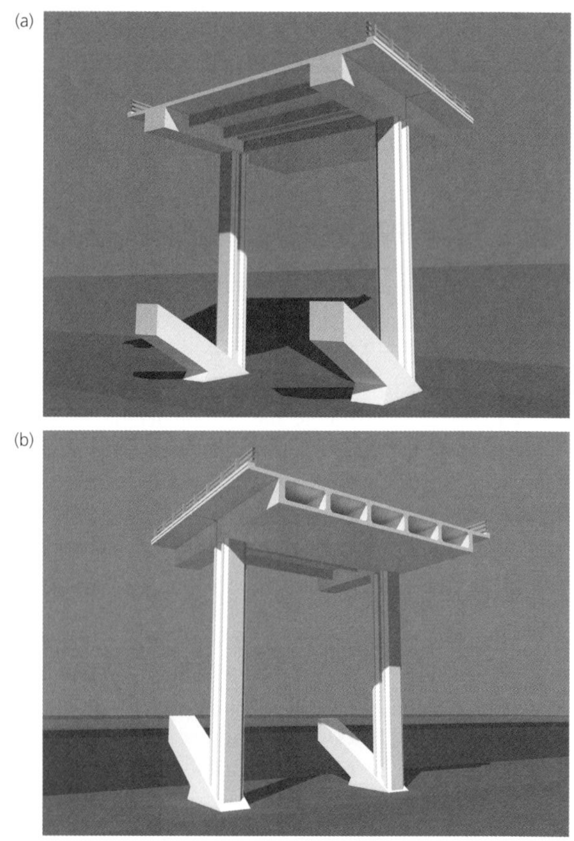

If the flood water level requires a slender deck, a through arch or cable-supported structure of fine dimensions that fits the scale of the landscape is more suitable (see Figures 4.67(b) and 4.67(c)). In these bridges, structural steel can be suitably combined with concrete to create lightweight and transparent structural systems. The bridge deck of these structures can be formed by a slender concrete slab (see Figure 3.5(a)), which for larger widths can be reinforced by concrete or steel floor beams (see Figures 3.5(d) and 3.12(e)).

For highway bridges, structures suspended on cables situated in vertical or inclined planes are suitable. However, this solution is not suitable for motorway bridges, because when the deck is suspended in four planes, the cables cross each other in an oblique view. Therefore, it is appropriate to suspend the structure in the motorway axis.

The bridge across the Ebro River in Zaragoza, Spain, designed by Casado Consulting Engineers, Madrid, is an excellent example of such a light and transparent structure (Figure 4.76) (Manterola, 2006). The bridge deck is composed of five spans (lengths 42 + 52 + 120 + 52 + 42 m) and is formed of a 31.9 m wide and 2.2 m deep two-cell box girder with long overhangs (Figure 4.77). In the main span the deck is suspended in the bridge axis on a steel arch. As the arch is fixed into the deck, the structure forms an economical self-anchored structural system (a tied arch). The deck was cast progressively: first, a spine girder was erected, then the overhangs were cast in the formwork supported on the already erected deck.

Figure 4.76 Bridge across the Ebro River, Zaragoza, Spain

Figure 4.77 Bridge in Figure 4.76: cross-section

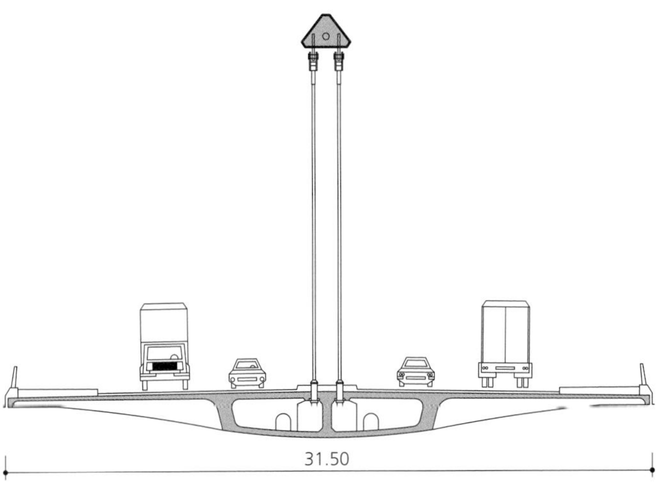

An excellent example of a structure for which architecture was derived from the true structural function and the technology of construction is the cable-stayed bridge on the Vienna airport motorway in Austria (Pauser and Beschorner, 1976). The three-span bridge crosses the Danube Canal at a skew angle of 45°. It has 119.0 m central span and 55.7 m side spans (Figure 4.78). The bridge deck is formed by a 15.8 m wide and 2.8 m deep trapezoidal three-cell box girder, which is suspended on 18.0 m tall pylons. The bridge superstructure was built on the banks and consequently rotated into the design position (see Figure 5.31).

Another example of a cable-stayed bridge is the one on the D11 motorway that crosses the River Elbe in the Czech Republic (Figure 4.79) (Stráský, 1993). This type of bridge was selected for this location because of the limited clearance available. The bridge deck is composed of three spans (lengths 61.60 + 123.20 + 61.60 m), and is formed by a spine box girder that is widened by overhangs supported by precast slab struts (Figures 4.80 and 5.88). The deck is 32.30 m wide and 2.50 m deep, and is suspended in the bridge axis on single 28.00 m tall pylons. The spine girder is assembled from precast, match cast segments (Stráský, 1993).

The same design approach used for the Ebro River bridge in Zaragoza, Spain, was also used in a design of the Brotonne Bridge, which was built across the River Seine in Normandy, France, in 1977 (Figure 4.81) (Mathivat, 1983). This cable-stayed bridge has a total length of 1278.4 m, and consists of the main bridge and two approach viaducts. The main crossing has a span length of 320 m, and the approaches have typical span lengths of 58.50 m. The bridge deck, which carries four lanes, is formed by a one-cell trapezoidal box girder with interior stiffening struts (Figure 4.82). The deck is 19.20 m wide and 3.80 m deep. The girder was progressively assembled from 3.0 m long segments formed by cast-in-place top and bottom slabs and precast webs. Both the main bridge and the approach spans were erected as symmetrical cantilevers starting at the piers (see Figure 11.28).

Short- and medium-span motorway viaducts are usually formed of two separate bridge structures. For bridging large obstacles, it is possible to connect the decks of these structures by transverse members and suspend them on arches or pylons. The vertical component of the cable force is then transferred from the cross-beam into both bridge decks, while the horizontal component is transmitted by the deck slab. The transverse members can be formed by a beam of a solid cross-section, by a truss or by part of the deck connected by ties and struts (see Figure 3.5(d)).

Figure 4.78 Bridge across the Danube Canal, Vienna, Austria: (a) cross-section; (b) partial elevation; (c) elevation

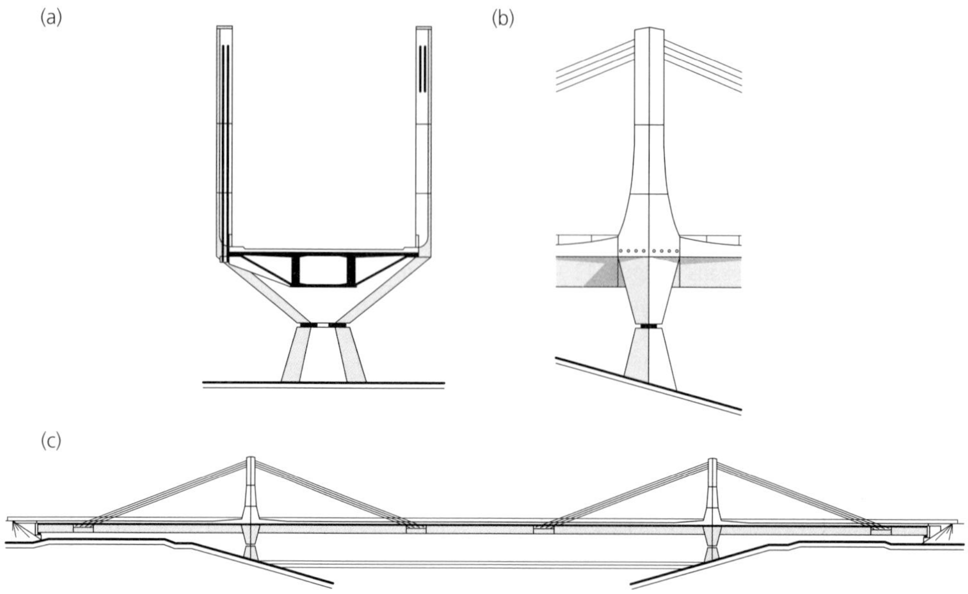

Figure 4.79 Bridge across the River Elbe, Czech Republic

Figure 4.80 Bridge in Figure 4.79: progressive deck assembly

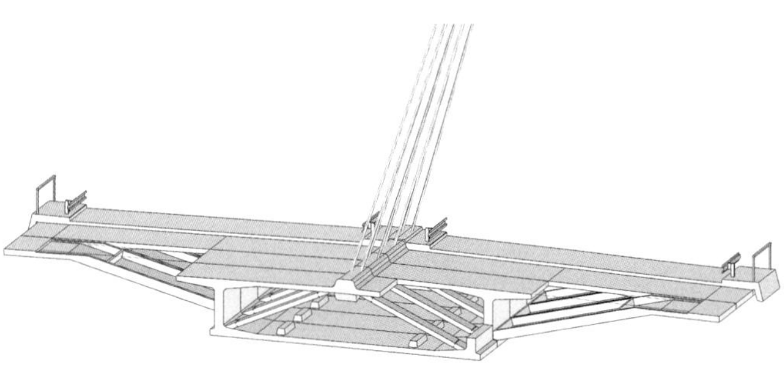

Figure 4.81 Brotonne Bridge across the River Seine, Normandy, France

The last of these arrangements is illustrated by the cable-stayed bridge across the River Odra and Antošovice Lake on the D1 motorway near Ostrava in the Czech Republic (Figure 4.83) (Stráský, 2009). The bridge is formed by a continuous structure of 14 spans, with span lengths of 21.50–105.00 m. The main span bridging the Odra River is suspended on a 46.81 m high single pylon situated in the bridge axis. As the stay cables have a symmetrical arrangement, the back-stays are anchored in two adjacent spans situated on the land between the river and the lake. The stay cables have a semi-radial arrangement; in the deck they are anchored at a distance of 6.07 m, while at the pylon they are anchored at a distance of 1.20 m.

Figure 4.82 Bridge in Figure 4.81: (a) pylon elevation; (b) cross-section of the pylon; (c) cross-section of the deck

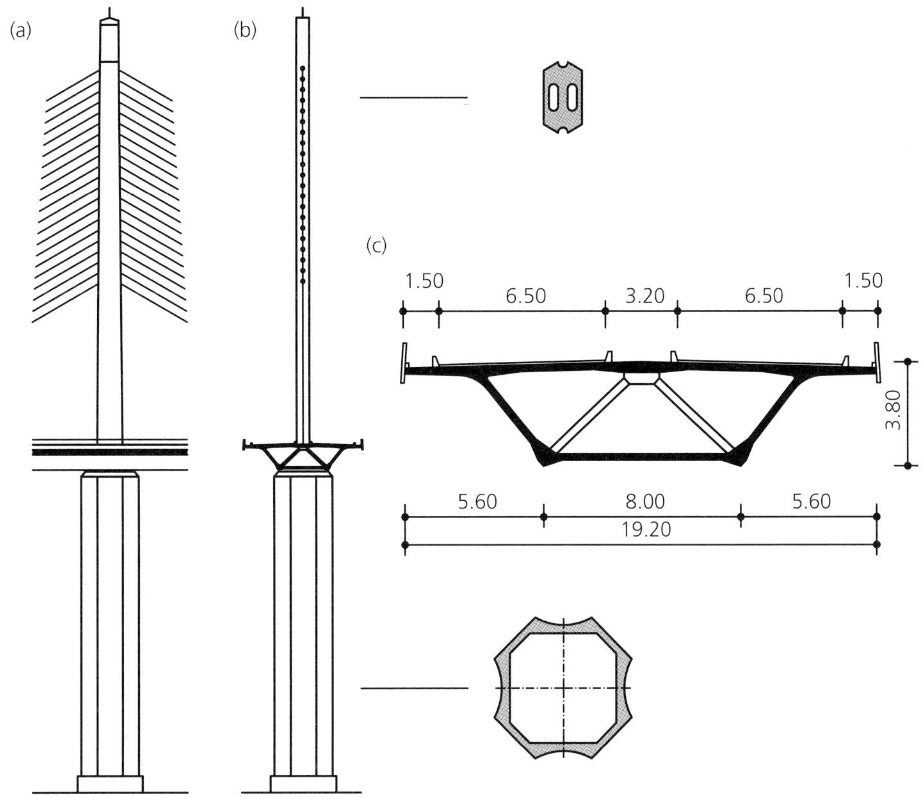

The decks are formed by two-cell box girders of depth 2.20 m, without traditional overhangs (Figures 4.83 and 4.84). The bottom slabs of both cells are inclined, and are curved in the middle of the girder. In the suspended spans the box girders are mutually connected by a top slab cast between the girders and by individual struts situated at a distance of 6.07 m. The stay cables are anchored at anchor blocks situated on the connected slab. The struts connect the curved bottom of the girders and, together with the inclined slabs, create a simple truss system that transfers the force from the stays into the webs. There are circular openings at the connected slab between the stay anchors (Figures 3.15 and 4.85).

Similar approach, but on a much larger scale, was used in the replacement of the eastern span of the Bay Bridge between San Francisco and Oakland, California, USA (Figure 4.86) (Brown, 2007). The main crossing is formed by a signature self-anchored suspension bridge linked to the 1.3 mile long Skyway viaduct. The twin decks of both bridges have a trapezoidal box section, and these sections have the same depth and perimeter at their connection (Figure 4.87). The 5.50 m deep decks of the suspension bridge are mutually connected by transverse box girders and are suspended on inclined suspension cables on their outer edges.

Figure 4.83 Bridge across the River Odra and Antošovice Lake, near Ostrava, Czech Republic

The concrete decks of the Skyway viaduct are strengthened by haunches. These are divided into four frames separated by expansion joints, which resist both the moment and the shear. Three frames have four piers and one frame has two piers. The typical span

Figure 4.84 Bridge in Figure 4.83: (a) cross-section of main span deck; (b) cross-section of approach spans; (c) cross-section at pylons

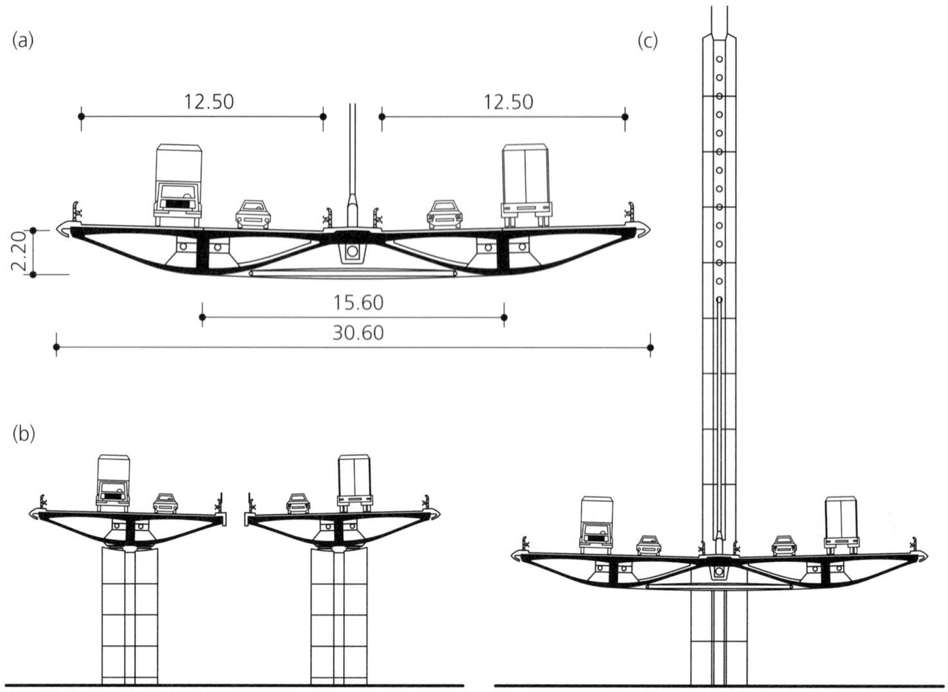

length is 160 m, the deck width is 26.7 m, and the depth varies from 9.0 m at the piers to 5.5 m at the midspans.

The typical span consists of a cast-in-place pier table and progressively erected precast segments. The segments, which each weigh 300–800 t, have a one-cell box section with large overhangs supported by slab struts. Struts made from lightweight concrete are inserted into the forms as precast members, and they are not mutually connected in the longitudinal direction. The piers are formed of four columns of pentagonal cross-section, which are mutually connected by transverse and longitudinal shear webs (see Figures 11.33–11.35).

A self-anchored suspension structure was also employed on the River Ebro crossing between the small cities of Deltebre and Sant Jaume D'Enveja in Spain. The bridge is situated close to the river estuary at the Mediterranean Sea. The arrangement of the bridge is the result of an architectural and structural competition (Stráský *et al.*, 2014). The client required a signature structure that corresponds to the scale of the cities it links (Figure 4.88). The bridge crosses the river at a skew angle, and has a crest elevation.

The bridge forms a self-anchored suspension structure composed of three spans (lengths 69.00 + 112.00 + 69.00 m). The 19.30 m wide deck is suspended on four suspension cables situated in the bridge axis. The torsionally stiff deck is formed by a composite four-cell box girder assembled from three steel webs, a curved bottom flange and a concrete deck slab (Figure 4.89). The central web is of variable depth and protrudes above the deck slab and substitute suspenders of the classical suspension structures, and so naturally divides the local highway from the pedestrian and bicycle routes. The steel structure is stiffened by transverse cross-beams at a distance of 3.00 m that support the composite deck slab. At the abutments the deck is stiffened by the end cross-beams, which transfer the load from the central webs into the bearings. The deck is connected via the frame to slender piers and low pylons. At the abutments the deck is supported by pairs of multidirectional bearings supplemented by stoppers (shock transition units). Guide bearings in the bridge axis resist the transverse effects.

The superstructure of the side spans and the sections situated close to the piers were assembled on temporary towers, and the central portion of the main span was floated and lifted into the design position (see Figure 7.34).

4.7. Bridges over deep valleys

The shape of arch bridges expresses the nature of the obstacles being bridged, so they are very suitable for bridging deep valleys (Figure 4.90(a)). However, they require a soil that is capable of resisting the arch horizontal forces. Recently, many interesting bridges have been built, although many of them are formed by stiff arches to ensure that instability does not occur at any stage of the construction or during the service life of the bridge.

Figure 4.85 Bridge in Figure 4.83: (a) cross-section of main span; (b) cross-section of approach spans

Figure 4.86 Bay Bridge between San Francisco and Oakland, California, USA

Figure 4.87 Bridge in Figure 4.86: (a, b) cross-sections of Skyway spans; (c) cross-section of main span

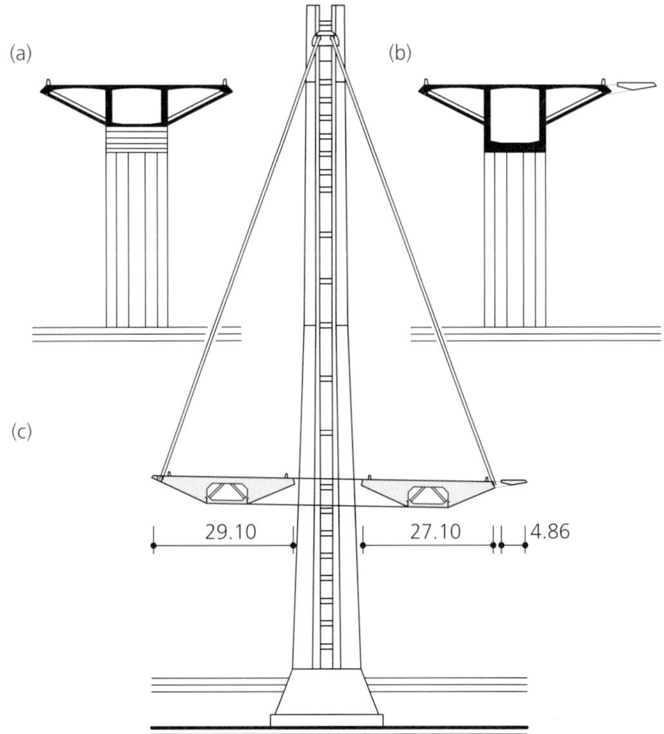

Figure 4.88 Bridge over the River Ebro between Deltebre and Sant Jaume D'Enveja, Spain

Figure 4.89 Bridge in Figure 4.88: cross-section at the pylon

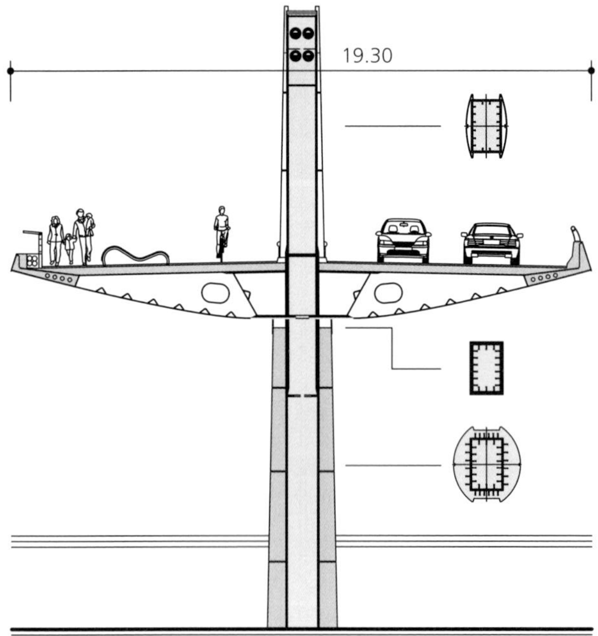

Figure 4.90 Bridges over deep valleys

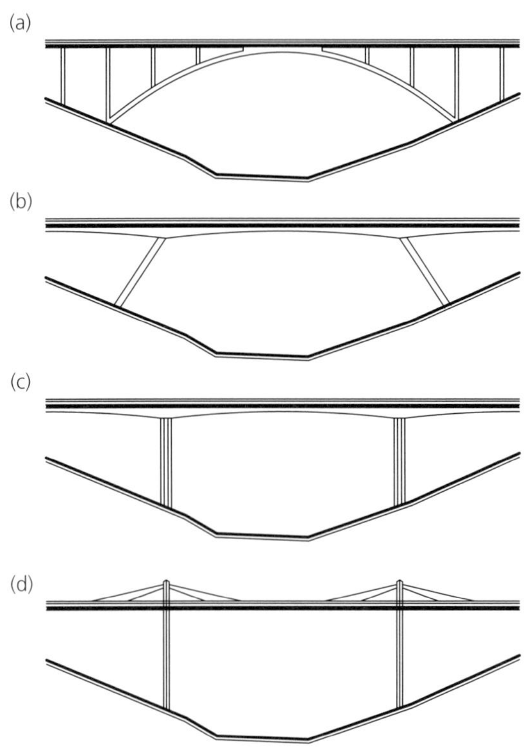

Beautiful slender arches with a stiff deck were designed by Robert Maillart in the first half of the 20th century. Later, in the second half of the 20th century, Christian Menn also designed some outstanding slender arches using prestressed concrete deck (Figure 4.91). However, the Infant Dom Henrique Bridge built in 2003 across the Douro River in Porto, Portugal (Figure 4.92), represents the most striking example of a beautiful and bold structure that utilises prestressing technology for both its construction and its service (Adão-da-Fonseca and Mato, 2005). The bridge is formed by a slender arch with a span of $L = 280$ m and a rise of 25 m. The arch supports the deck, which is formed of a 4.5 m deep one-cell box girder (Figure 4.93). The arch has a constant thickness of 1.50 m. Although arches of longer spans have been built, the Infant Dom Henrique Bridge is the world's 'boldest structure', having a static coefficient of $L^2/f = 3136$ m.

Bridges formed of strutted frames are also very efficient (see Figure 4.90(b)). Unlike arch structures, strutted-frame bridges can

Figure 4.91 Cascella Bridge, Mesocco, Switzerland

Figure 4.92 Infant Dom Henrique Bridge, Porto, Portugal

be built with a plan curvature. An excellent example of such a structure is the Kylesku Bridge built in 1982 in north-west Scotland (Figure 4.94) (Ciampoli, 1987). The bridge deck has a total length of 275.80 m, and is formed of a five-span curved box girder that is supported by V-shaped supports. The span lengths of the deck vary from 48.64 to 85.40 m, while the span lengths of supports are 71.90 + 132.00 + 71.90 m. The bridge axis has a complex geometry, with a minimum plan radius of 143 m. The V-shaped supports, which are also inclined transversally, give the structure a feeling of stability (Figure 4.95). While most of the structure was built on stationary scaffolding, the central part (43 m long) was cast away from the bridge and consequently floated to the construction site and lifted into the design position (see Figure 5.29).

For the bridging of wide valleys, typical cantilever structures are commonly used (see Figure 4.90(c)). Examples are the bridges at Valy and Rieka (total lengths 592 and 502 m, respectively) on the D3 motorway in Slovakia (fib, 2018). The bridges form continuous structures with span lengths of 24.50–92.00 m (Figure 4.96). The

Figure 4.93 Bridge in Figure 4.92: (a) cross-section at the midspan, (b) cross-section at a quarter span

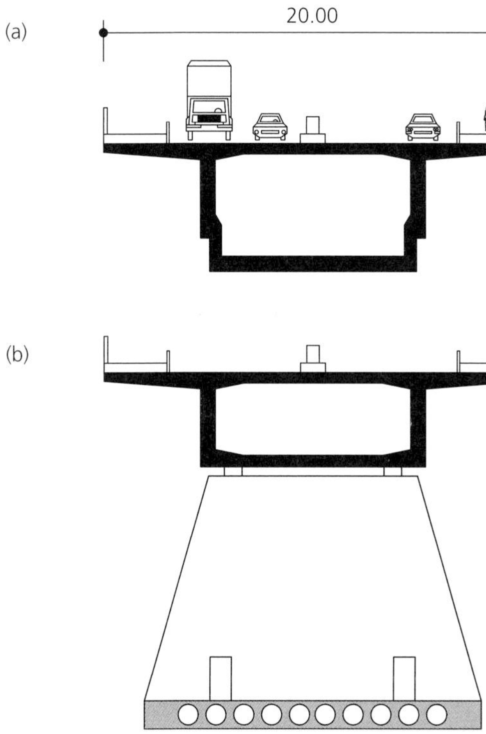

Figure 4.94 Kylesku Bridge, Sutherland, Scotland, UK

Figure 4.95 Bridge in Figure 4.94: (a) cross-section of the deck (b) cross-section of the bridge (c) elevation

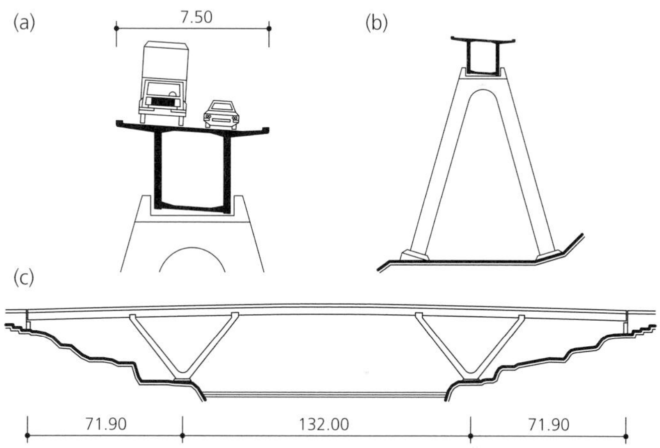

Figure 4.96 Bridge at Valy, Slovakia

11.40 m wide decks of both bridges are formed by single-cell box girders with a variable depth of 2.70–5.00 m (Figure 4.97). The decks are supported by slender piers with H cross-sections, which are formed of two slender transverse walls mutually connected by a longitudinal wall. The longitudinal walls are omitted at the top portion of the piers (see Figures 11.38 and 11.39).

If longer spans are needed, typical cantilever structures will have large depths, and the structures will become very heavy and look too robust in the landscape. Therefore, it is useful to suspend the deck on low pylons. The deck can be suspended via prestressed concrete walls (see Figure 4.90(d)) or multiple stay cables. Prestressed concrete walls were used in the spectacular Ganter Bridge at Ried-Brig in Switzerland, designed by Professor Menn and built in 1980 (Figure 4.98) (Menn, 1979). With its spectacular location, unique profile and innovative design the Ganter Bridge is considered to be one of the great bridges of the 20th century.

Figure 4.97 Bridge in Figure 4.96: (a) cross-section of deck; (b) cross-section of piers

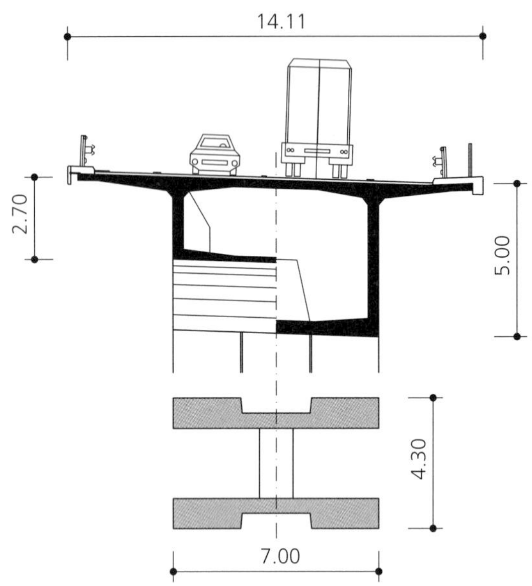

Figure 4.98 Ganter Bridge, Ried-Brig, Switzerland

In this bridge the 678 m long deck is formed of a box girder without traditional overhangs, and has eight spans of lengths varying from 35 to 174 m. The bridge is supported by two main towers and five piers. The main span between the towers is straight, while the other seven spans are curved. The 10 m wide box girder has a variable depth of 2.50–5.00 m, and is suspended on prestressed walls that follow the upward line of the girder webs (Figure 4.99). The height of the pylon above the deck is 17.53 m. The deck was segmentally cast in symmetrical cantilevers, starting at the towers and piers. During the construction of the main and neighbouring spans the deck was progressively suspended on the tendons situated in the partially cast walls. The tendons have dead anchors in the girder webs and overlap at the pylon heads, where they are anchored. After the midspan closures had been cast, the walls were prestressed.The prestressed concrete walls, which give the Ganter Bridge sufficient stiffness, can be substituted by multiple stay cables combined with a stiffer deck. This system, which is known as 'extradosed', was also used in the Mukogawa Bridge in Kobe, Japan, which was built in

Figure 4.99 Bridge in Figure 4.98: cross-section

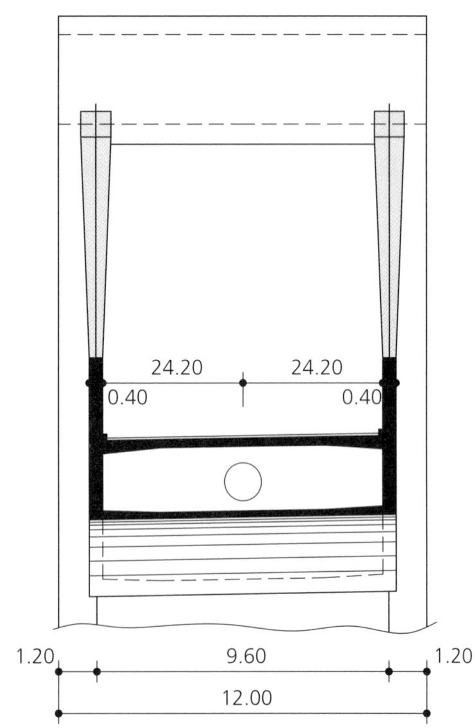

2016 (Figure 4.100) (Kasuga, 2017). The bridge has a total length of 442.20 m, and comprises five spans (lengths 72.80 + (3 × 100.00) + 72.80 m). The 24.20 m wide bridge deck, which is suspended in the bridge axis, is assembled from twin box girders mutually connected by the top slab (Figure 4.101). Close to the piers, where the box girders are suspended on low pylons, the girders are also mutually connected by the bottom slab, which is situated between them. The girder webs are formed of 'butterfly' precast panels, which were erected during the cantilever construction of the bridge deck. The stay cables are anchored at low pylons (8.50 m high, 1.35 m wide) formed of two inclined columns connected by steel plates.

Figure 4.100 Mukogawa Bridge, Kobe, Japan

Figure 4.101 Bridge in Figure 4.100: deck cross-section

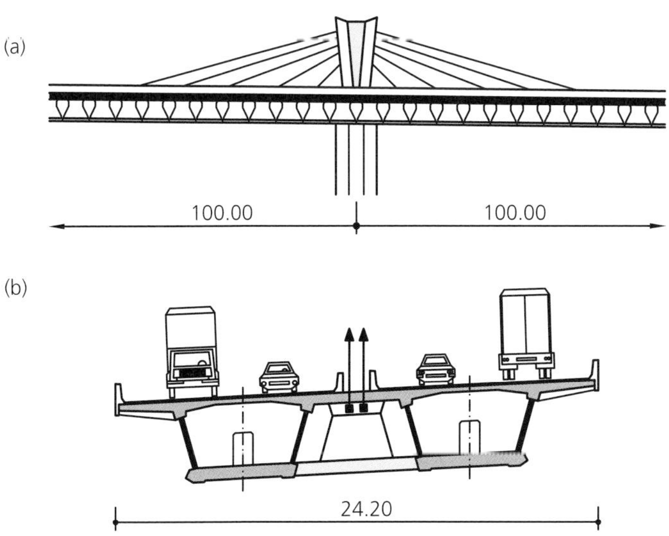

4.8. Footbridges

The design of footbridges requires special attention. Footbridge structures should be human in scale, lightweight, transparent and comfortable for pedestrians. Light, cable-supported footbridges are discussed in the author's book (Stráský, 2011), in which a comprehensive list of references can be found. Unfortunately, the limited scope of the present book does not allow for a more detailed description.

REFERENCES

Adão-da-Fonseca A and Mato FM (2005) Infant Henrique Bridge over the River Douro, Porto. *Structural Engineering International* **15(2)**: 85–85.

Brown DC (2007) Built to resist the big one. *ASPIRE* Winter: 12–17.

California Department of Transportation (1980) *California Bridges*. University Art Museum, Santa Barbara, CA, USA.

Ciampoli M (1987) The Kylesku bridge in Scotland: a prestressed concrete deck with partially precast central span. *Industria Italiana del Cemento* **7–8**: 458–471.

Dywidag (n.d.) Elevated highway railroad station Benrath. *Prestressed Concrete Bridges*. Report No. 7.

fib (2006) Structural concrete in the Czech Republic 2002–2005. *2nd fib International Congress, Naples, Italy*. National Report of the Czech Republic.

fib (2010) Structural concrete in the Czech Republic 2006–2009. *3rd International fib Congress*, Washington, DC, *USA*. National Report of the Czech Republic.

fib (2014) Structural concrete in the Czech Republic 2010–2013. *4th International fib Congress, Mumbai, India*. National Report of the Czech Republic.

fib (2018) Structural concrete in the Czech Republic 2014–2017. *5th International fib Congress, Melbourne, Australia*. National Report of the Czech Republic.

FIP (1998) Structural concrete in the Czech Republic 1994–1997. *13th International Congress of FIP, Amsterdam, The Netherlands*. National Report of the Czech Republic.

Kasuga A (2017) Butterfly web effects on bridge design. *Structural Concrete* **18(1)**: 128–142.

Leonhardt F (1984) *Bridges. Aesthetics and Design*. Deutsche Verlags-Anstalt, Stuttgart, Germany.

Manterola J (2006) *Puentes, Apuntes para su diseño, cálculo y construcción*, vols I and II. Colección Escuelas, Madrid, Spain.

Mathivat J (1983) *The Cantilever Construction of Prestressed Concrete Bridges*. John Wiley, New York, NY, USA.

Menn C (1979) The Ganter Bridge, Simplon. *IABSE Structures* **3**: 7–79.

Menn C (1990) *Prestressed Concrete Bridges*. Birkhäuser, Basel, Switzerland.

Michalek V and Klimes J (1970) *Bridge over the Nusle valley in Prague*. Práce, Prague, Czech Republic.

Muller J and Barker JM (1985) Design and construction of Linn Cove Viaduct. *PCI Journal* September–October: 38–53.

Murugesh G and Cormier K (2007) When lighter is better. *ASPIRE* Summer: 20–24.

Pauser A and Beschorner K (1976) Betrachtungen über seilverspannte Massivbrüken, ausgehend vom Bau der Schrägseilbrücke über den Donaukanal in Wien. *Beton- und Stahlbetonbau* **71(11)**: 261–265.

Schlaich J and Bergermann R (1992) *Fußgängerbrücken. Ausstellung und Katalog*. ETH, Zurich, Switzerland.

Smith DA, Hewson NR and Hendy CR (2009) Design of the Dubai Metro light rail viaducts – superstructure. *Proceedings of the Institution of Civil Engineers – Bridge Engineering* **162(2)**: 55–62.

Stráský J (1993) Design and construction of cable-stayed bridges in the Czech Republic. *PCI Journal* November-December: 24–43.

Stráský J (2009) Cable-stayed bridge across the Odra River, Czech Republic. *26th Annual International Bridge Conference 2009: Meeting Bridge Challenges in Challenging Times*, pp. 352–360.

Stráský J (2011) *Stress Ribbon and Cable Supported Pedestrian Bridges*, 2nd edn. Thomas Telford, London, UK.

Stráský J (2015) Viaducts with progressively erected decks. In *Multi-Span Large Bridges* (Pacheco P and Magalhaes F (eds)). CRC Press, London, UK, pp. 27–36.

Stráský J (2016) *Innovation in Bridge Design and Construction. Ponte da Arrábida*. University of Porto, Porto, Portugal.

Stráský J and Husty I (1999) Arch bridge crossing the Brno–Vienna Expressway. *Proceedings of the Institution of Civil Engineers – Civil Engineering* **132(4)**: 156–165.

Stráský J, Maťaščík M, Novák R and Táborská K (2010) Multi-span extradosed viaduct in Povazska Bystrica, Slovakia. *3rd fib International Congress 2010*. Washington, D.C.

Stráský J, Cobo del Arco D, Novotny P and Raventós I (2014) Construction of "Lo Passador": a self-anchored suspension bridge at the Ebro River delta. *Structural Engineering International* **24(1)**: 114–117.

Stráský J, Romportl T, Kocourek P and Rayor G (2013) Integral arch bridges. *ARCH 2013*. 7th International Conference on Arch Bridges, Split, Croatia.

Troyano LF (2003) *Bridge Engineering. A Global Perspective*. Thomas Telford, London, UK.

Stráský and Nečas
ISBN 978-0-7277-6385-3
https://doi.org/10.1680/dcpb.63853.083
ICE Publishing: All rights reserved

Chapter 5
Bridge construction

Concrete bridges are either cast-in-place or they are assembled from prefabricated elements. Composite structures, which combine steel, precast and cast-in-place structural members, are also used.

5.1. Cast-in-place structures

Cast-in-place structures are concreted on stationary or movable scaffolding, and incrementally launched or cast progressively, segment by segment. Structures can also be cast outside the obstacles and consequently rotated or lifted into the design position.

5.1.1 Stationary scaffolding

Only relatively short bridges (up to about 100 m) are concreted all at one time on a stationary scaffolding located along the entire length of the structure. These structures are usually prestressed by continuous tendons that are tensioned from both ends. In the case of longer bridges, the structures are usually cast incrementally, span by span, with short cantilevers protruding into the next span. The joints between the sections are usually chosen at sections where the dead-load bending moments are zero. In the case of a constant cross-section, the section width is approximately 1/5 of the span length. The bridge decks are prestressed by continuous cables tensioned and coupled at the joint. Because of the possibility of cracks (see Chapter 10), only one half of the cables can be coupled at the joints.

Over the years, inventory systems have been developed to enable economical construction. Stationary scaffolding is made up of either light columns (posts) that directly support a formwork (Figures 5.1(a) and 5.2) or the towers and beams (Figures 5.1(b) and 5.3). In both cases, the supports must be properly braced to ensure proper stability of the scaffolding. When designing scaffolding, it is necessary to devote attention to the design of a proper foundation. In addition, the possible scaffolding deformation and the corresponding camber of the scaffolding must be considered. A scaffolding design has to solve not only its assembly, but also its dismantling, which will be carried out under the finished bridge.

In the case of longer spans, which are assembled from several sections that are progressively prestressed, it is necessary to carefully check not only the state of the stresses in the sections but also the corresponding reactions that load the scaffolding. Figure 5.4 shows a three-span bridge in which the side spans with parts of the deck above the supports and the central part of the main span were cast and prestressed advance. After the closures had been cast, the whole bridge was prestressed by continuity tendons. Prestressing of the central section can cause lifting of the deck from the scaffolding and transfer of the whole dead load into the side towers. If these towers are not sufficiently strong, the scaffolding will fail.

Figure 5.1 Stationary scaffolding: (a) short-span structure; (b) long-span structure

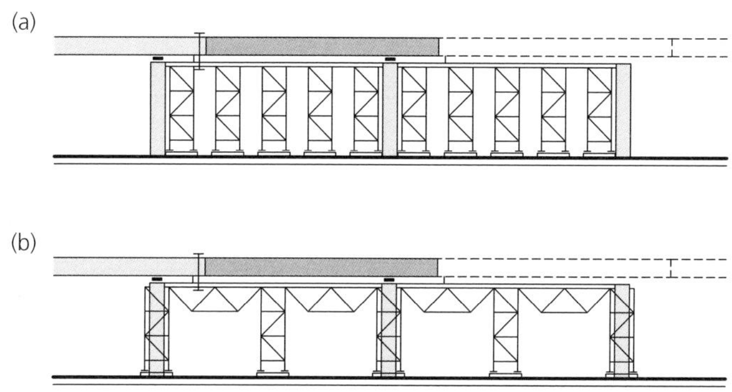

Figure 5.2 Místecká Bridge, Czech Republic

Figure 5.3 Olse River bridge, Czech Republic

Figure 5.4 Progressive construction of the main span: (a) casting; (b) prestressing

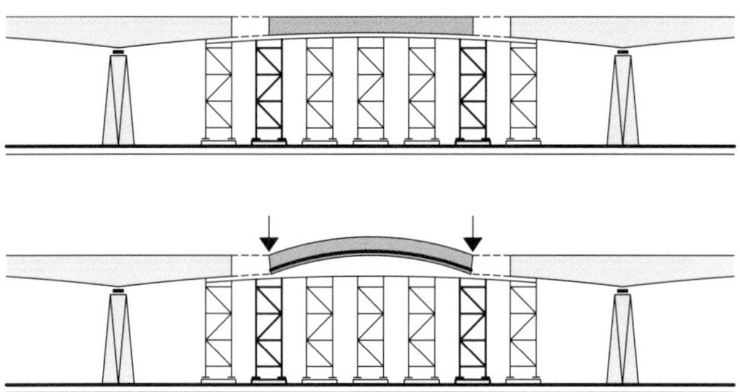

The function of the scaffolding must also be carefully checked for the construction of bridge decks that are supported by struts or arches. It is necessary to create structural models of the scaffolding and the erected structure, and to check the stresses during progressive casting, prestressing and dismantling of the scaffolding.

Stationary scaffolding is also used in construction of arch bridges. The design of the scaffolding will depend on the range of span lengths and the local conditions. For smaller spans, scaffolding similar to that used in the construction of girder structures is used (Figures 5.5 and 5.6), while for larger spans and deep valleys ingenious structures formed by struts and beams are constructed (Figures 5.7(a) and 5.8) (Ministry of Public Works, 1982). Scaffolding made of timber truss arches (Figure 5.7(b)) braced by a space

Figure 5.5 Bridge in Redmond, Oregon, USA

Figure 5.6 Bridge over the Willamette River, Eugene, Oregon, USA

net of cables has also been used. Scaffolding formed of a steel box (Figure 5.9) was used in the construction of the Arrábida Bridge in Porto, Portugal (Appleton, 2019). This concrete bridge has a beam deck, which is supported by two parallel arches. When the first arch had been cast, the steel box was moved transversely for the casting of the second arch.

To reduce the effects of concrete shrinkage and to guarantee uniform loading of the scaffolding, arches are usually cast symmetrically

Figure 5.7 Stationary scaffolding of arch bridges

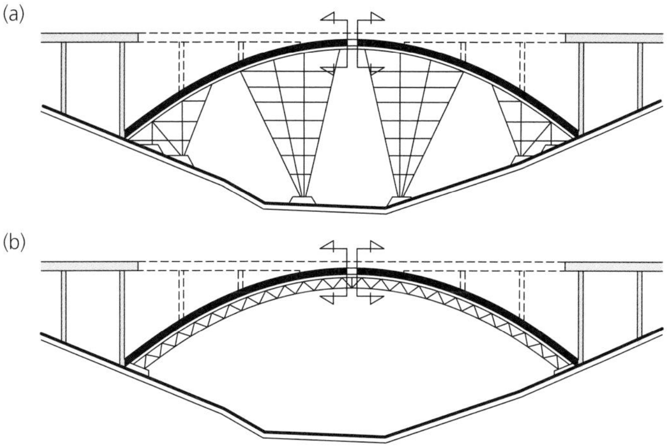

Figure 5.8 Houffalize Viaduct, Belgium

Figure 5.9 Arrábida Bridge, Porto, Portugal

Figure 5.10 Jacking frame for the bridge over the Willamette River, Eugene, Oregon, USA

segment by segment. A very economical progressive construction of arches having a box cross-section has also been used in the construction of several arch bridges. In this case, the scaffolding was designed to carry only the weight of the bottom slab, and the arches were cast progressively. First, the bottom slab was cast, and then the scaffolding was partially unloaded. Finally, the remaining members of the cross-section were cast. In this way the bottom slab resists the stresses originating from its own weight and from the weight of the webs and the top slab. The stability of the slender slab was guaranteed by the unloaded scaffolding that was tied to the bottom slab.

Arches are often stripped by the jacking of hydraulic jacks located at the arch crowns (Figure 5.10). The jacks are usually inserted in jacking frames, in which strong bolts resist the compression force before the joint is cast. To eliminate the effects of the elastic deformations of the subsoil, the jacking force is usually repeated after a couple of days. The bolts are then fixed, and joint is cast.

5.1.2 Movable scaffolding

If a bridge is longer than 400 m, or if a structure needs to pass over an inaccessible obstacle, it is appropriate to cast the structure in a formwork supported or suspended on movable scaffolding (Dyckerhoff and Widman, 1973; Wittfoht, 1972). Over the years, sophisticated scaffolding systems have been developed, and these have subsequently influenced bridge design. Movable scaffolding can be used in the construction of bridges of constant or variable cross-section that are led in a space curve or in a super-elevation.

With respect to the design of structures, scaffolding can be divided into scaffolding used for the concreting of the whole span (Figures 5.11(a) and 5.11(b)) and scaffolding used for the symmetrical concreting of

Figure 5.11 Movable scaffolding systems

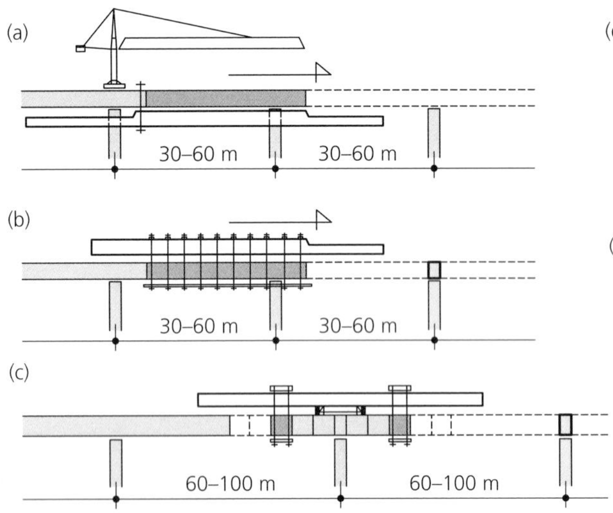

segments (Figure 5.11(c)). The first type of is optimal for span lengths in the range 30–60 m, while the second type, which combines the technology of cantilevering with the technology of movable scaffolding, is suitable for lengths in the range 60–100 m.

Scaffolding for casting the whole span can be situated either under the deck (Figures 5.11(a), 5.12 and 5.13) or above the deck (Figures 5.11(b), 5.14 and 5.15). These are usually referred to as underslung movable scaffolding systems and overhead movable scaffolding systems, respectively. The formwork is either supported on or suspended from the scaffolding. The advantage of under-deck scaffolding is that the working space is kept free. The disadvantage is that the weight of the scaffolding carrying the wet concrete has to be transferred into the piers by a temporary steel structure (see Figure 5.13), and it is necessary to use a crane to transport the material.

Conversely, overhead scaffolding can be used for the transport of materials. However, the working space will be limited by the bars carrying the formwork. The scaffolding is supported above the piers by support segments (see Figure 5.15), which have to be cast in advance. If these segments are supported by bearings, it is necessary to temporarily fix the bearings to ensure the stability of the pier segments. The pier segments are sometimes substituted by steel structures which are, during casting of the span, embedded in the pier diaphragms.

Figure 5.12 Bridge near Olomouc, Czech Republic

Figure 5.13 Viaduct near Bitarová, Slovakia

A complication when using movable scaffolding is that it is necessary to laterally shift the formwork when moving the scaffolding onto the next span (see Figures 5.11(e) and 5.11(g)). Therefore, cross-sections that require minimum lateral movement have been developed for bridges with smaller spans. If the scaffolding is situated under the deck, a double-T cross-section without any diaphragms is used (see Figures 5.11(d) and 5.12). A spine girder with large overhangs that supported by narrow piers is also very economical. Similar bridge types also offer an optimum solution for structures where the scaffolding is situated above the deck (see Figure 5.11(f) and 5.14) (Zorzi *et al.*, 1978). In the past, 'mushroom' decks with widths of up to 30 m were commonly built (Dyckerhoff and Widman, 1973).

Figure 5.14 Viaduct on the Palermo–Sciacca highway, Sicily, Italy

Figure 5.15 Bridge over the Váh River, Slovakia

In order to ensure a continuous connection between the newly cast span and the previously cast structure, the scaffolding is suspended at the end of the cast part. As the scaffolding deforms during the casting, it is usual to delay the setting of the concrete by the addition of retarders, which postpone the setting until the whole span has been casting.

As the length of the span increases, the weight of the scaffolding increases disproportionately. It is for this reason that segmental casting of decks is usually selected. The deck is incrementally cast and prestressed, segment by segment, in symmetrical cantilevers, starting from the piers (Figures 5.11(c) and 5.16) (Matyassy and Palossy, 2006). The formwork for segments up to 10 m long is hung on scaffolding supported by the pier segments and previously constructed cantilevers. When the last segments have been cast, the structure is post-tensioned by continuity tendons, and the scaffolding is moved into a new position.

When designing structures that will utilise movable scaffolding, it is necessary properly to determine the camber of the scaffolding. This depends not only on the deformation of the main load-bearing steel girders, transverse frames and ties, but also on the deformations of the concrete cantilevers supporting the scaffolding. The latter deformation depends on the structural arrangement and layout of the prestressing cables in the previously cast spans and on the modulus of elasticity of the concrete, which depends on the concrete's age. To determine the camber, it is necessary to carefully survey the constructed spans and scaffolding and to undertake tests on the concrete. It is usually necessary to survey two or three spans if satisfactory results are to be obtained.

If the deck is formed by a box girder that is cast in two steps (e.g. bottom slab and webs first, top slab later), it is necessary to consider

Figure 5.16 Kőröshegy Viaduct, Hungary

Figure 5.17 Underslung movable scaffolding system with organic prestressing

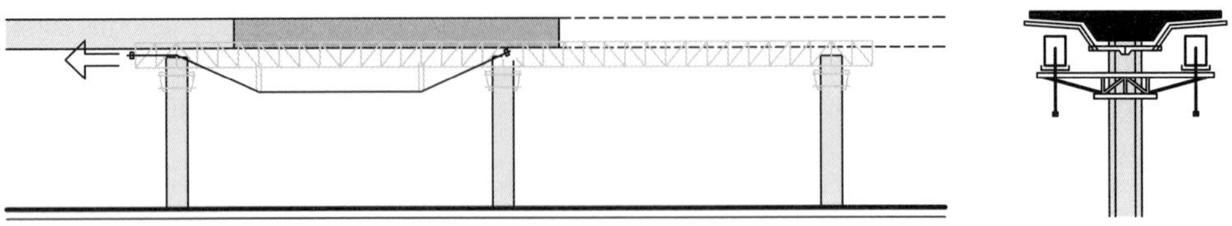

Figure 5.18 Bridge over the Corgo River, Portugal

Figure 5.19 Overhead movable scaffolding with organic prestressing

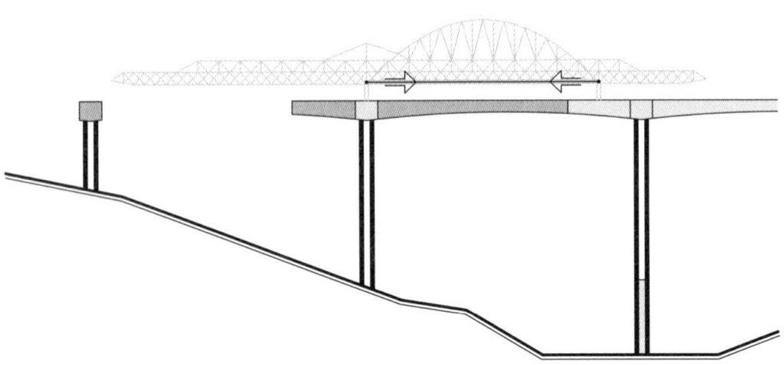

that first casting participates in resisting the weight of the second casting (see Chapter 10).

The above problems can be easily solved by using scaffolding that has been 'organically prestressed' (Pacheco, 2015). The scaffolding is prestressed by cables that balance the dead load. During casting, the deformation of the scaffolding is monitored automatically. When the deformation of the scaffolding reaches certain limits, the tension in the cables automatically increases, and in this way the deformation is eliminated.

Figure 5.20 Bridge on the D1 motorway, near Janovce, Slovakia

In the case of underslung movable scaffolding systems, the scaffolding is formed of beams prestressed by polygonal external cables situated under the structure (Figures 5.17 and 5.18). In the case of overhead movable scaffolding systems, the scaffolding is formed by a tied arch in which the ties are formed of cables (Figures 5.19 and 5.20) The use of cables allows to the weight of the scaffolding to be reduced and its span length to be increased. To date, scaffolding of this type has been used for spans up to 70 m.

5.1.3 Incremental launching

Incrementally launched structures are cast segment by segment in lengths of 15–40 m in forms that are usually situated behind the abutments (Rosignoli, 2002). After casting and prestressing a new segment to the already cast structure, the whole structure is launched into a new position to allow the casting of a new segment (Figure 5.21). It is possible to launch bridge decks the profile and alignment of which have a constant curvature.

When launching, the webs of the cross-sections must be evenly supported. Therefore, launched decks usually have a one-cell box section. However, multi-cell or double-T sections have also been used.

During launching the structure is stressed by large cantilever bending moments. To reduce these, a lightweight steel nose is usually

Figure 5.21 Incremental launching: (a) with a steel nose; (b) with a temporary pylon and stays

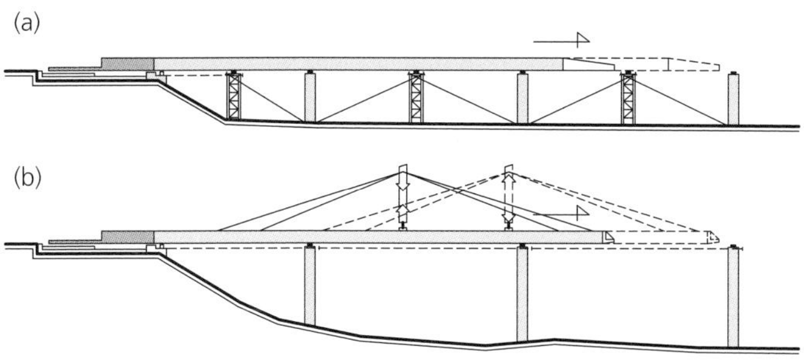

Figure 5.22 Incremental launching of the bridge over Rybny Creek, Czech Republic

Figure 5.23 Incremental launching of the bridge over Rybny Creek, Czech Republic

attached to the end of the cantilever (Figures 5.21(a), 5.22 and 5.23). The length of the nose is typically 60–80% of the length of a typical span. The nose is usually prestressed to the launched structure by means of prestressing bars.

The cantilever moments can also be reduced by means of a temporary pylon with stay cables (see Figure 5.21(b)). In this case only a short steel nose sufficient in length to get onto the pier is attached to the cantilever. Therefore, the pylon is usually situated at a distance that is equal to the length of the typical span of the launched structure. During launching, the pylon loads the structure with variable vertical forces, so it is necessary – depending on the pylon position – to adjust these forces. The pylon is usually supported by hydraulic jacks that allow lifting or sagging of the pylon. By this vertical movement, the forces in the pylon, and in the cables, are regulated. Of course, it is possible to combine the typically used steel nose with a temporary pylon with stays (Figure 5.24) (Cezard and Servant, 1988).

There are two types of a launching equipment: pushing and pulling. The pushing device consists of vertical and horizontal hydraulic jacks (Figure 5.25(c)). During launching the structure is first raised vertically, then shifted horizontally. The pulling device utilises prestressing jacks and cables that are anchored at the end of the launched structure (Figure 5.25(d)). The jacks are supported by the end abutment.

The form is usually constructed in two steps. In the first, the bottom slab is cast (Figure 5.25(a)), and in the second the webs and top slab are cast (Figure 5.25(b)). Because the geometry of the launched structure depends on the accuracy of the execution of the bottom plate, its shape must be carefully defined. In the construction of several newly launched bridges, the bottom slab with webs was cast first, followed by the top slab (Figure 5.26).

Figure 5.24 Charix Viaduct, France

The deck is launched across the concrete blocks, the top surface of which is provided with stainless steel plates. When launching, Teflon plates are inserted between the blocks and the bottom slab of the girders (Figure 5.27(a)). Guide elements that maintain the transverse direction of the launched structure are situated on the supports (Figure 5.27(b)). Modern launched structures also utilise special blocks formed of steel plates supported by neoprene layers that allow the plates to rotate (Figure 5.28).

Figure 5.25 Form and launching: (a) cross-section A–A; (b) cross-section B–B; (c) longitudinal section A–A with pushing equipment; (d) longitudinal section A–A with pulling equipment

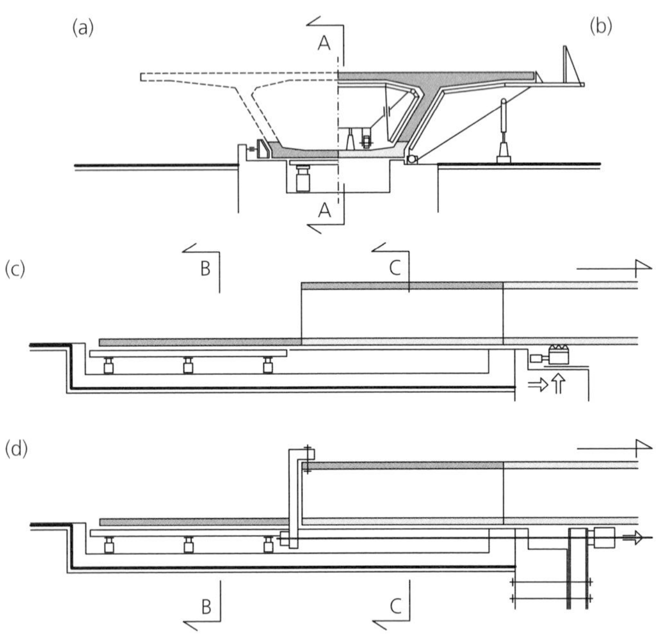

Figure 5.26 Bridge over Rybny Creek, Czech Republic

When designing a launching device, it is necessary to remember that the launching blocks are positioned parallel to the girders, and therefore it is necessary to overcome the corresponding component of the self-weight during launching. It is also necessary to overcome the friction between the Teflon plates and the structure, which is in the range 0.01–0.03. Launching is usually done down and along the slope, and therefore it is not necessary to overcome the component of the self-weight, but rather it is necessary to ensure effective braking of the structure. However, launching devices are conservatively designed for a friction value of 5%. This value is also considered when designing the temporary and definitive supports.

The deck depth of the launched structures is usually relatively large – from 1/10 to 1/14 of the span length. Therefore, in many cases it is possible to launch structures without the need for temporary supports. If it is necessary to design a smaller construction depth, one to two temporary supports are used in the spans. Because the supports are stressed by a large horizontal force due to friction, they must be anchored (see Figure 5.21(a)). Anchoring is sometimes used for the definitive piers too.

Pier diaphragms are usually cast after the entire structure has been launched. During launching the deck is centrically prestressed. After

Figure 5.27 Launching concrete blocks: (a) elevation, (b) cross section

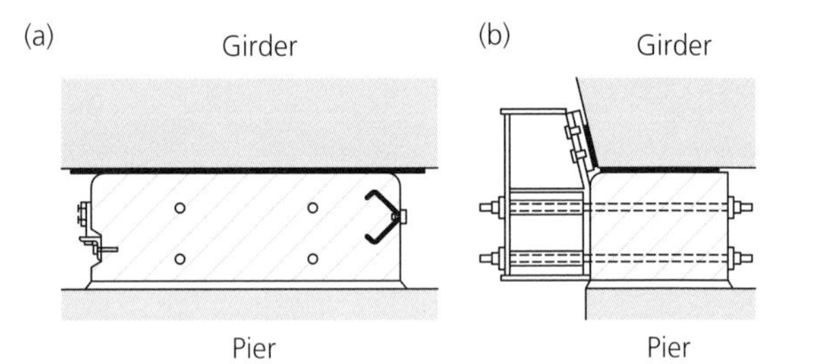

Figure 5.28 Bridge over Rybny Creek, Czech Republic

launching it is necessary to post-tension the continuous cables along the course that corresponds to the course of the bending moments of the continuous beam. The cables are tensioned after the structure has been supported by the definitive bearings and after the removal of the launching blocks. As it is difficult to produce the concrete deck with a camber, the radial forces from the continuous cable should balance the bridge dead load.

The first incrementally launched bridge had a total length of 480 m (Wittfoht, 1972) and was assembled from prefabricated segments that were centrically prestressed by external cables after casting of the joints. After the entire structure had been launched, the cables were moved to a position corresponding to the course of the bending moments. It is evident that precast segments incrementally assembled and prestressed beyond the abutments can also be used. As shown in Chapter 10, the deck can be prestressed by temporary and final external cables.

Structures can be launched not only longitudinally but also transversely. Usually the bridge deck is cast in a position parallel to the projected position, and is then moved transversely on temporary supports situated next to the definitive supports.

5.1.4 Heavy lifting and rotating

Whole bridge spans or their parts may also be produced outside of the bridge and then transported under the structure. The structure is then suspended on the prestressing cables and lifted into the design position by the tension in the hydraulic jacks (Figures 5.29 and 4.94) (Ciampoli, 1987). When the joints between the lifted and already erected structure are cast, the whole structure is prestressed by means of continuity tendons.

The structure or part of the structure can also be constructed alongside the obstacle and then rotated into the design position (Figure 5.30). The rotated structure must be supported at three

Figure 5.29 Kylesku Bridge, Sutherland, Scotland, UK

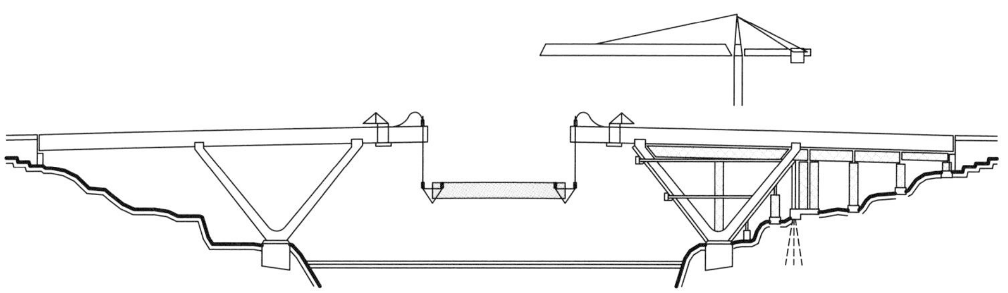

Figure 5.30 Rotation: (a) elevation; (b) plan

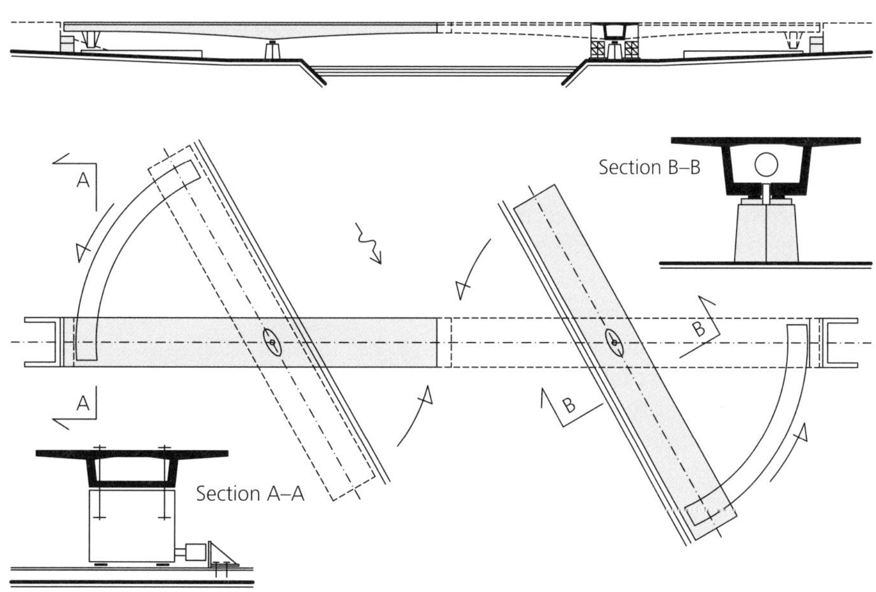

Figure 5.31 Danube Canal bridge, Vienna, Austria: (a) elevation; (b) plan

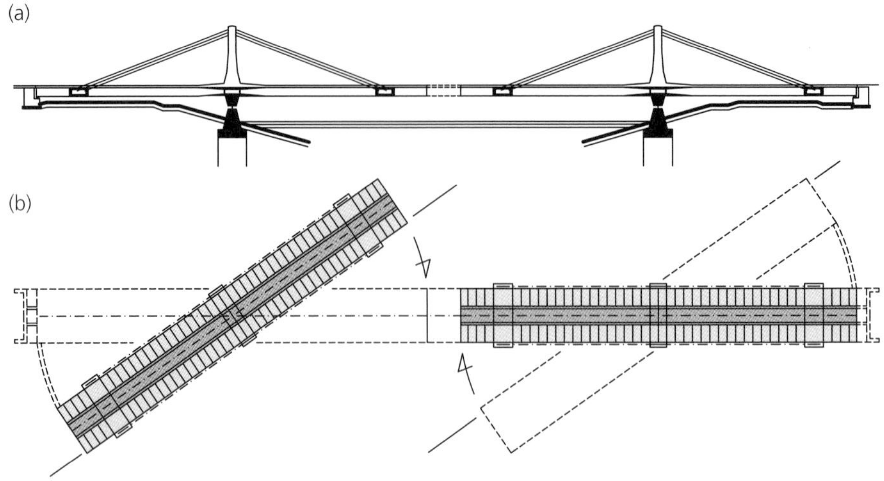

points. The first point forms a support around which the structure rotates. When rotating, the structure is usually supported by elastomeric bearings that have a Teflon surface. The position is guaranteed by a vertical steel pin. The second two points are supported by a rear support that moves along a circular path. The support is provided by a temporary or definite pier; the circular pass is made of a concrete footing with strips of stainless steel on its surface. When rotating, Teflon plates are inserted between the stainless steel and the rear support, similarly to the procedure for launched structures. The rotation is caused by a force created by hydraulic jacks inserted between the rear support and brackets anchored into the footings.

One of the first prestressed concrete structures for which rotation was used is Danube Canal bridge, which was described in Chapter 4 (see Figure 4.78) (Pauser and Beschorner, 1976). Because construction was not allowed to interfere with navigation on the canal, the structure was built in two 110 m halves on each bank and parallel to the canal (Figure 5.31). Upon completion, the two halves were rotated into their final position and a cast-in-place closure joint was made. During rotation of the two half-bridges, the deck and pylon sat on a bearing consisting of five epoxy-glued circular steel plates. The top plate was coated with Teflon, and sat on a reinforced concrete block, which sat on a sand box. After rotation the structure was lowered onto the permanent bearings by emptying the sand box. At the canal-bank end, the deck had a concrete wall on its underside, bearing on a circular concrete sliding track. After rotation, the two halves of the structure were connected by a cast-in-place closure joint and continuity tendons were placed and stressed.

5.1.5 Cantilever construction

The deck of a cantilever bridge is cast segment by segment into a formwork supported by travellers that are fixed into the already cast and prestressed part of the structure (Figures 5.32 and 5.33). The travellers can be positioned above (Figure 5.34(a)) or below (Figure 5.34(b)) the bridge deck. Once the segment has been cast, the prestressing cables are post-tensioned. In this way the new segment is prestressed to the previously completed part of the structure. Then the traveller moves into a new position. The procedure is repeated until the cantilever reaches the centre of the bridge span (Dyckerhoff and Widman, 1973; Mathivat, 1983; Podolny and Muller, 1982). Here, between the neighbouring cantilevers, the middle segment is cast (Figure 5.35), and the structure is prestressed by continuity and span tendons.

The technology used for the cantilever construction greatly influences the bridge design. As the structures are usually concreted in cantilevers extending from the supports to the middle of the bridge spans, a substantially variable cross-section corresponding to the

Figure 5.32 Bridge at Valy, Slovakia

Figure 5.33 Cantilever construction: (a) one-span bridge; (b) three-span bridge; (c) multiple-span bridge erected with a steel launching beam

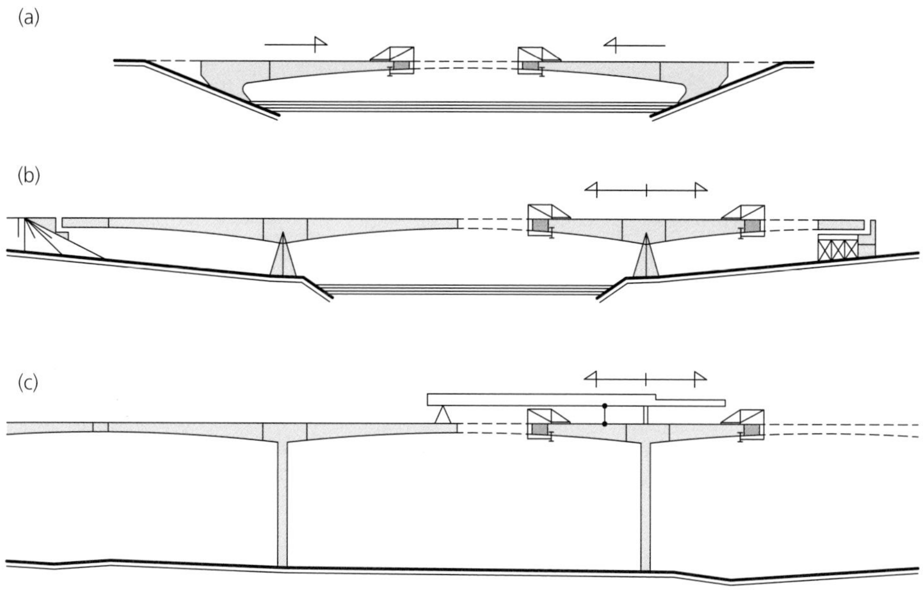

Figure 5.34 Traveller: (a) above the deck; (b) below the deck slab

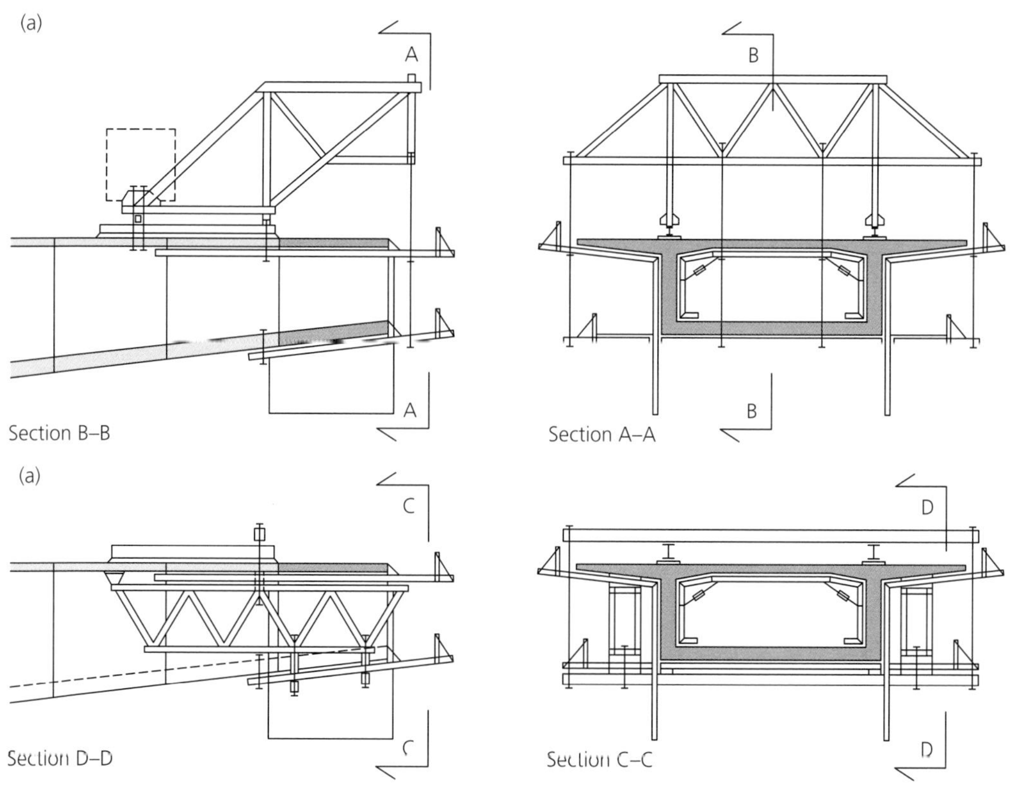

Figure 5.35 Casting of the mid-span segment

Figure 5.37 Casting of typical segments

course of bending moments is suitable. One-span bridges are usually cast in cantilevers extending from the end abutments to the middle of the bridge (see Figure 5.33(a)), while multi-span bridges are usually cast in cantilevers extending from the intermediate supports to the middle of the bridge spans (see Figure 5.33(b)). With respect to the weight limitation of the travellers and with respect to the stresses of the preceding segments, the segment length is usually 3–5 m.

During construction, pier tables of length 7–10 m, on which the travellers are assembled, are cast first (Figure 5.36). They are usually cast on the stationary scaffolding. As the pier tables are stiffened by pier diaphragms in which continuity tendons are usually anchored, they are heavily reinforced. Therefore, their construction takes several weeks.

Figure 5.36 Pier table

A typical segment is usually cast all at one time (Figure 5.37). Casting begins with casting of the bottom slab, and then the webs and the top plate are cast. In order to ensure proper compaction, several holes for vibration are usually provided in the webs. Two symmetrical segments are usually concreted within one week. The deck is usually prestressed when the concrete reaches a strength of about 80% of the design strength.

The structure of the travellers consists of a truss structure that is anchored in the previous segment. The traveller must be stiff enough to avoid cracks in the joint between the segments. For this reason, it is quite heavy – usually weighing 50–80 t. This weight represents a considerable load, which must be included in the analysis of the erection stages.

During construction it is necessary to ensure the stability of the build structure. This is done by permanent or temporary fixing into the substructure (see Chapter 11). Stability can also be provided by a steel launching beam linking the end of the already concreted cantilever to the pier table (see Figure 5.33(c)), a method that has been used in the construction of several large bridges (Dywidag, n.d.). These beams also serve to transport material and workers, and to move the travellers.

Progressive cantilever construction has also been used, where the cantilever is constructed progressively from one end abutment to the other. To reduce static effects in the cantilevers, either temporary supports (Figure 5.38(a)) or temporary pylons with stay cables (Figure 5.38(b)) have been used. Progressive construction is commonly used when the length of the side spans is larger than one-half of the length of the main span. The temporary support is activated after the segment above has been cast, and the traveller is then moved into the next position.

Figure 5.38 Progressive cantilever construction: (a) with temporary supports; (b) with a temporary pylon and stays

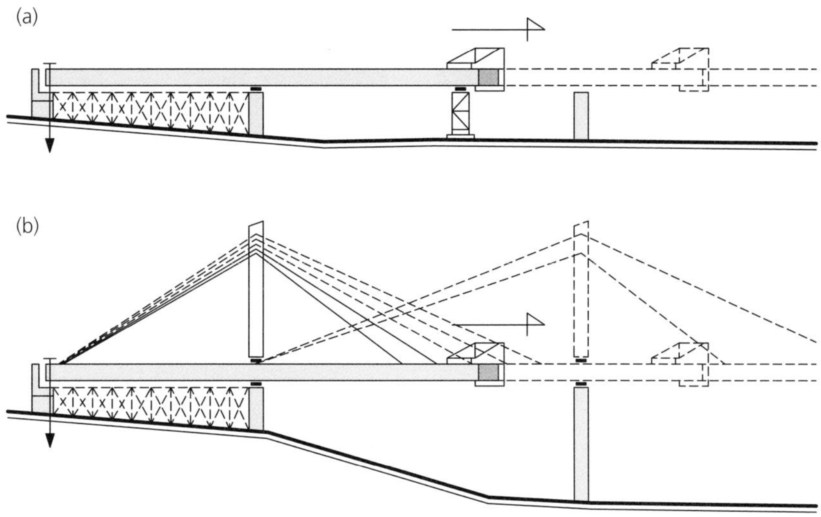

Figure 5.39 Viaduct at Považká Bystrica, Slovakia

Cantilever construction is also utilised in the construction of cable-stayed bridges. For structures with a stiff deck, conventional travellers are used (Figure 5.39); for structures with a slender deck, the travellers are usually suspended on the pylons (Figures 5.40 and 5.41). The travellers can be advantageously suspended on the definitive stay cables that are anchored in precast members, which are connected to the previously cast structures and to the travellers (Walther *et al.*, 1998). The precast members transfer the horizontal components of the stay forces. To guarantee the linear behaviour of the stay cables, it is usually necessary to temporarily load the travellers. Once the segments have been cast, this loading is progressively removed. This can be easily done using tanks filled with water that are then drained.

Cantilevering is also used in the construction of arch bridges (Adão-da-Fonseca and Mato, 2005; Arenas *et al.*, 2016). The arches are either temporarily suspended on a pylon located on supports situated above the arch footing (Figures 5.42(a) and 5.43), or above the

Figure 5.40 Cantilever construction of bridges with a slender deck: (a) elevation; (b) partial elevation; (c) cross-section

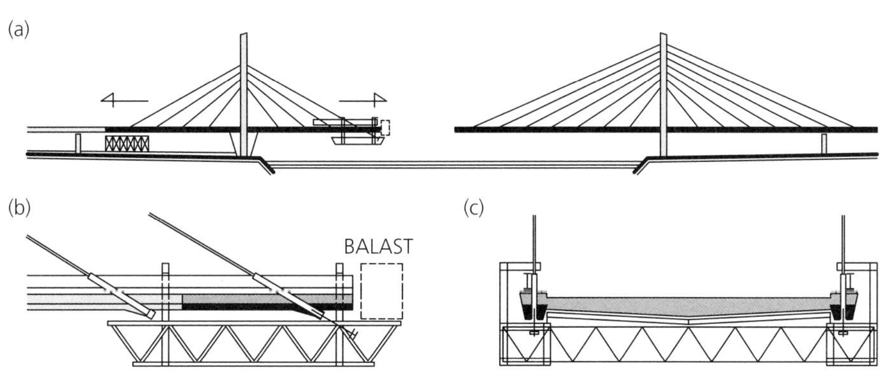

Figure 5.41 Helgeland Bridge, Norway

Figure 5.42 Construction of arch bridges

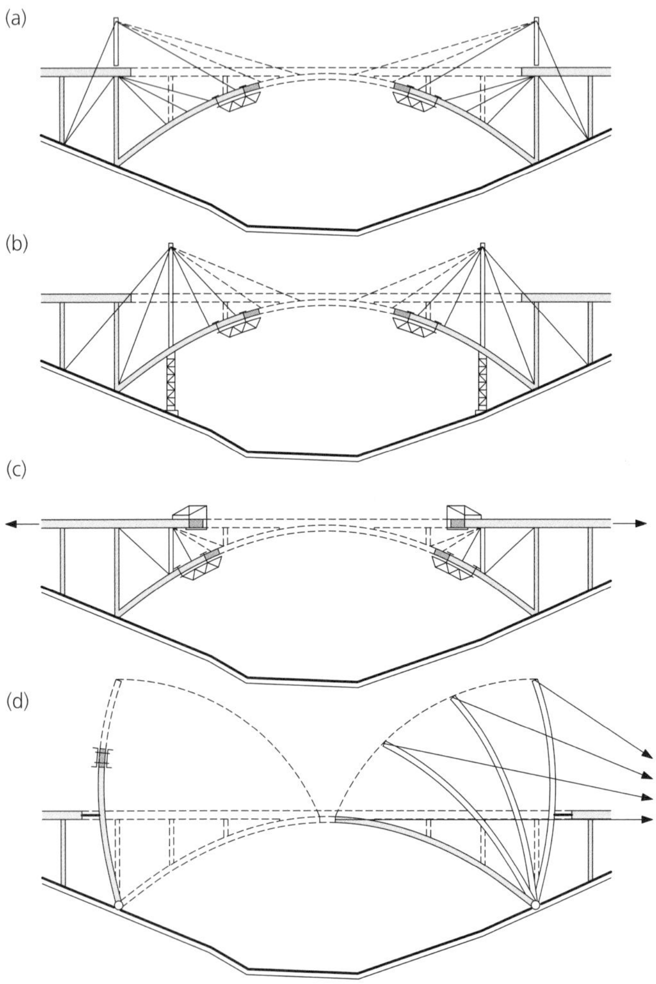

Figure 5.43 Almonte River Viaduct, Garrovillas de Alconétar, Spain

Figure 5.44 The Arco de la Regenta bridge over the Cabo River, Spain

temporary supports located under the arches (Figure 5.42(b)). Temporary suspension can be achieved with truss structures that use the tension capacity of the bridge deck and temporary stay cables and the compression capacity of the already cast piers and arches (Figures 5.42(c) and 5.44). To reduce the deformation of the erected structure, the diagonal cables are usually formed of prestressing bars. Bridge arches can also be concreted vertically, and subsequently rotated into the design position (Figures 5.42(d) and 5.45).

5.2. Precast structures

It is undisputed that industrial-scale production in stationary plant makes it possible to produce high-quality prefabricated elements. However, the prefabricated elements must be transported to the

Figure 5.45 Bridge over the Kobara valley, Japan

construction site, and assembled there. The bridge structure can be erected from longitudinal precast elements (Figure 5.46(a)) or from transverse precast members (segments) (Figure 5.46(b)) (Mathivat, 1983; Podolny and Muller, 1982). For wider bridges, segmental structures can also form longitudinal elements that need to be cross-connected.

5.2.1 Longitudinal precast beams

Bridge decks are usually assembled from precast beams that have an open cross-section (inverted-T, I, T, U and V girders – see Chapter 9) and cast-in-place composite slabs. The beams are mostly erected by means of crane trucks. Depending on the span length, they are erected using either one or two cranes (Figures 5.47(a), 5.48 and 5.49). If the girders are suspended on one crane, an erection beam should be used to resist the compression forces, or the girders must be checked for the compression force originating from the inclined suspenders. As the girders usually have a low transverse stiffness, it is necessary to guarantee their vertical position during their erection, and their stability must be carefully checked.

If the deck is to be erected across an inaccessible obstacle (see Figure 5.47(a)) an erection beam is usually used. The girder is placed on this beam, and then moved into the erected span, where it is moved transversely by a couple of cranes.

The precast girders can be also erected by means of a launching gantry, which allows erection of the girders independently of the terrain under the bridge (Figures 5.47(b) and 5.50). The launching gantry has a length that corresponds to the length of two spans, and is situated both in the erected span and in the already assembled span. The beams allow the girders to be lifted and moved longitudinally and transversely into the design position.

Figure 5.46 Precast structures: (a) precast beams; (b) precast segments

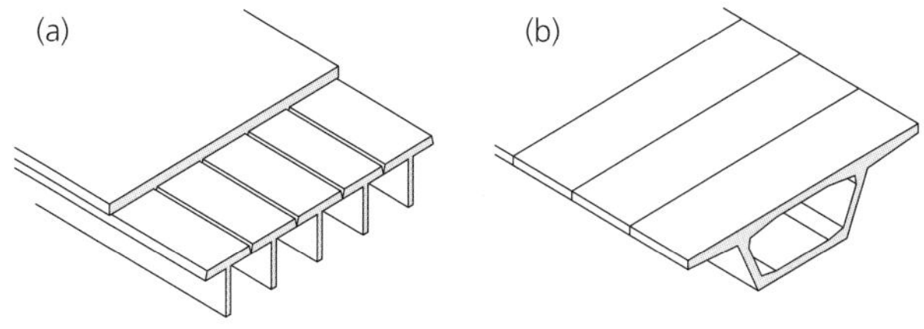

Figure 5.47 Erection of precast girders: (a) using cranes; (b) using a launching gantry

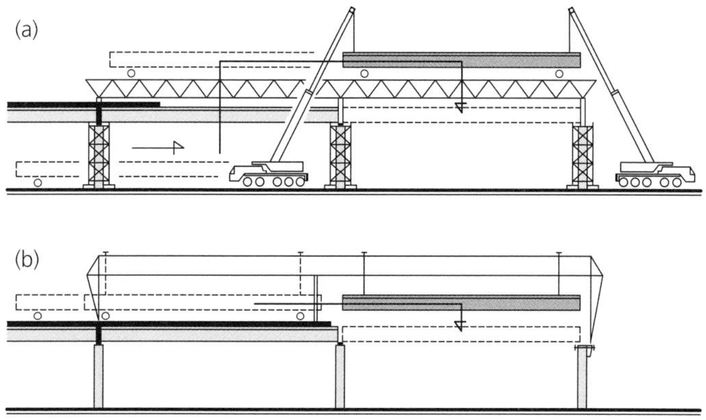

Figure 5.48 Bridge on the D1 motorway, Slovakia

Figure 5.49 Bridge on the D1 motorway near Višňová, Slovakia

Figure 5.50 Bridge on the D3 motorway, Slovakia

Figure 5.51 Viaduct on the Rome–Florence railway, Italy

Figure 5.52 Bridge on the D5 motorway, Czech Republic

In the construction of long railway bridges, the decks are formed of a chain of simply supported girders having a box or channel section. Launching gantries have been developed that allow the erection of girders weighing up to 600 t (Figure 5.51) (Gallo, 1990).

5.2.2 Precast segments

The bridge decks are assembled from prefabricated elements – segments of length 1.5–4.0 m. The segments usually have a single-cell box section (Figure 5.52), although multi-cell box sections, or double-T and channel sections have also been used. In the early segmental structures, the joints between the segments were formed by concrete. Modern segmental structures have epoxy joints; however, on several structures that have been built in areas with favourable climatic conditions the joints have no filler material.

Concrete joints are now only used in the case of closure joints between individual assemblies. In epoxy joints, the epoxy resin forms only a thin layer, filling the tiny irregularities between the adjacent segments. To ensure good contact at the joints, the segments need to be manufactured in a contacting fashion. This means that the face of the already cast segment becomes the formwork for the adjacent face of the newly cast segment.

Currently, there are a number of commercially available epoxy resins. They are usually prepared from two-components: a self-adhesive and a hardener. For proper curing of the epoxy sealant, a uniform compression of 0.2–0.4 MPa is required in the joint. Although it has been verified in laboratory tests that the tensile strength of the epoxy joint is greater than the tensile strength of concrete, this factor is not relied on in the construction of structures. Rather, the strength of a joint depends mainly on the quality of the workmanship. In common segmental structures, the unhardened epoxy sealant acts as a lubricant for the erection of the contact segments, and ensures watertightness of the joints in the completed structure. During construction the epoxy resin has no shear resistance when the segments are erected, so it is necessary to design shear keys in the joints.

5.2.2.1 Production of segments

Due to the fact that the geometry of modern bridges has to correspond to a three-dimensional layout of the highways, each segment has different dimensions. Two methods are used to produce segments: the long-line method and the short-line method.

Long-line method

When casting the segments, a casting bed that provides the formwork for the soffit of the deck is constructed first. The geometry of the bed corresponds to the geometry of the soffit of the bridge deck. The length of the line built usually corresponds to the length of one-half of the constructed spans. The segments are progressively cast into a form that moves along the line, with one face being formed by the already cast segment (Figures 5.53 and 5.54). The advantage of the long-line method is that the soffit geometry can be easily checked; its disadvantage is that it requires considerable working space.

Figure 5.53 Long-line casting of segments

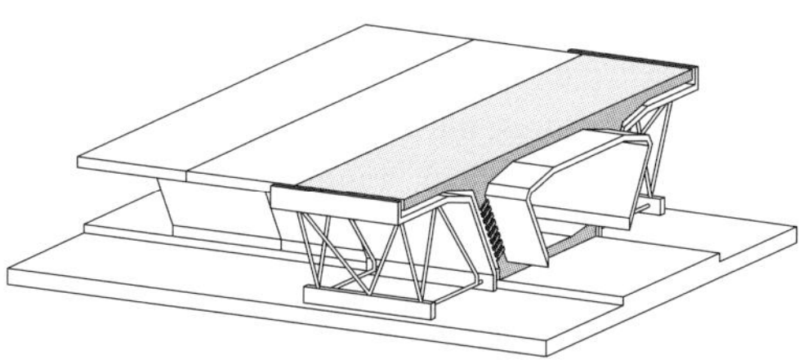

Figure 5.54 Casting of the segments for the Skyway section of the Bay Bridge between San Francisco and Oakland, California, USA

Short-line method

In the short-line method, the segments are cast in a production cell that has a length corresponding to the length of two segments. The production cell usually consists of two parts: moulds, and rectifying frames that serve to adjust the position of the segment already produced (Figure 5.55 and 5.56). The form itself comprises a bulk face, a mould bottom, side walls and a core. The production process is as follows. The already cast (contact) segment is moved, on its mould bottom, into the contact position, where it is placed on the rectifying frames to adjust its position to the design position. Then, the mould bottom and the side walls of the formwork for the next segment are attached to the contact segment, the reinforcing cage and the formwork of the production cell is inserted in the formwork, and the new segment is cast. When the concrete has sufficient strength, the contact segment is taken to stockpile, and the side walls and core are removed. The new segment on the mould bottom is moved into to the position of the contact segment, and the procedure is repeated.

Figure 5.55 Short-line casting of segments

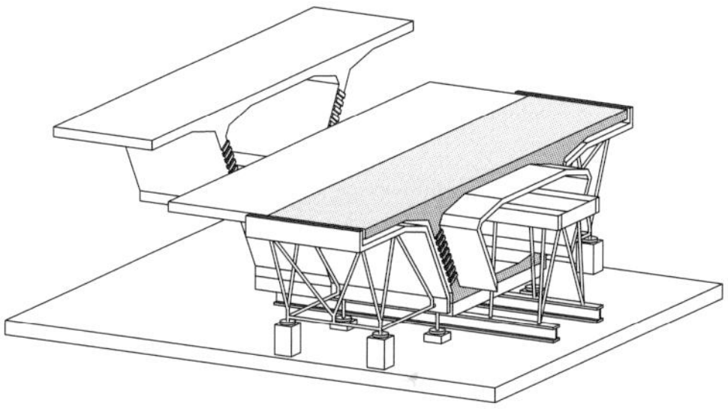

Figure 5.56 A segment on its rectifying frames

Figure 5.57 Segment geometry: (a) elevation; (b) plan

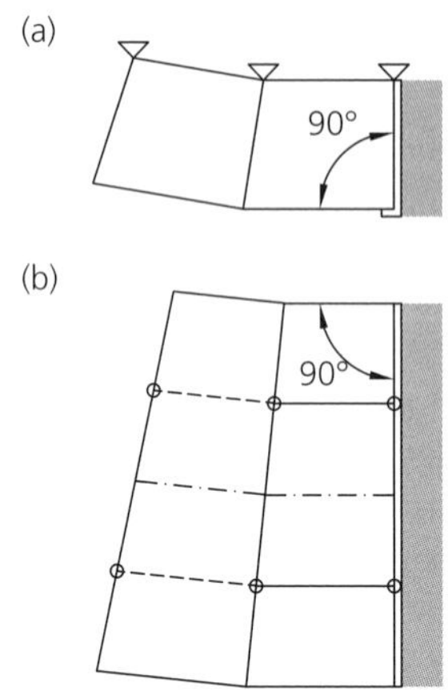

Figure 5.58 Segment erection: (a) on scaffolding; (b) as cantilevers

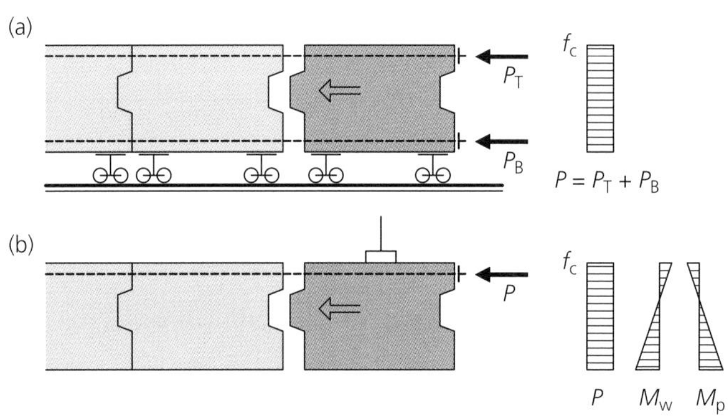

Successful segment production requires exact positioning of the contact segment, and the values shown in Figure 5.57 are necessary. In addition, during the casting of the segments, it is necessary to pay attention to the development of the heat of hydration. The advantage of a short-line method is that it requires a relatively small space; the disadvantage of the method is that it requires a detailed programme that determines the positioning of the contact segment.

5.2.2.2 Erection of segments

The construction of bridges formed from precast segments is similar to the construction of monolithic structures. Bridges with span lengths up to 50 m are usually erected on scaffolding (Figure 5.58(a)). Bridges with longer spans are erected as cantilevers (Figure 5.58(b)), either as symmetrical cantilevers from the piers, or as a progressive cantilever with construction moving from one abutment to the other.

Structures erected on scaffoldings

In the early days, the segments were assembled in a similar way as for monolithic structures, span by span, with short cantilevers protruding into the next span. The joints between the sections were usually chosen to be at points where the dead-load bending moments were zero. The bridge decks were prestressed by continuous cables tensioned and coupled at the joints.

Modern structures, which are mostly prestressed by external cables, are erected span by span. The external cables usually overlap at the pier diaphragms, where they are also anchored (see Chapter 10). The scaffolding supports the segments only in the erected spans, and is moved into the next span when the cables have been post-tensioned. As for monolithic structures, the scaffolding can be located below or above the assembled bridge deck (Figure 5.59). In the first case, the segments are supported by rectifying trolleys, in the latter case the segments are suspended.

Figure 5.59 Erection on scaffolding

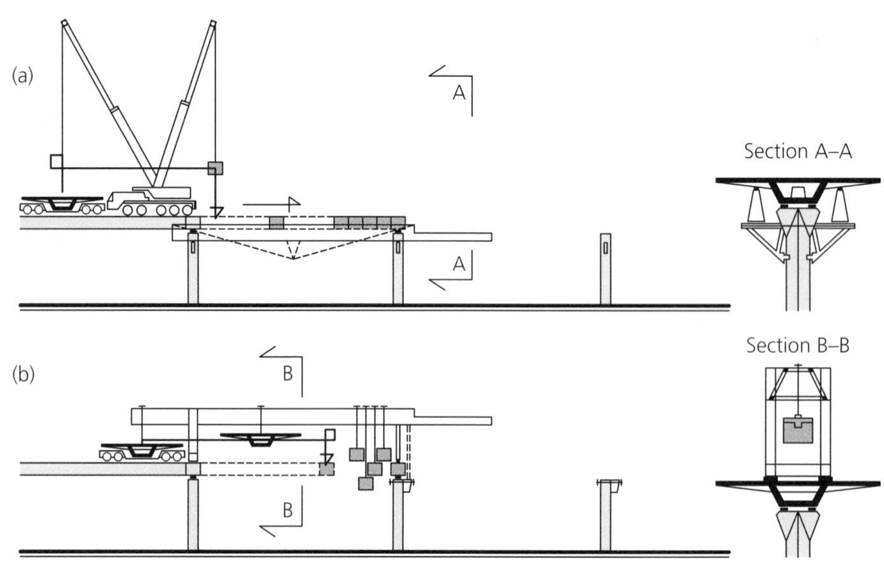

If the scaffolding is located below the bridge, the segments can be erected using mobile cranes (Figure 5.60), portal cranes moving on crane tracks located along the assembled structure, or cranes situated on the already erected spans close to the supports (see Figure 5.59(a)). In the latter case, which allows construction independently of the terrain, the segments are shifted into the designed position by winches.

If the scaffolding is located above the erected bridge deck, it also serves as erection equipment. The front support is usually located on a pier segment, which has to be erected first, or on piers. The rear support is situated on the already assembled structure. If the front support is located on the pier caps (Figure 5.61), the piers have to be designed for the corresponding bending.

The segments can be lifted from the ground or they can be transported on the already the assembled structure. In the latter case, it is necessary to allow a segment to rotate and to design the rear support to allow the passage of the segment through the support (see Figure 5.59(b)). To enable rotation of the segment in the erected spans, the segments are usually temporarily placed close together and suspended at different heights. When all the segments of the erected spans are suspended, they are moved into their design positions.

Figure 5.60 Long Key Viaduct, Florida, USA

Figure 5.61 Anita Garibaldi Bridge, Santa Catarina, Brazil

Before the segments are prestressed, it is necessary to adjust their positions and tie them together so that the geometry of the erected span corresponds to the design position, and so that no additional stresses arise in the joints between the segments.

Cantilever erection

Bridges with span lengths longer than 50 m are usually progressively erected as cantilevers, either as balanced cantilevers symmetrically from the piers, or as progressive cantilevers directed from one abutment to the other.

Balanced cantilevers. The majority of segmental bridges are usually erected progressively as balanced cantilevers starting at the piers. Support segments forming pier tables are erected first. These segments are either connected to the substructures via the frame (permanently or temporarily), or supported by temporary towers. Then, typical segments are erected symmetrically. First, epoxy resin is applied to the face of the erected segment, and then the segment is prestressed to match the already erected structure by means of prestressing rods. The position of the rods in the cross-section is determined in such a way that, together with the bending moment due to the segment's self-weight, uniform compression is created in the joint. The value of the compression required for the hardening of the resin is usually in the range 0.2–0.4 MPa. After a symmetrical segment has been erected, the prestressing cables are placed and tensioned. This procedure is repeated until all segments of the erected span have been assembled. The adjacent cantilevers are then temporarily connected and the gap between them is cast. Finally, the span tendons and continuity tendons are placed and post-tensioned.

As in the case of cast-in-place concrete structures, the stability of the erected structure must be ensured during construction (see Chapter 11).

Segments can be erected using mobile cranes, portal cranes situated on the crane tracks along the erected structure, or by winches supported by erection beams anchored into the already assembled part of the structure (Figure 5.62) or supported by the neighbouring cantilevers (Figure 5.63(a)).

Figure 5.62 Bay Bridge, California, USA

Figure 5.63 Cantilever erection

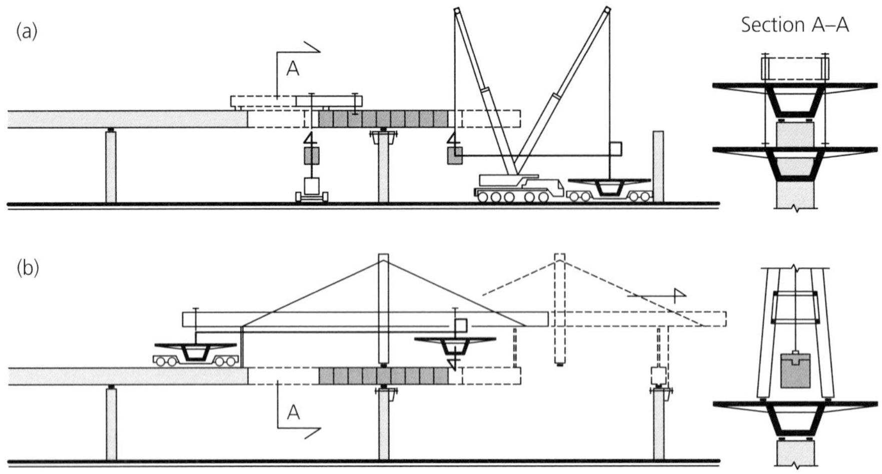

Independently of the terrain, the segments are erected using launching gantries (Figures 5.63(b) and 5.64). The gantry is formed by a horizontal beam that is supported by a central tower located above a pier. The stability of the bridge is secured by an erection support located at the end of the already assembled cantilever. The central tower and the end support allow the passage of segments. During assembly, the erected segment is hung on the horizontal beam and moved into the assembled span, where it is turned and moved into the design position (Figure 5.65(a)).

After all the segments of the assembled balanced cantilever have been erected, the launching gantry is moved along the structure to the position where the central tower is situated at the end of the cantilever (Figure 5.65(b)). Subsequently, the front end of the beam is temporarily supported by an erection support, and a pier segment for the next erected balanced cantilever is erected (Figure 5.65(c)). Once the pier segment has been fixed into the substructure, the erection support is substituted by a front tower, and the launching gantry moves into a new erection position, allowing the erection of typical span segments (Figure 5.65(a)).

Figure 5.64 Bridge on the D1 motorway near Považká Bystricaia, Slovakia

The launching gantries described have a length corresponding to the length of the assembled spans. Extremely long bridges, which need to be built in a short time, use a launching gantry that allows the simultaneous erection of two symmetrical segments. The length of the launching gantry then corresponds to the length of the two spans (Figure 5.66).

Formerly, sophisticated gantries stiffened by erection cables were used. The erected decks had to be fixed into the substructures that had to guarantee the stability of the erected cantilevers. Nowadays, rather robust truss structures are used. Their stiffness can be utilised as a guarantee of the stability of the erected structure (Figure 5.67).

Progressive cantilevers. Bridge decks can also be erected in progressive cantilevers, moving from one abutment to another. The static effects in the erected cantilevers can be reduced by temporary supports (Figure 5.68(a)) or by erection pylons with stay cables (Figure 5.68(b)).

The temporary supports must be designed so that they can load the erected cantilevers by controllable forces, and the pylons with the

Figure 5.65 Erection positions of the launching gantry

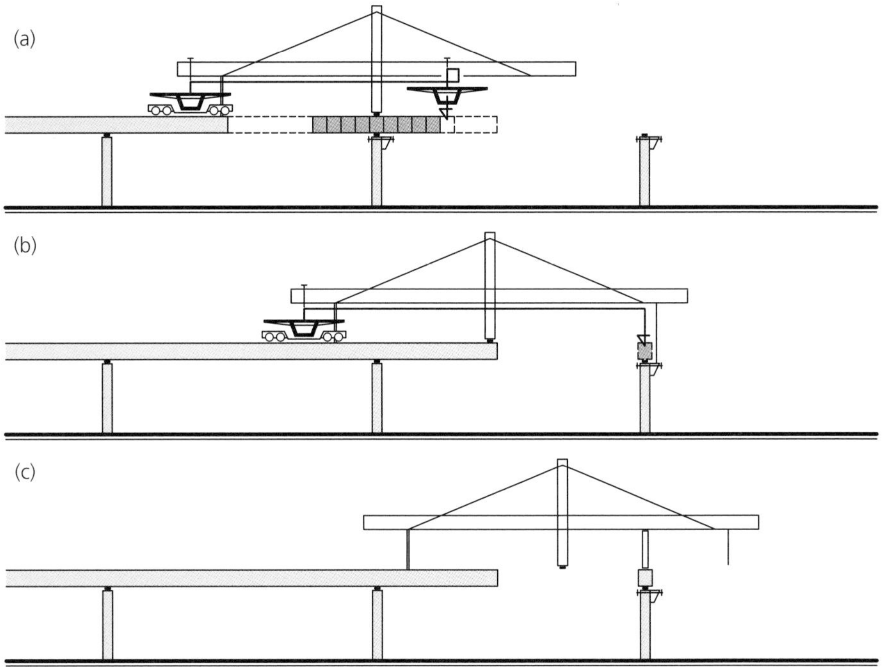

Figure 5.66 Ré Island Bridge, France

Figure 5.67 Crown Princess Mary's Bridge, Frederikssund, Denmark

stay cables must be designed to allow adjustment of the forces in the stays. Segments can be erected using mobile cranes, portal cranes moving on the crane tracks situated along the erected structure, or special cranes. In the Czech Republic, a crane that moves on the already assembled structure (see Figures 5.68(a) and 5.69) has been used, and in France a swivel crane anchored in the erected structure has been used (see Figure 5.68(b)). The segments of the Linn Cove Viaduct, North Carolina, USA were erected using a stiff-leg derrick (Figure 5.70), and a portal crane and temporary pylon with stays were used in the construction of the side spans of the cable-stayed bridge across the Víšovice railway switching yard in Prague, Czech Republic (Figure 5.71).

Figure 5.68 Erection in progressive cantilevers

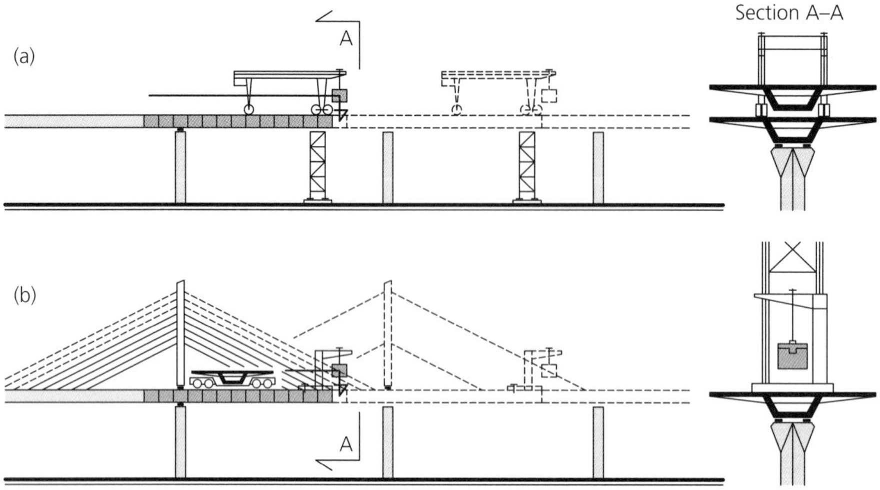

Figure 5.69 Viaduct du Claix, on the TGV South-Europe-Atlantic high-speed rail route, Claix, France

It is evident that not only decks, but also piers, pylons and arches can be assembled from precast segments. An example of where pylons have been assembled in this way is the recently completed Lunda Footbridge in Umeå, Sweden (Figure 5.72).

5.3. Hybrid structures

The previously described procedures can also be used in the construction of hybrid structures, which combine steel with prefabricated and monolithic concrete (Stráský *et al.*, 2001). These structures can be erected progressively (Figure 5.73). A basic load-bearing element in the form of an arch (Figure 5.74), a stay and suspension cable (Figure 5.75), or a beam is erected in advance, and is later used as a support member for the construction of the other parts of the structure. Prefabricated elements or a part of a cast-in-place structure serve as scaffolding and formwork for other parts of the structure (Figure 5.76).

Figure 5.70 Linn Cove Viaduct, North Carolina, USA

The advantage of the hybrid solution is that the structure can be assembled from lightweight elements that can be easily transported, and assembled using available erection equipment, independently of the terrain ground.

Figure 5.71 Bridge across the Vršovice railway switching yard, Prague, Czech Republic

Figure 5.73 Progressive construction of hybrid structures: (a) arch; (b) cable-stayed; (c) suspension

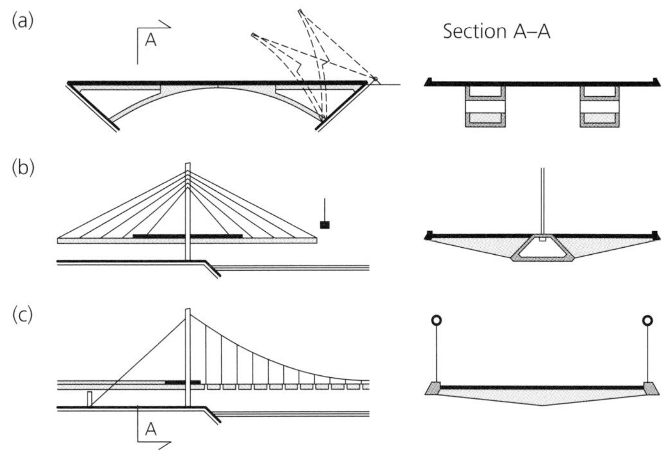

Figure 5.72 Lunda Footbridge, Umeå, Sweden: (a) structural arrangement of pylons; (b) erection of a pylon segment

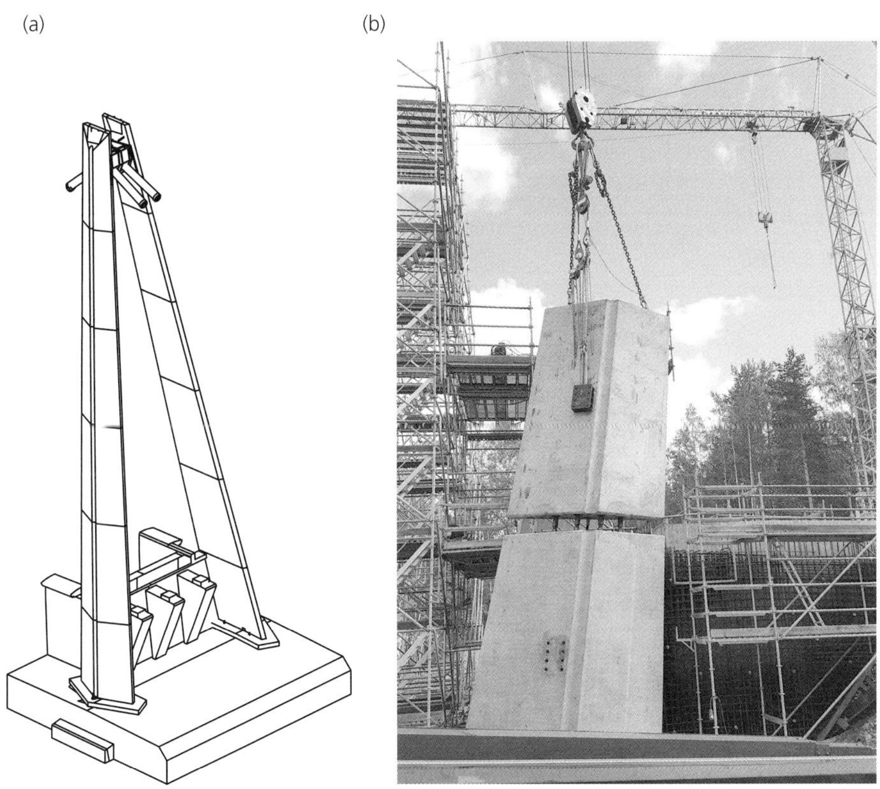

Figure 5.74 Wisconsin Avenue Viaduct, Milwaukee, Wisconsin, USA: casting of the arch fill

Figure 5.75 Vranov Lake pedestrian bridge, Czech Republic: deck segments suspended on the cables

Figure 5.76 Progressive construction of a girder structure

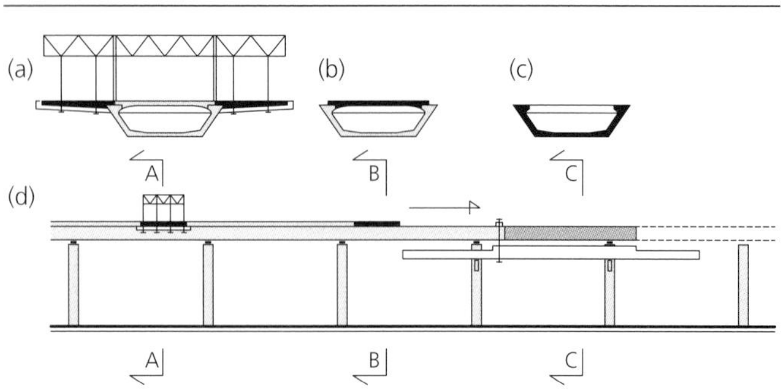

Progressive construction is most often used for girder structures in which the basic support element consists of a prefabricated or steel beam (see Chapter 9). However, the girder may also be made of cast in situ concrete or be erected from precast segments. Figure 5.76 shows an example of a progressively cast girder structure. First, the base trough section formed by the bottom plate and the ribs is cast in a formwork supported by movable scaffolding. Then, after it has been prestressed, the slab between the ribs is cast, and finally the overhangs are completed.

Figures 5.77 and 5.78 show another example of a progressively erected deck of an urban viaduct, this one in the city of Plzeň in the Czech Republic (see also Chapter 11), where deck is composed of precast segments and a cast-in-place deck slab. The precast segments

Figure 5.77 Plzeň Bridge, Plzeň, Czech Republic: erection of segments

Figure 5.78 Plzeň Bridge, Plzeň, Czech Republic: casting of the composite deck slab

Figure 5.79 Furukawa Bridge, Tokyo, Japan: erection of the segments of the trough section

were erected in balanced cantilevers and, after casting and prestressing of the midspan joint, the deck slab was cast. The segments were progressively prestressed by internal cantilever and span tendons and, after the deck slab had been cast, the external cables and continuity cables were post-tensioned.

Progressive erection of the deck has also been used in the construction of several viaducts in Japan (Kasuga, 2017). The bridge decks are formed of a box girder, the top slab of which is stiffened by transverse ribs. First, the precast segments of a trough section were erected span by span (Figure 5.79). Then, precast panels supported by transverse ribs were assembled, and, finally, the deck was cast in place (Figure 5.80). The deck was prestressed by external cables once the erection of the segments had been completed and after the casting of the deck slab.

In several new Japanese bridges, the webs of the precast segments are formed by precast 'butterfly' panels made from high-strength concrete (Figure 5.81; see also Figures 3.7(c), 3.11(c) and 4.100) (Kasuga, 2017). These panels (Figure 5.82) are also used in the construction of cast-in-place structures erected on scaffolding or at symmetrical cantilevers.

Progressively erected decks are mainly used in the construction of the wide decks of motorway bridges (see Figure 3.12(a)). The decks of these bridges are usually formed of a spine box girder that is further widened by large overhangs supported by precast struts. Usually, the struts are erected, and then the overhangs are cast on a scaffolding moving on the already erected spine girders (Figures 5.83 and 5.84).

Figure 5.80 Furukawa Bridge, Tokyo, Japan: cast-in-place deck

Figure 5.81 Okegawa Viaduct, Tokyo, Japan

It is also possible to suspend the precast overhangs on the spine girder (Figures 5.85, 5.86(b) and 5.87) and place precast slab members on them (Figure 5.86(c)) (Riberio, 2015). Then, the reinforcement and transverse prestressing tendons are placed, and the composite

Figure 5.82 Takubogawa Bridge, Hyuga, Japan

Figure 5.83 West viaducts to the Kayak Bridge, Bratislava, Slovakia

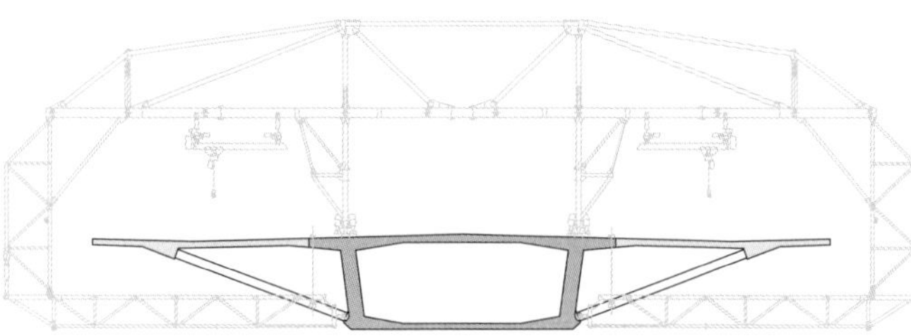

Figure 5.84 West viaducts to the Kayak Bridge, Bratislava, Slovakia

Figure 5.85 Anita Garibaldi Bridge, Santa Catarina, Brazil: progressive construction

Figure 5.86 Anita Garibaldi Bridge, Santa Catarina, Brazil - progressive erection: (a) spine girder, (b) precast struts, (c) precast slabs memers, (d) composite slab

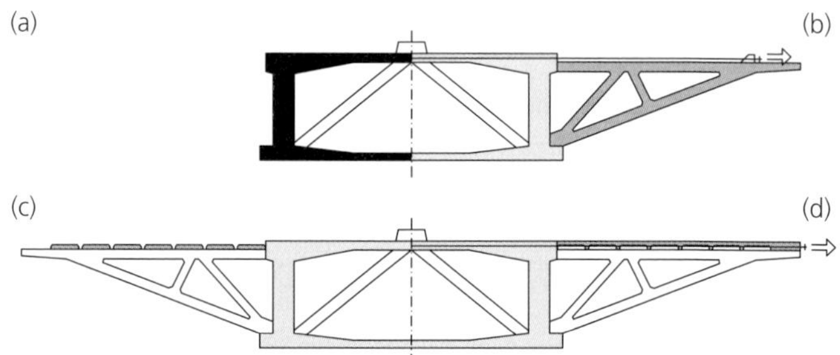

Figure 5.87 Anita Garibaldi Bridge, Santa Catarina, Brazil: precast overhangs suspended on the spine girder

Figure 5.88 Bridge across the River Elbe, Czech Republic: progressive construction

Figure 5.89 Viaduct across the Doliansky Creek Valley, Slovakia

Figure 5.90 Viaduct across the Hostovsky Creek Valley, Slovakia - progressive erection: (a) spine girder, (b) precast struts, (c) overhangs, (d) carriageway

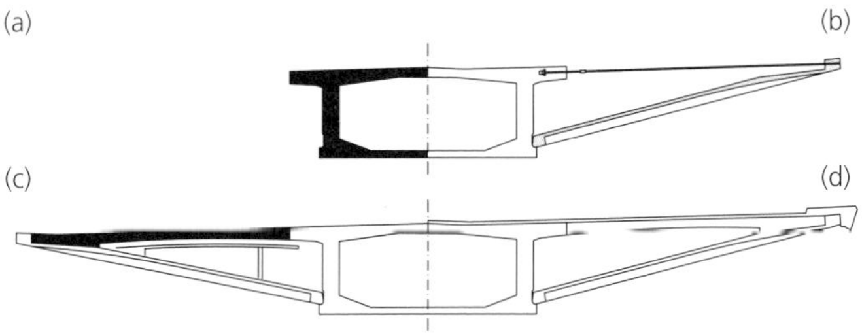

Figure 5.91 Viaduct across the Hostovsky Creek Valley, Slovakia

slab is cast. When the concrete of the overhangs has sufficient strength, the deck is transversally prestressed (Figure 5.86(d)).

Another very efficient type of bridge construction is where the overhangs are supported by precast slab members that are not mutually connected (see Chapter 3, Figure 3.16(c)). The author used this approach in the design of his first cable-stayed bridge built across the River Elbe in the Czech Republic (see Chapter 4, Figures 4.79 and 4.80). The bridge deck is formed by a spine box girder assembled of precast–match cast segments, and is further widened by overhangs supported by precast slab struts (Figure 5.88).

This solution has been successfully used recently in the construction of several viaducts in Slovakia (Figure 5.89). The slab members are supported by short corbels protruding from the webs, and are suspended on prestressing bars anchored at the top slab of the spine girders (Figures 5.90(b) and 5.91). The deck slab is then cast in a very simple formwork supported by the precast struts. Then the deck slab is transversally prestressed (Figure 5.90(d)). The deck is longitudinally prestressed in two stages, at first, when the spine girder is constructed and later, when the overhangs are cast and transversally prestressed (see Chapter 10).

REFERENCES

Dywidag (n.d.) Cantilever construction method with launching truss and form follower. *Prestressed Concrete Bridges*. Report No. 7.

Adão-da-Fonseca A and Mato FM (2005) Infant Henrique Bridge over the River Douro, Porto. *Structural Engineering International* **15(2)**: 85 85.

Appleton J (2019) History of arch bridges in Portugal. In: *Structural Integrity. Proceedings of ARCH 2019* (Arêde A and Costa C (eds)). Springer, Cham, Switzerland, pp. 31–50.

Arenas J, Capellán G, Martínez J *et al.* (2016) Viaduct over the River Almonte. Design and analysis. Challenges in Design and Construction of an Innovative and Sustainable Built Environment, 19th IABSE Congress Stockholm, Sweden, pp. 2336–2343.

Cezard C and Servant C (1988) Charix Viaduct. *13th IABSE Congress, Helsinki, Finland*. Curran, Red Hook, NY, USA, pp. 63–70.

Ciampoli M (1987) The Kylesku bridge in Scotland: a prestressed concrete deck with partially precast central span. *Industria Italiana del Cemento* **7–8**: 458–471.

Dyckerhoff and Widman (eds) (1973) *Festschrift. Ulrich Finsterwalder 50 Jahre für Dywidag*. G. Braun, Karlsruhe, Germany.

Gallo P (1990) Railway viaducts for the Rome – Florence express line in the Arrezzo Sud–Figline Valdarno section. Prestressed Concrete Structures in Italy 1986/1990. *Industria Italiana del Cemento* **4**: 423–428.

Kasuga A (2017) Butterfly web effects on bridge design. *Structural Concrete* **18(1)**: 128–142.

Mathivat J (1983) *The Cantilever Construction of Prestressed Concrete Bridges*. John Wiley, New York, NY, USA.

Matyassy L and Palossy M (2006) Koroshegy Viaduct. *Structural Engineering International* **16(1)**: 36–38.

Ministry of Public Works (1982) *Realisations of Ministry of Public Works*. Ministry of Public Works, Belgium.

Pacheco P (2015) Multi-span large decks – the organic prestressing impact. In *Multi-Span Large Bridges* (Pacheco P and Magalhaes F (eds)). Taylor & Francis, London, UK, pp. 103–124.

Pauser A and Beschorner K (1976) Betrachtungen über seilverspannte Massivbrüken, ausgehend vom Bau der Schrägseilbrücke über den Donaukanal in Wien. *Beton- und Stahlbetonbau* **71(11)**: 261–265.

Podolny W and Muller J (1982) *Construction and Design of Prestressed Concrete Bridges*. John Wiley, New York, NY, USA.

Riberio CF (2015) The Octavio Frias de Oliveira and Anita Garibaldi cable-stayed bridges. *Proceedings of the International Conference on Multi-Span Large Bridges, 1–3 July 2015, Porto, Portugal*.

Rosignoli M (2002) *Bridge Launching*. Thomas Telford, London, UK.

Stráský J, Navrátil J and Suský S (2001) Applications of time-dependent analysis of in the design of hybrid bridge structures. *PCI Journal* **46(4)**: 56–74.

Walther R, Houriet B, Walmar I and Moïa P (1998) *Cable Stayed Bridges*. Thomas Telford, London, UK.

Wittfoht H (1972) *Triumph der Spannweiten*. Beton, Düsseldorf, Germany.

Zorzi S *et al.* (1978) Viaducts for the Palermo–Sciacca highway. Prestressed Concrete Structures in Italy 1974–1978. *Industria Italiana del Cemento* **4**: 228–233.

ICE Publishing

Stráský and Nečas
ISBN 978-0-7277-6385-3
https://doi.org/10.1680/dcpb.63853.111
ICE Publishing: All rights reserved

Chapter 6
Structural concrete

Previous bridge design specifications and guidelines separated the design of reinforced and prestressed concrete structures. They were based on an outdated view of the different behaviours of the two types of concrete. In the past, the main advantage of prestressed structures was the elimination of tensile stresses and cracks. Over the years, however, it was found that the requirement to reduce cracking in some structures is uneconomical and leads to the design of bridges that are susceptible to side-effects such as deflections of supports and non-uniform temperature changes. Also, it has been found that some structures cannot be designed as fully prestressed. This applies, in particular to structures in which the ratio of the permanent to the live loads is small, or structures subjected to a bending moment of the opposite sign.

Therefore, the requirement for full prestressing was abandoned, and a uniform approach to the design of concrete structures was developed. When designing bridge structures, plain, reinforced or prestressed concrete (collectively 'structural concrete') is used, which is effectively combined in the structure according to the nature and function of the structural elements.

Modern concrete structures combine reinforced concrete with different levels of post-tensioning to obtain the most appropriate behaviour of the structure at both the service and the ultimate load. In this way so-called 'structural concrete' is developed.

6.1. Behaviour of structural concrete

Figure 6.1 shows the results of tests done on a two-span prestressed beam to understand the function of prestressing and reinforcing steel, performed by Professor Walther at the University of Lausanne (Walther, 1982). The beam was designed with different degrees of prestressing, from full to zero. The degree of prestressing λ, which was taken as the ratio of the ultimate tensile force in the prestressed reinforcement to the total ultimate tensile force, was $\lambda = 0.9$ for full prestressing (element A1) to $\lambda = 0.0$ for elements A4 and B4. Element A1 was prestressed by three tendons, elements A2 and B2 by two tendons and elements A3 and B3 by one tendon; elements A4 and B4 were reinforced only by reinforcing bars.

The sections of the beams were reinforced for the static effects obtained by a linear analysis of the structure (B) and for the static effects where the redistribution of moments (A) was considered

Figure 6.1 Tests on a two-span prestressed beam: (a) elevation; (b) bending moments; (c) maximum deflection y as a function of the ratio Q/Q_u; (d) maximum crack width over the intermediate support w as a function of the ratio Q/Q_u

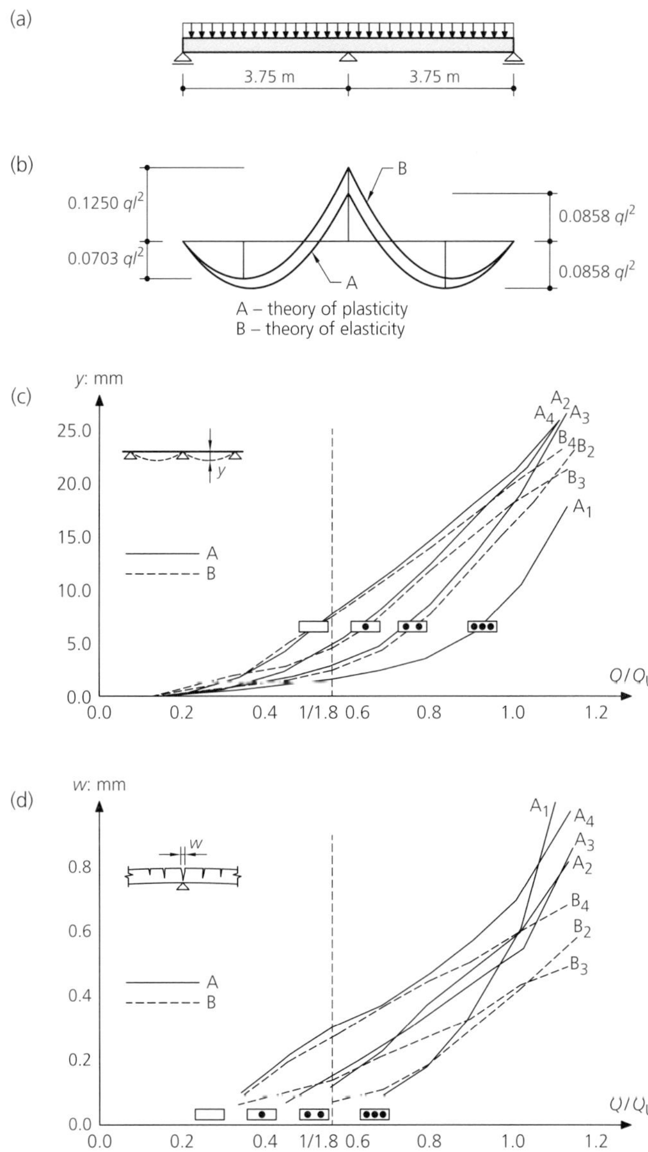

(Figure 6.1(b)). In the latter case, the sections were designed to have the same moments at the spans and above the support, in accordance with the theory of plasticity.

Each beam was gradually loaded until the ultimate capacity of the structure was reached. Figure 6.1(c) shows the development of the deformations in the middle of the span, and Figure 6.1(d) shows the crack width above the inner support, according to the ratio of the achieved and ultimate loads (Q/Q_U).

It is clear from Figure 6.1(c) that the method of distribution of reinforcement along the length of the structure, whether the structure is designed according to the theory of elasticity or the theory of plasticity, has no significant effect on the deformation. The course of the deformation of the structures show that prestressing significantly reduces the deformation under the service load. From Figure 6.1(d) it is clear, that the crack width does not depend on how the reinforcement is distributed. Under the service load, prestressing limits the width of the cracks or delays their occurrence. However, at full prestressing, the crack width increases very quickly.

The results show, therefore, that prestressing limits the crack width and deformation of the structure, while reinforcing steel increases the ductility (plastic deformation) of the structures. They also confirm that a reasonable redistribution of static effects according to the principles of plasticity theory can be used to design structures.

As described in Chapter 2, bridge structures can be prestressed by internal bonded tendons, internal unbonded tendons, or external cables situated inside or outside the perimeter of the cross-section. The different function of these elements is illustrated in Figure 6.2.

Figure 6.2(a) shows a simple T-beam that is prestressed by either an internal bonded or an unbonded tendon. The bonded tendon can be formed of strands grouted in the steel duct. The unbonded tendon can be formed by monostrands that are cast in the beam, or by monostrands grouted in the duct. The number of the strands, the geometry of the tendon and the prestressing force are the same for both tendons. The function of both tendons is identical for the service load. However, the beams behave differently under the increased load when the first crack in the concrete occurs.

The elongation of the strands in the bonded tendon in the cracked section corresponds to the increase in the crack width, while the elongation of the strands in the unbonded tendon corresponds to the difference in length of the beam before and after the loading: $\Delta L = L_{T,U} - L_T$. As the force in the tendon corresponds to its elongation, the force in the bonded tendon is much larger than the force in the unbonded tendon. Therefore, the ultimate capacity of the structure prestressed by bonded tendons is larger than that of the structure prestressed by unbonded tendons.

Figure 6.2(b) shows a simple T-beam that is prestressed by external cables situated outside the cross-section. The cables are anchored at the end diaphragms, and are deviated at deviator blocks situated at the bottom fibres close to a quarter of the beam's span. The cables are formed by strands grouted in polyethylene pipes. It is assumed that the friction at the deviator prevents movement of the cable. The number of strands is the same as in the beam in Figure 6.2(a). After the first crack originates, the elongation of the cable corresponds to the difference in the length of the deviators before and after the loading: $\Delta L = L_{C,U} - L_C$.

Therefore, the force in the cable corresponds to this elongation. When checking the ultimate moment, a possible change in the eccentricity should be considered. If slippage of the cables at the

Figure 6.2 Types of prestressing: (a) internal; (b) external, situated within the perimeter of the cross-section; (c) external, situated outside the perimeter of the cross-section

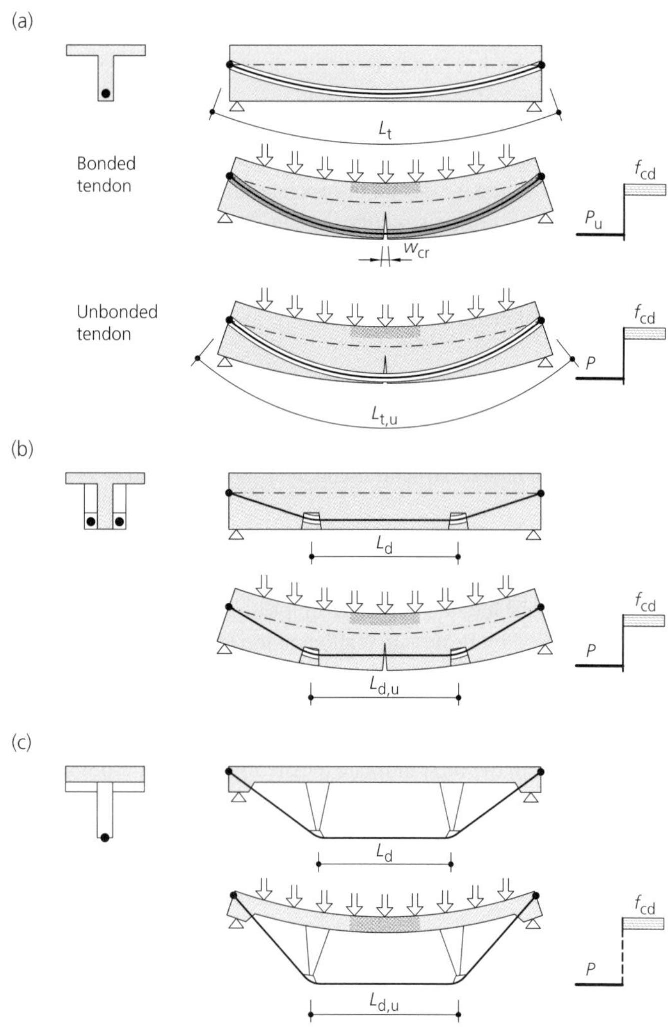

deviator is possible, the corresponding elongations and forces at the cables should be taken into consideration.

Figure 6.2(c) shows a simply supported structure composed of a slender beam of rectangular cross-section which is carried and prestressed by external cables situated underneath it. It is evident that the analysis of this structure is more complex. It is shown in Chapters 7 and 8 how the initial forces in the cable have to be determined with respect to concrete shrinkage and creep, and with respect to the non-linear behaviour of the cable. At the ultimate state, a large deformation of the structural system has to be considered. Also in this case, the elongation of the cable corresponds to difference in the length of the deviators before and after the loading: $\Delta L = L_{C,U} - L_C$.

6.2. Design philosophy

The uniform approach to the design of concrete structures was the basis of the FIP Recommendations 1996 – *Practical Design of Structural Concrete* (FIP, 1998). The FIP (International Federation for Pre-stressing) merged with the CEB (Euro-International Committee for Concrete) in 1998 to create the *fib* (International Federation for Structural Concrete).

As the FIP Recommendations 1996 were prepared for practicing engineers and the rules are clearly written, they are repeated in the following text.

The design of concrete bridges should be done in following steps

- check the ultimate limit state (ULS)
- check the serviceability limit state (SLS)
- comply with the technological requirements
- comply with the durability requirements.

In all cases, detailing practice and the technological requirements should be carefully observed, because these are as important to the serviceability and durability of concrete structures as checks done by calculations. The overall structural integrity is particularly dependent on adequate dimensioning and proper detailing, especially at regions of geometrical or local discontinuity (D regions, see Chapter 8, Section 8.9).

The primary focus when designing a structure should be directed towards a careful consideration of the overall or global structural behaviour, and the achievement of an efficient flow throughout the structure. The effects of potentially damaging restraints and aggressive environmental factors should also be considered. In general, the various load-carrying members in a structure should be interlinked in order to ensure a satisfactory overall performance in terms of structural stability and robustness. In particular, it should be ensured that the structure cannot be subjected to progressive collapse as a consequence of localised damage due to abnormal use or accident.

6.2.1 Ultimate limit state (ULS)

(*a*) The ULS verification should be based on clear and realistic models of structural behaviour that stimulate the correct failure mechanisms under the ultimate loads. For these calculations, the theory of plasticity provides a simple and efficient approach in many cases. A non-linear analysis method of a more general character may also be used, particularly in the case of instability.

(*b*) In all cases, it must be ensured that the structure is sufficiently ductile, allowing for restraint effects, to be able to reach the assumed ULS without premature brittle failure. In a structure is sufficiently ductile, the effects of temperature, creep, shrinkage and foundation settlement will, in general, have an insignificant influence on the ultimate load carrying capacity.

(*c*) The ULS conditions are satisfied if the following symbolic equation is respected for all relevant combinations of actions.

$$F_{\text{act,d}}\left(G;\ \gamma_q \sum Q\right) \le F_{\text{res,d}}\left(\frac{f_{ck}}{\gamma_c};\ \frac{f_{sk}}{\gamma_s}\right) \tag{6.1}$$

Design value of load or actions — Design value of load-carrying capacity

where (with the associated partial safety coefficients in parenthesis) G is the permanent action (mean values) (γ_g), Q is the variable action (nominal values) (γ_q), f_{ck} is the characteristic concrete strength (γ_c) and f_{sk} is the characteristic strength of the reinforcing or prestressing steel (γ_s).

(*d*) As it very unlikely that the maximum values of all variable actions Q will occur at the same time, representative values (as defined later) may be introduced. In the case of two or more variable actions, combined values characterised by the coefficient ψ_0 may be used.

(*e*) The effect of prestressing may be considered as either the external forces on the action side or the internal forces on the resistance side.

(*f*) The above general criterion (*c*) is satisfied if all significant sections of the structure fulfil the following condition.

$$S_d\left(\gamma_g G;\ \gamma_q \sum Q\right) \le R_d\left(\frac{f_{ck}}{\gamma_c};\ \frac{f_{sk}}{\gamma_s}\right) \tag{6.2}$$

Design value of critical combination of action effects — Design value of resistant action effects

However, the overall structural integrity should also be ensured by checking the arrangement and the anchorage of the reinforcement, especially for regions of discontinuity.

(*g*) The distribution of internal forces in the structure must satisfy the conditions of equilibrium. This is always the case for elastic distributions. Thus, if the structure has already been designed for serviceability conditions (SLS), the same distribution multiplied by an appropriate load factor can often be used to check the required capacity at the ULS.

(*h*) For structures in which the equilibrium is affected by the deformations of the members, the equilibrium conditions should be formulated on the deformed structure. However, Equation 6.2 may be used if the second-order effects are included in the terms S_d. It should be noted that the ultimate capacity of the section is not always attained.

(*i*) Fatigue problems are normally not critical in reinforced and prestressed concrete, as long as severe cracking under the appropriate actions γ_q is prevented.

6.2.2 Serviceability limit state (SLS)

(*a*) The SLS verification should be based on clear and realistic models of structural behaviour, including, where relevant, cracking and time-dependent effects.

(*b*) Normally, the SLS calculations are based on the theory of elasticity. In certain cases, a non-linear analysis method may be used.

(*c*) Depending on the particular case, the SLS check should be done using one or more of the following three methods:

- by limiting the stresses $\sigma_d \leq \sigma_{lim}$
- by limiting the deformations (deflections or angles) $a_d \leq a_{lim}$
- by limiting the crack width $w_d \leq w_{lim}$

The limit values should be established on the basis of the functional requirements of the structure. Other requirements, such as watertightness, tolerances and vibrations, should also be checked where appropriate.

(*d*) In the SLS calculations, the effects of prestressing should be considered either on the action side, or in the material characteristics as an imposed deformation. However, it is usually considered as an external action. The prestressing force should be considered, using its mean value.

(*e*) The check by calculation can sometimes be omitted by respecting minimum reinforcement or detailing regulations.

6.2.3 Design specifications

At present, the bridges built in Europe are designed according to the system of Eurocodes, which use the partial factor method (CEN, 2002, 2003, 2005). ICE Publishing has produced a series of designers' guides to the Eurocodes, which provide comprehensive guidance on the form of design aids, indications for the most convenient design procedures and worked examples (Croce, 2010; Gulvanessian *et al.*, 2002; Hendy and Smith, 2007).

The bridges built in the USA are designed according to load and resistance factor design (LRFD) methodology (AASHTO, 2004).

REFERENCES

AASHTO (American Association of State Highway and Transportation Officials) (2004) LRFD Bridge Design Specifications. AASHTO, Washington, DC, USA.

CEN (European Committee for Standardization) (2002) EN 1990:2002 + A1:2005. Eurocode – Basis of structural design. Annex A2: Application for bridges. CEN, Brussels, Belgium.

CEN (2003) EN 1991-2:2003. Eurocode 1 – Action on structures. Part 2: Traffic loads on bridges. CEN, Brussels, Belgium.

CEN (2005) EN 1992-2:2005. Eurocode 2 – Design of concrete structures. Part 2: Concrete bridges: design and detailing rules. CEN, Brussels, Belgium.

Croce P (2010) *Design of Bridges*. Klokner Institute, Prague, Czech Republic.

FIP (1998) FIP Recommendations 1996. *Practical Design of Structural Concrete. FIP Congress, Amsterdam, The Netherlands.*

Gulvanessian H, Calgaro JA, Holický M and Gulvanessian H (2002) *Designers' Guide to Eurocode: Basis of Structural Design. EN 1990.* Thomas Telford, London, UK.

Hendy CR and Smith DA (2007) *Designers' Guide to EN 1992-2 Eurocode 2: Design of Concrete Structures. Part 2: Concrete Bridges.* Thomas Telford, London, UK.

Walther R (1982) Partial prestressing. In *Prestressed Concrete of Switzerland.* 9th FIP Congress, Stockholm, Sweden, 1982. Technische Forschungs- und Beratungsstelle der Schweizerischen Zementindustrie, Wildegg, Switzerland.

ice Publishing

Stráský and Nečas
ISBN 978-0-7277-6385-3
https://doi.org/10.1680/dcpb.63853.115
ICE Publishing: All rights reserved

Chapter 7
Effects of concrete creep and shrinkage

During the construction of modern prestressed concrete bridges there are a number of different static systems in play: boundary conditions change, new structural members are added or cast, post-tensioning is applied, and temporary support elements are erected and subsequently removed. Structural elements of various ages are combined, and the concrete is gradually loaded. Therefore, when designing a concrete structure, account must be taken of the creep and shrinkage of the concrete, both during construction and throughout the service life of the structure.

7.1. Creep and shrinkage functions

The deformation of concrete due to shrinkage and creep may vary considerably with the type of cement and aggregate, the climate (temperature and humidity), the member size and the time of loading.

Modern creep functions (Figure 7.1) combine the theory of delayed elasticity (wherein the final value of creep does not depend on the age of concrete) and the rate-of-creep theory (Dischinger, Mörsch and Ross) (Leonhardt, 1964; Navrátil, 2014; Smerda and Kristek, 1988).

For final design, a time-dependent analysis using the CEB-FIP Model Code 90 (MC 90) (CEB, 1993; fib, 2013) or specific tests should be used. MC 90, which replaced the earlier MC 78 (CEB, 1986), uses the following function for shrinkage (cs) and creep (cc) strains.

$$\varepsilon_{cs}(t, t_s) = \varepsilon_{cs0}\beta_s(t - t_s) \tag{7.1}$$

$$\varepsilon_{cc}(t, t_0) = \frac{f_c(t_0)}{E_c}\varphi(t, t_0) \tag{7.2}$$

where

β_s is the coefficient that describes the development of shrinkage over time, $\varphi(t, t_0)$ is the creep coefficient, f_c is the normal stress, E_c is the modulus of elasticity at age 28 days, t_0 is the time of loading (days), t_s is the age of the concrete (days) at the beginning of shrinkage or swelling, and t is the age of the concrete (days).

For the preliminary considerations, the final values of shrinkage strain and the creep coefficient can be taken from Tables 7.1 and 7.2 (FIP, 1998). The tables present mean values, and apply to concrete

Figure 7.1 Creep functions: (a) delayed elasticity; (b) rate of creep; (c) modified rate of creep

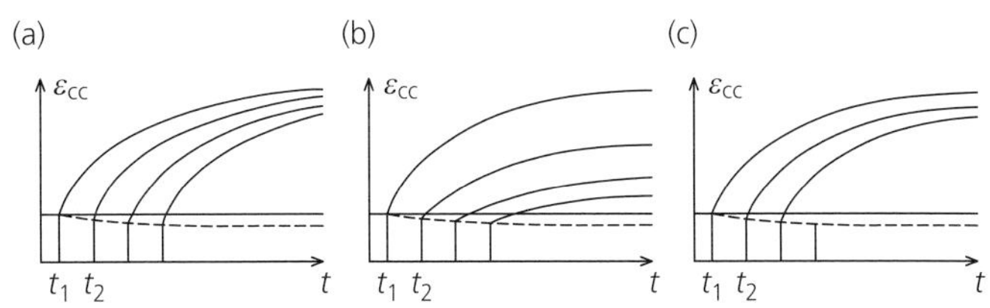

Table 7.1 Final value of shrinkage strain, ε_{cs} (10^{-3})

Atmospheric conditions	Effective member size, $2A_c/u$: mm		
	50	150	600
Dry; indoor (RH = 50%)	−0.53	−0.51	−0.36
Humid; outdoor (RH = 80%)	−0.30	−0.29	−0.20

A_c, cross-sectional area of concrete; u, exposed perimeter of A_c.

Table 7.2 Final value of the creep coefficient, φ

Age at loading, t_0: days	Effective member size, $2A_c/u$: mm					
	Dry; indoor (RH = 50%)			Humid; outdoor (RH = 80%)		
	50	150	600	50	150	600
1	5.6	4.6	3.7	3.7	3.3	2.8
7	3.9	3.2	2.6	2.6	2.3	2.0
28	3.0	2.5	2.0	2.0	1.8	1.5
90	2.4	2.0	1.6	1.6	1.4	1.2
365	1.9	1.5	1.2	1.2	1.1	1.0

A_c, cross-sectional area of concrete; u, exposed perimeter of A_c.

Figure 7.2 Shrinkage strain $\varepsilon_{cs}(t)$ and creep coefficient $\varphi(t)$ at time t days, divided by the ultimate shrinkage or creep coefficients given in Tables 7.1 and 7.2

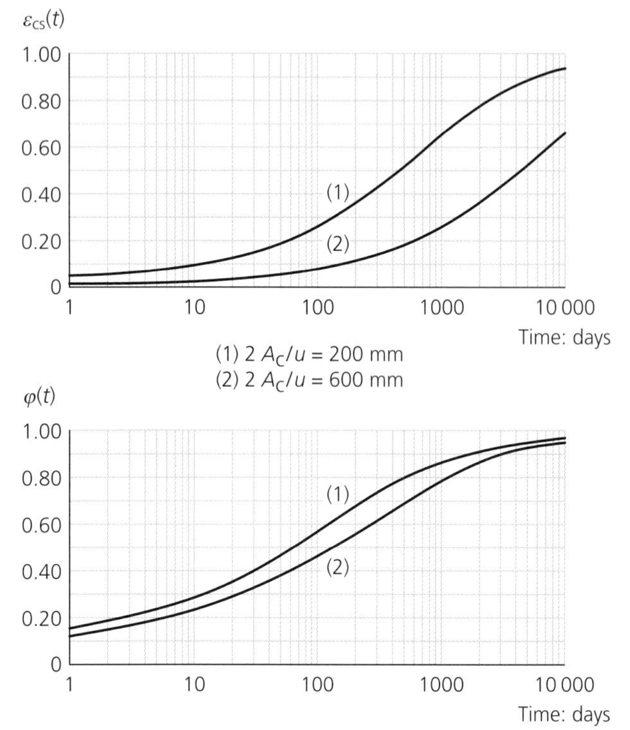

of characteristic cylinder strength 20–50 MPa subjected to a stress not exceeding $0.4f_{c,t0}$ at age t_0 of loading.

The development of the shrinkage strain and creep coefficient with age may be estimated from Figure 7.2.

7.2. Time-dependent analysis

As modern bridge structures are assembled of members of different ages and the bridges utilise different static systems during construction, the effects of concrete creep and shrinkage have to be considered in their designs (CEB, 1993; Stráský *et al.*, 2001). Formerly, only special problems using the rate-of-creep theory could be analysed by solving the differential equations. Nowadays, however, step-by-step time-dependent analyses can be done using comprehensive software. The software programs can usually use any creep function. However, since Model Code 90 was accepted internationally (CEN, 2005), the majority of programs have used it.

The methods of analyses are based on a step-by-step computer procedure in which the time domain is subdivided at discrete times (time nodes) into time intervals. The finite-element analysis is performed at each time node. The software assumes a linearity between stresses and strains in order to allow the use of the linear superposition procedure.

The numerical solution is based on the replacement of the Stieltjes hereditary integral by a finite sum. The general creep problem is thus converted into a series of elasticity problems.

The total strain $\varepsilon_m(t)$ at time t is given by

$$\varepsilon_m(t) = \sum_{j=0}^{n}\left([1 + \varphi(t, t_j)]\frac{\Delta f(t_j)}{E(t_j)}\right) \tag{7.3}$$

The shrinkage and creep of the structural members is predicted using the mean properties of a given cross-section, taking into account the average relative humidity and member size. The development of the modulus of elasticity due to ageing (i.e. with time) is also considered.

Although software has also been developed for the analyses of three-dimensional structures assembled of solid or shell elements, the majority of the programs analyse three-dimensional structures assembled of beam elements. The elements represent, for example, precast segments, composite slabs, stay or external cables, prestressing tendons or reinforcement (Figures 7.3 and 7.4). Within the software, elements can be installed or removed according to the construction scheme, and the influence of concrete of different ages in both the longitudinal and transverse directions of the structure is taken into account. The various operations used in construction, such as the addition or removal of segments and prestressing cables, changes in boundary conditions, loads and prescribed displacements, can be modelled.

Considering the extensive amount of input and output data, the software has a user-friendly graphic interface for pre- and post-processing. The pre-processor guides the user through the process of

Figure 7.3 Time-dependent modelling of a progressively erected segmental structure: (a) partial elevation; (b) cross-section; (c) detail A

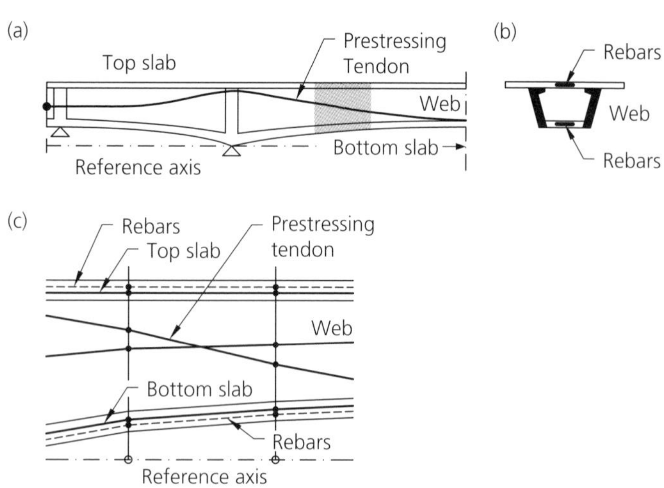

Figure 7.4 Time dependent modelling of a cable-supported structure: (a) partial elevation; (b) cross-section; (c) detail A

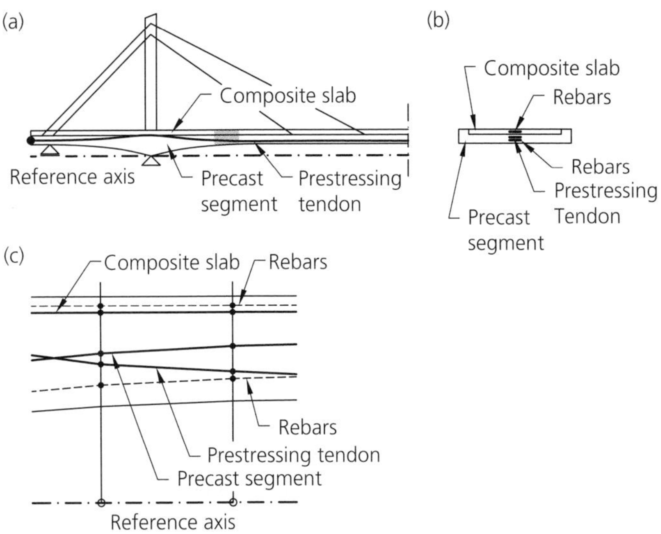

Figure 7.5 Modelling of a progressively erected and prestressed viaduct

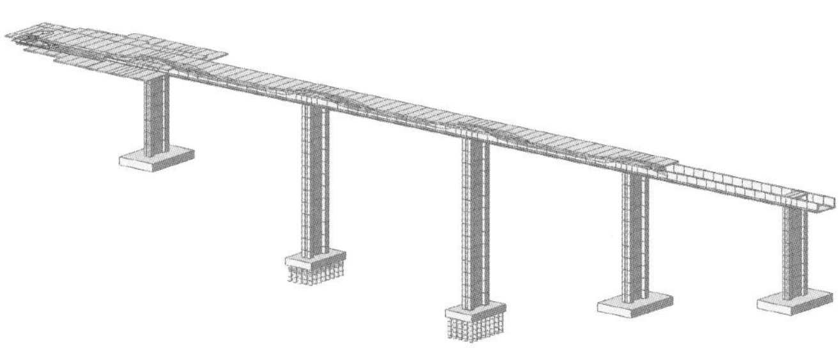

preparing and inputting data. Post-processors allow the user to select and view only the data they need. In a graphical environment it is possible to filter the output data, to switch on and off individual parameters (internal forces, stresses, deformations), to zoom in and out of the structure, to choose the time node to view and other operations. The possibilities of modern software are illustrated by Figure 7.5, which shows the calculation model for the progressively erected structure shown in Figures 5.89–5.91.

The importance of the time-dependent analysis will be demonstrated in following chapters.

7.3. Redistribution of the stresses between members of different age

It is well known that a significant redistribution of stresses occurs in structural members composed of elements of different ages, and in structures formed by steel and concrete (Figure 7.6). The significance

Figure 7.6 Progressively cast concrete member: (a) precast segment and composite slab; (b) precast segment, composite slab and prestressing tendon

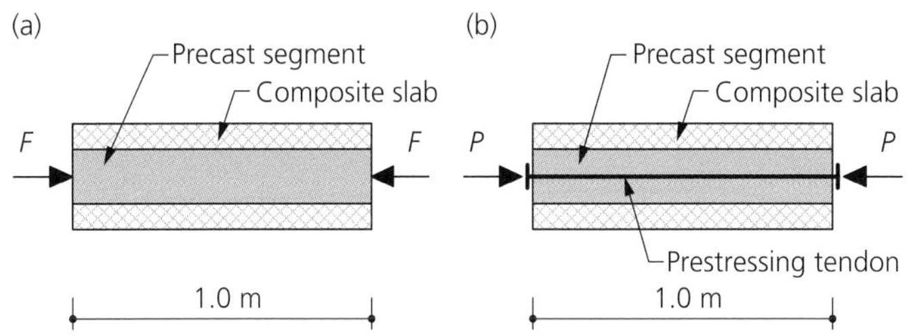

Figure 7.7 Redistribution of the forces in the member shown in Figure 7.6(a)

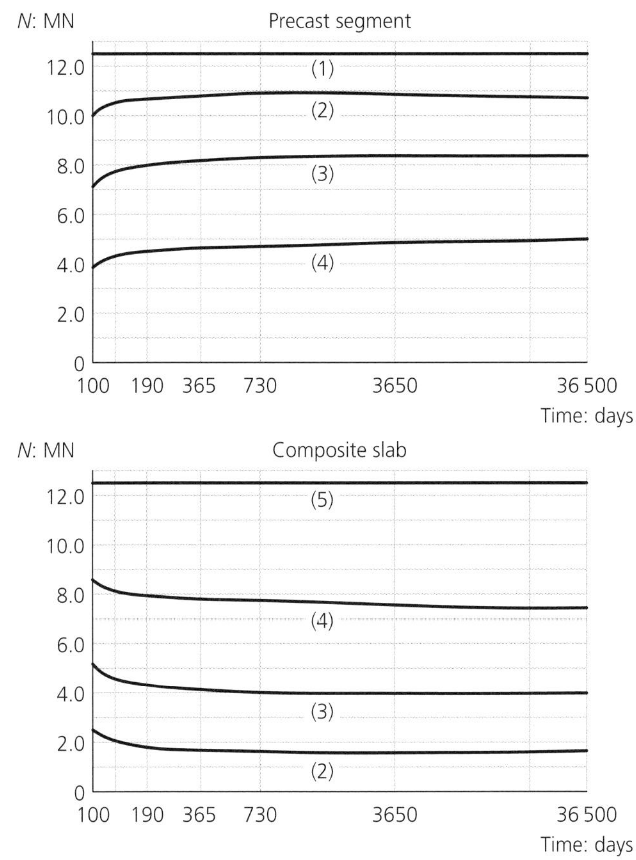

of this phenomenon is evident from Figures 7.7 and 7.8, which show the redistribution of stresses between two parts of a structural member of different ages. The cross-section of the member is rectangular (5.00 × 0.25 m). The analysis was performed for five cases that differ in the proportion of the area of the precast segment (A_{PS}) to the area of the composite slab (A_{CS}) (Table 7.3). The precast segment was

Figure 7.8 Redistribution of the forces in the member shown in Figure 7.6(b)

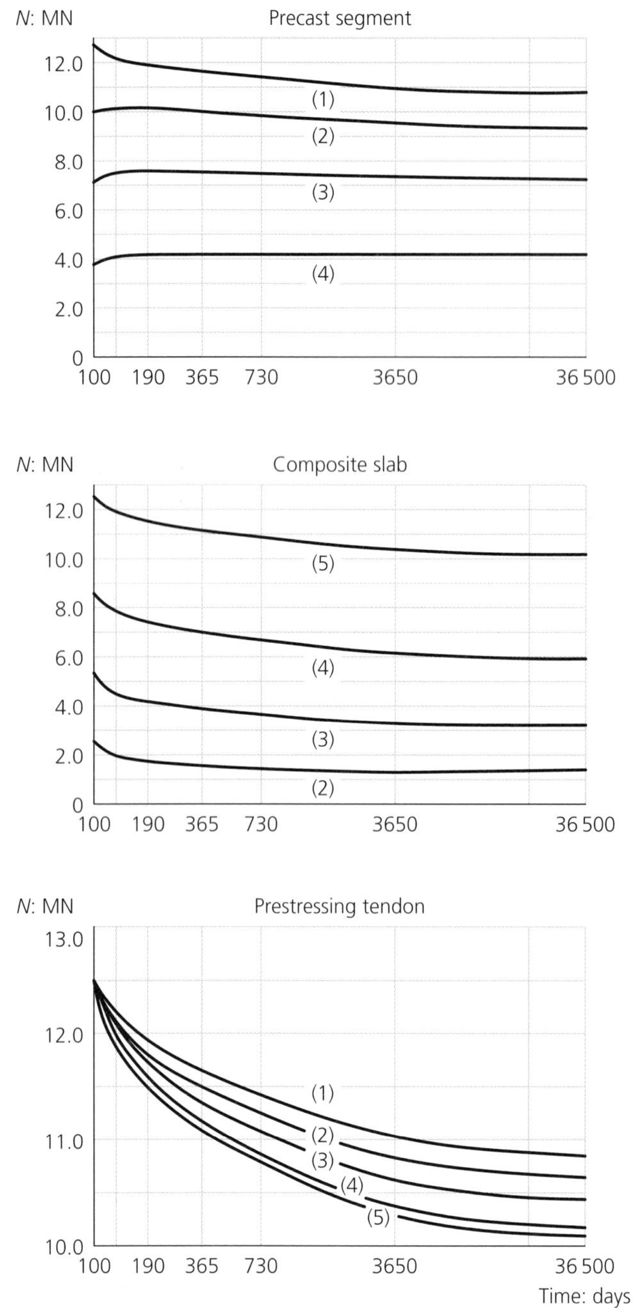

assumed to have a characteristic strength f_c = 50 MPa ($E_c(28)$ = 38.5 GPa); the composite slab was assumed to have a characteristic strength of f_c = 40 MPa ($E_c(28)$ = 36.50 GPa). The redistribution of stresses was determined using software that utilises MC 90 rheological functions.

Table 7.3 Progressively cast concrete member

Case	Area: m^2		
	Precast segment	Composite slab	Total
1	1.2500	0.0000	1.2500
2	0.9375	0.3125	1.2500
3	0.6250	0.6250	1.2500
4	0.3125	0.9375	1.2500
5	0.0000	1.2500	1.2500

Figure 7.7 shows the redistribution of stresses between the two parts of the progressively cast concrete member when it was loaded by a constant force. The process of casting and loading was as follows. The first part of the member was cast and, after 3 days of curing, it was placed on a support that allowed it to shrink. After 87 days (time 90 days) the second part of the member was cast and cured for 3 days. After an additional 7 days (time 100 days) the element was loaded with a constant force F = 12.5 MN. From the figure it is evident that at the time of loading the force was distributed between the two parts of the member proportionally to their stiffness, given by the product of the area $A_c(t)$ and the modulus of elasticity $E_c(t)$. Over the course of time, the stresses (or portion of the force) were redistributed from the younger to the older part of the compressed member.

Figure 7.8 shows the redistribution of stresses between the two parts of the progressively cast concrete member when it was prestressed by a prestressing tendon. The tendon has an area $A_{PT} = 0.01512$ m^2, with a modulus of elasticity E_s = 195 GPa. The process of casting and loading was the same as in the previous example. The prestressing force applied at day 100 was P = 12.5 MN. At the time of the post-tensioning the tendon had zero stiffness ($A_{PT}E_s = 0$); after post-tensioning the tendon was incorporated in the structure.

Similar to the previous example, at the time of loading the prestressing force was distributed among the components of the member proportionally to their stiffness, as given by the product of the area A_c at the modulus of elasticity $E_c(t)$. With time, the prestressing force was redistributed. Due to the shortening of the member caused by the creep and shrinkage of the concrete, the tension force in the tendon was reduced, and the compression stresses were redistributed from the younger to the older concrete.

The reduction in the tension forces in the prestressing tendons is usually referred to as the 'time-dependent loss' of the prestressing force. Figure 7.8 shows the simplest example of the losses in a straight, centrically compressed member prestressed at one time. However, usually, structures are subjected to normal stresses originating from a combination of the normal forces and the bending moments caused by the dead load and the prestressing.

7.4. Redistribution of the stresses due to the deformations of supports

The influence of the deformation of the supports is demonstrated here using the example of a three-span structure that was cast in formwork supported by stationary scaffolding (Figure 7.9(a)). The dimensions of the structure are shown in Figure 8.5(b). After 14 days, the formwork was removed and the bearings were activated. The intermediate supports were then subjected to vertical deformations of Δ = 100 mm. These deformations were done immediately after the bearing activation (14 days), or after 3 months, 1 year, 3 years, 10 years, 30 years and 60 years.

The effects of the deformation of the supports, which were calculated using the MC 90 creep function in the MIDAS Civil software, are given in Table 7.4 and Figure 7.9(b). The figure shows the course of the bending moments when the deformation occurs, and after 100 years (time t_∞).

Formerly, using the old rate-of-creep theory, the influence of the deformation of a support was given by (Leonhardt, 1964)

$$M_b(t) = M_b e^{-\varphi(t)} \tag{7.4}$$

where $M_b(t)$ is the bending moment above the intermediate support b at time t, M_b is the bending moment in the same section determined by elastic analysis and φ is the creep coefficient.

Figure 7.9 Redistribution of stresses: (a) analysed structure; (b) bending moments

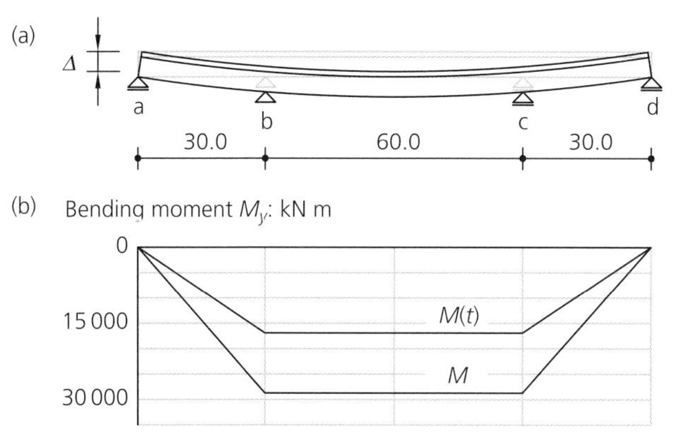

To compare the old and present creep functions, Table 7.4 gives the multiples of the bending moments that stress the structure at time t_∞, which were determined using MC 90 and Equation 7.4. It is evident that, according to MC 90, the early reduction in the static effects is smaller than the reduction determined according to the rate-of-creep theory (Equation 7.4). However, if the older structure is subjected to the deformation of the supports, there is still a significant reduction in the static effects.

7.5. Redistribution of stresses in structures in which the static systems change

To understand the problem, several basic examples are discussed here. All the calculations were done using the MC 90 function in the MIDAS Civil software.

7.5.1 Redistribution of the stresses in structures assembled of simple beams

The first example looks at the influence of creating of continuous structures from simple beams. The problem is illustrated using a two-span structure assembled of two 40 m long simple beams. The beams have the cross-section shown in Figure 8.5(a) and are made from concrete 35/45. They were cast in a formwork supported by stationary scaffolding. After 14 days the formwork was removed, and the beams were supported by bearings. The beams were then connected above the intermediate support.

Time-dependent analyses were undertaken for several different times of making the connection: immediately after placing the beams on the bearings (14 days), or at 3 months, 1 year, 3 years, 10 years, 30 years and 60 years.

As the new connection prevents free rotation of the beams at the intermediate support, the negative bending moments originate there. Figure 7.10(b) shows the bending moments that originate at the structure at time t_∞ if the connection was made after 1 year. The bending moment diagrams of the structure cast at one time are also presented for comparison.

Table 7.5 gives, for all the connection times considered, multiples of the bending moments originating in the structure cast on the scaffolding at one time. It is evident from the table that a significant redistribution occurs if the connection is made early after beam casting.

Table 7.4 Multiples of bending moments

Theory	Time of deformation, t						
	14 days	3 months	1 year	3 years	10 years	30 years	60 years
Rate of creep	0.0990	0.2252	0.4405	0.6900	0.9180	0.9917	0.9991
Model code	0.2404	0.4243	0.5201	0.5888	0.6577	0.7142	0.7307

Figure 7.10 Redistribution of stresses: (a) analysed structure; (b) bending moments

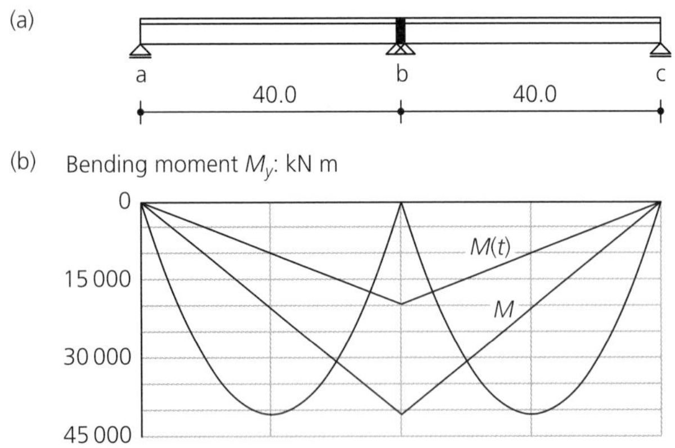

This well-known problem, which is connected with the construction of bridge decks assembled from precast beams, was formerly solved by using the old rate-of-creep theory. In this theory, the bending moment originating at the intermediate support b is given by (Leonhardt, 1964)

$$M_b(t) = M_b\left(1 - e^{-\varphi(t)}\right) \tag{7.5}$$

where $M_b(t)$ is the bending moment at time t, M_b is the bending moment determined by elastic analysis of the structure cast one time, and φ is the creep coefficient. To show the difference between the new and old creep theories, the multiples of the bending moments that stress the structure at time t_∞ determined using Equation 7.5 are also presented in Table 7.5.

7.5.2 Redistribution of the stresses in cantilever structures

The redistribution of stresses during the construction of cantilever structures is illustrated in Figure 7.11(a) for the example of a three-span structure made from concrete 35/45 and assembled incrementally. First, a structure formed by two symmetrical cantilevers was cast on stationary scaffolding. At structure age 14 days, the scaffolding was removed, and the intermediate bearings were activated. From this time on, the structure functions as two symmetrical cantilever structures (the stability of the cantilevers is guaranteed by temporary supports). After 14 days or 3 months, the midspan closure between the cantilevers was created, the outer ends of the cantilever were supported (a very small force was created at the supports), and the bearings were placed.

Figure 7.11 Redistribution of stresses: (a) analysed structure; (b) deformations; (c) bending moments

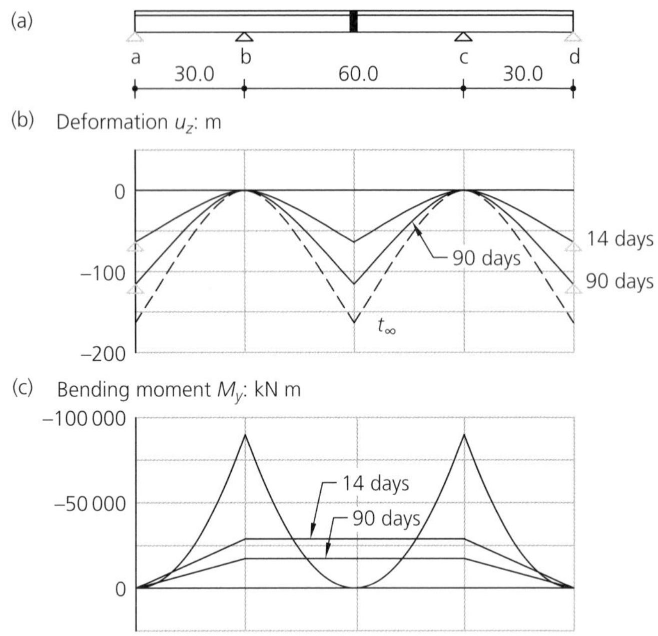

The results of the time dependent analysis are shown in Figures 7.11(b) and 7.11(c), which give the deformation curves and bending moment diagrams for the time when the static system changed and at time t_∞. It is evident that the progressively erected structure has a tendency to function as a structure cast at one time.

The next example is of the same structure, but in this case in the end supports were vertically deformed immediately after the closure was made. The value of the vertical deformation corresponds to the deflection of the end of the cantilever. Figure 7.12 shows the deformation curves and bending moment diagrams at the time when the static system changed and at time t_∞. It is evident that, if the deformation occurs early, the structure behaves similarly to the structure cast at one time.

Table 7.5 Multiples of bending moments

Theory	Time of connection, t						
	14 days	3 months	1 year	3 years	10 years	30 years	60 years
Rate of creep	0.9010	0.7748	0.5595	0.3100	0.0820	0.0083	0.0009
Model code	0.7315	0.4519	0.2780	0.1396	0.0187	0.0001	0.0000

Figure 7.12 Redistribution of stresses: (a) analysed structure; (b) deformations; (c) bending moments

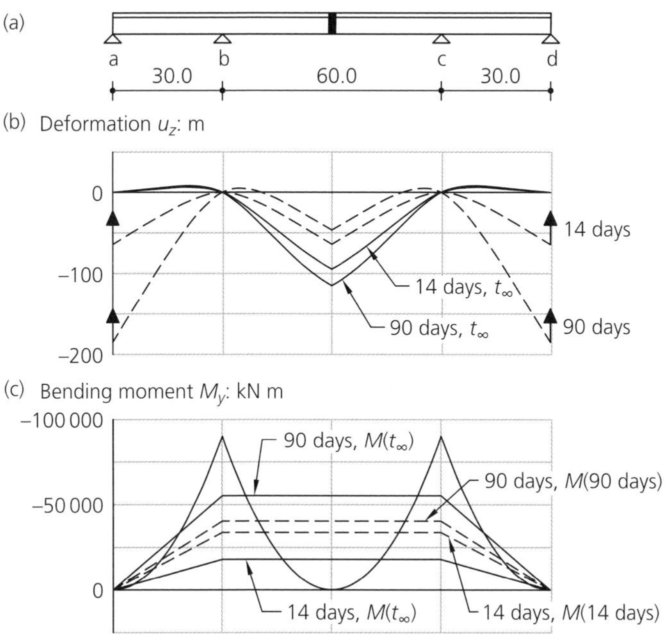

Figure 7.13 Redistribution of the bending moments in a two-span beam

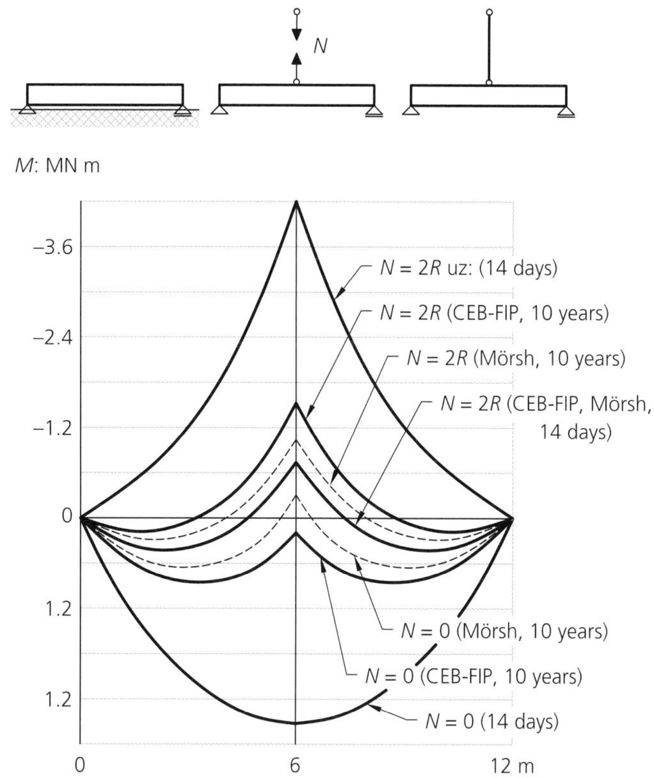

7.6. Balancing the dead load

Figure 7.13 shows a simple beam of span length 12 m. After 14 days of curing, the beam was suspended at its midspan on a very stiff stay cable ($E_sA_s = \infty$). Before the beam was suspended, a force N was introduced in the cable. The values of N were 0, R and $2R$, where R is the reaction at the intermediate support of a two-span continuous beam under uniform dead load.

The time-dependent analysis was performed using the Mörsch and MC 90 creep functions. With time, a significant redistribution of bending moments occurs for $N = 0$ and $N = 2R$. The bending moment diagram becomes more like the diagram for a two-span beam. Concrete is a natural material, and therefore the structure tries to behave naturally, that is, as a continuous beam. In this case, a larger redistribution of stresses is obtained when the Mörsch creep function is used.

It is important to realise that for the force $N = R$ there is no redistribution with either creep function. The structure keeps its shape and the stresses are constant over time. Their values do not depend on the creep function adopted.

As it is difficult to design a structure in which the stresses change over time, it is very important that structures are designed in such a way that the redistribution of stresses is minimal in the initial stage. This means that the geometry and forces in the internal prestressing tendons or external cables (situated inside or outside the perimeter of the deck) have to be determined in a way that their effects in conjunction with the dead load create zero deflection in the structures.

In beam structures, the dead load deformation and bending moments can be balanced by prestressing tendons arranged in one of two ways

- draped tendons
- straight tendons.

To quantify the different effects of these two arrangements, a detailed analyses was undertaken of a three-span structure loaded by its self-weight and by prestressing. The structure is composed of a one-cell box girder (with the dimensions given in Figure 8.5(b)) that has been prestressed by draped and straight prestressing tendons (Figure 7.14).

Both the draped and the straight tendons balanced the self-weight bending moments (Figure 7.14(b)). The draped tendons were led into the webs, and the straight cables were anchored in the corners of the box section and were led into the top and bottom slabs as is done in common cantilever structures. Due to the progressive anchoring of the tendons, the compression stresses arising from the short tendons

Figure 7.14 Balancing of the dead load: (a) layout of the prestressing tendons; (b) bending moments; (c) shear stresses due to the dead load; (d) shear stresses due to prestressing

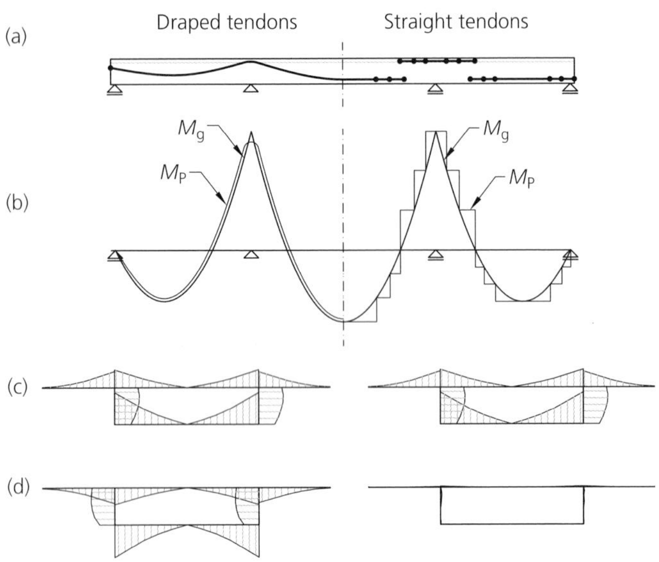

were not distributed across the whole width. For this reason, larger compression stresses originated above the webs, and these balanced the non-uniformly distributed self-weight tension stresses.

In this way, the self-weight bending stresses were well balanced by the prestressing tendons of both arrangements. As the straight tendons do not create shear forces, the shear stresses due to prestressing are zero (Figure 7.14(d)). However, as the draped tendons create radial forces, the box girder is stressed by shear forces of the same value, but of opposite sign, as the self-weight shear forces, and so the resulting shear forces are zero.

It is evident that the draped tendons reduce the shear stresses both in the webs and in the top and bottom slabs. Therefore, the prestressing should also balance the dead load shear forces. If structures are prestressed by external cables, the vertical components of the prestressing should ensure zero deflections at the deviators (Figure 7.15). Then the structure will be loaded only by the normal force, and will keep its shape over time. This approach, which was developed by Professors F. Leonhardt and T. Y. Lin (Leonhardt, 1964; Lin and Burns, 1981)), guides us to use partial, limited or full prestressing. The importance of load balancing was also demonstrated later by Professor R. Favre (Favre and Markey, 1994).

In cases where the deck is suspended on arches or pylons, the initial forces in the stay or suspension cables have to be determined for the condition of zero deck deflection at the anchor points. The forces should be determined from the reaction of the continuous beam supported at the points of the cable anchors (Figure 7.16).

Figure 7.15 Dead-load balancing in a two-span beam by means of: (a) internal tendons; (b) external tendons situated within the depth of the cross-section; (c) external tendons situated outside the depth of the cross-section

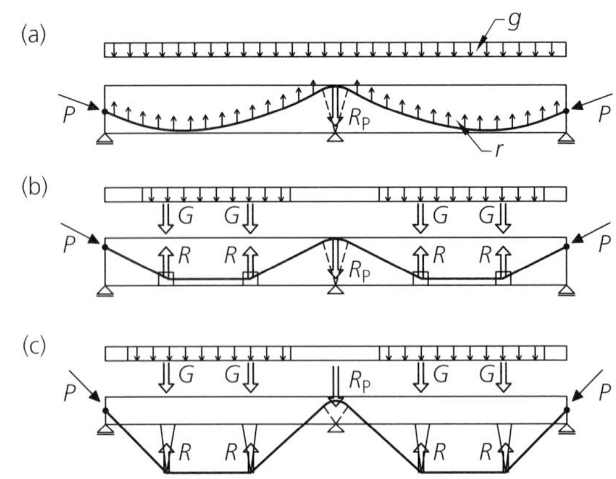

Figure 7.16 Dead-load balancing: (a) arch structure; (b) cable-stayed structure; (c) suspension structure; (d) equivalent continuous beam

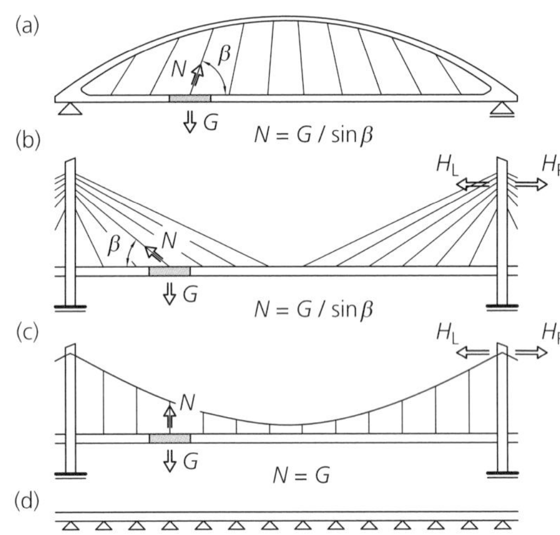

As an example, a time-dependent analysis was performed for an arch structure, with a span of 99 m and a rise of 14.5 m, composed of a steel arch and a concrete deck (Figures 7.17 and 7.18). The structures in Figures 7.17 and 7.18 both have the same geometry and the same sectional and physical properties. Both structures were loaded by their self-weight at time 14 days, but the structure in Figure 7.18 was also loaded by forces acting in the suspenders, which were prestressed. The forces in the suspenders correspond to the weight of the deck acting on a length equal to the distance between neighbouring suspenders (i.e. by the forces determined according to recommendation specified in the previous paragraph).

Figure 7.17 Bending moments without prestressing of the suspenders

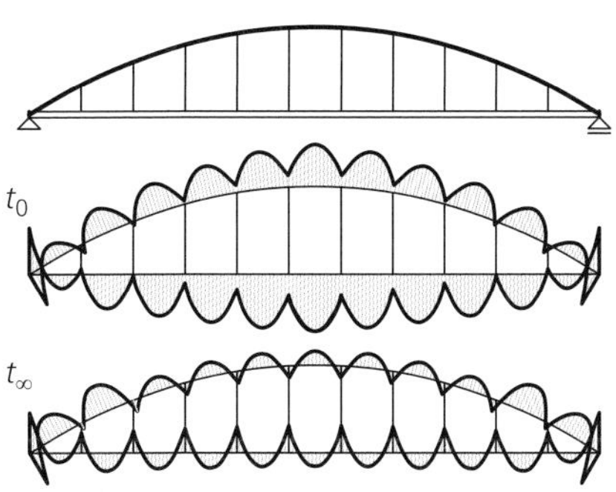

Figure 7.18 Bending moments with prestressing of the suspenders

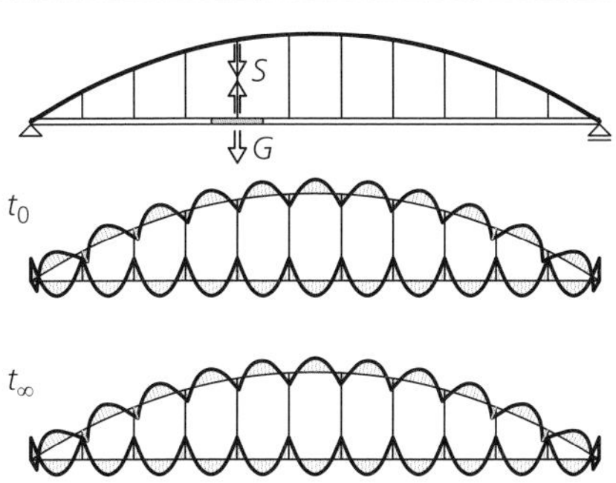

It is evident from Figures 7.17 and 7.18 that the structure, for which the forces in the suspenders were not actively determined, is subjected to a large redistribution of the bending moments originating in both the deck and the arch. However, the structure for which the forces in the suspenders were correctly chosen is stressed by constant bending moments.

7.7. Balancing the dead load in cable-stayed structures

Cable-stayed structures are suspended on inclined stay cables, which stress the beam by normal forces. The normal stresses create horizontal deformations that, due to the creep of concrete, increase over time. In addition, shrinkage of the concrete causes horizontal deformations of the deck. Therefore, the stay anchor moves horizontally, and its initial force is reduced. The change in the stay force causes a redistribution of the stresses in the structure.

Extensive studies have been done to quantify this phenomenon. Here only few results are presented. Figure 7.19(a) shows the same beam as in Figure 7.13, which, after 14 days of curing, was suspended at its midspan on a very stiff inclined stay cable ($E_sA_s = \infty$). Before the beam was suspended, a force

$$N = R/\cos\beta \tag{7.6}$$

was created at the cable, where R is the reaction at the midspan support of a continuous two-span (2 × 6 m) beam.

Figure 7.19 Redistribution of bending moments

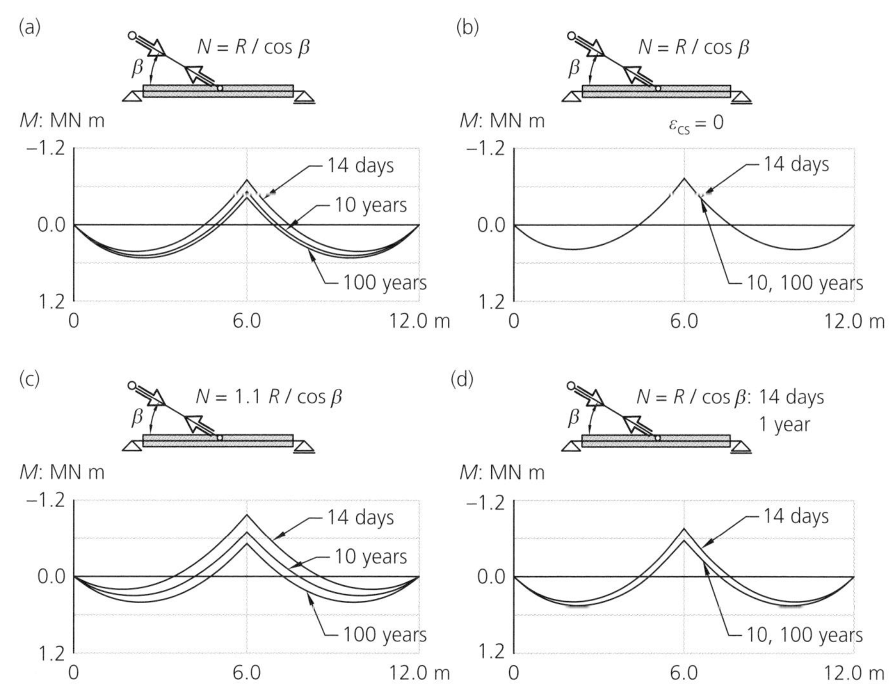

A time-dependent analysis was performed using the CEB-FIP (MC 90) creep function. Over time, the bending moments are redistributed from the anchor point to the span; at time $t_\infty = 100$ years, the moment at the anchor point is approximately 60% of the initial moment.

Another analysis was done, this time assuming that there is no shrinkage of the beam. In this case, the redistribution is very small (Figure 7.19(b)).

The time-dependent analysis was also done for the case where the initial force in the structure was increased by 10% (Figure 7.19(c)). The degree of redistribution of the bending moments was similar to that in Figure 7.19(a), but the final value was closer to the bending moment that corresponds to the force $N = R/\cos \beta$.

Figure 7.19(d) shows the results of the analysis done for the case where the force in the stay cable is adjusted after 1 year. The redistribution of the moments was significantly reduced; at time $t_\infty =$ 100 years the moment at the anchor point is approximately 80% of the initial moment.

It is evident that the redistribution of moments can reach significant values. However, it is necessary to realise that, in structures where the deck is suspended on multiple stay cables, the bending moments due to the dead load are very small compared with those due to the live load and temperature changes. However, it is wise to specify an adjustment of the stay forces when the deck deflection reaches specified values.

The deck of many cable-stayed bridges is not suspended on the stay cables in the area close to the pylons (Figures 7.20 and 7.21(a)).

Figure 7.20 Bridge across the Vršovice railway switching yard, Prague, Czech Republic

Figure 7.21 Dead-load balancing: (a) without suspension near the pylon; (b) without suspension near the pylon or the midspan; (c) an equivalent continuous beam

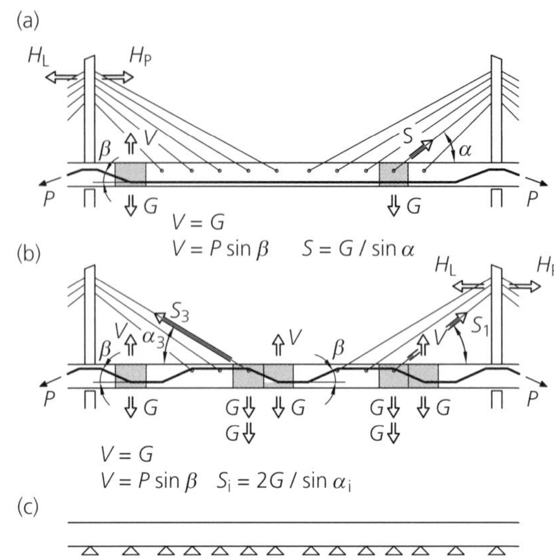

To balance the dead load, it is necessary to add prestressing cables that resist the bending moment that originates in that area. In cases where the deck is also not suspended in the area close to the midspan, it is necessary to design the layout of the prestressing cables that transfer the dead load into the area supported by the stay cables (Figure 7.21(b)).

The above solutions are illustrated by the following examples. The bridge across the River Elbe (see Figure 4.79) and the bridge across the Vršovice railway switching yard in Prague (see Figure 7.20), both in the Czech Republic (Stráský, 1993), have similar decks composed of a one-cell box girder with large overhangs supported by slab struts. The decks, which are suspended in the bridge axis on single pylons, are assembled from precast, match-cast segments. The segments are reinforced by prestressed concrete rods that transfer the cable force into the bottom corners of the box girder.

The River Elbe bridge is 248 m long and has three spans (lengths 61.60 + 123.20 + 61.60 m) (Figure 7.22(a)); the Vršovice bridge is 396 m long and has nine spans (lengths 26.40–101.20 m) (see Figure 7.22(b)). Due to the local conditions, the main piers of the Vršovice bridge are formed by single columns (Figure 7.23). Therefore, the torsion is resisted only at four points: the abutments, at pier No. 6 and at pylon pier No. 8. With regard to torsion, the structure acts as a three-span structure with span lengths 167.20 + 145.20 + 81.40 m.

Figure 7.22 Elevation: (a) the bridge across the River Elbe, Czech Republic; (b) the bridge across the Vršovice railway switching yard, Prague, Czech Republic

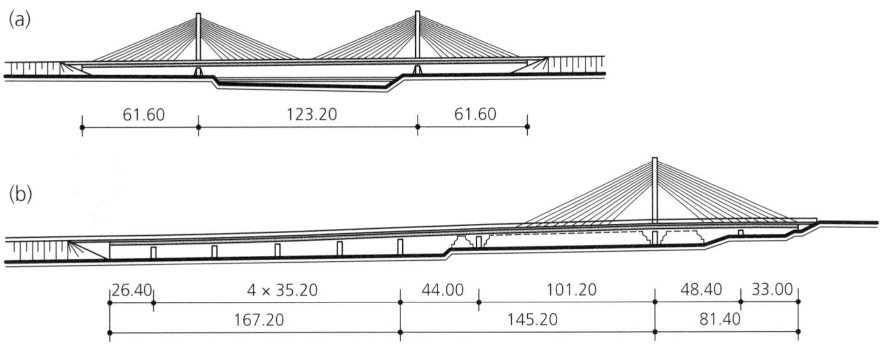

Figure 7.23 Cross-section: (a) typical; (b) at pier No. 6; (c) at pier Nos. 2, 3, 4, 5 and 9; (d) at pylon No. 8

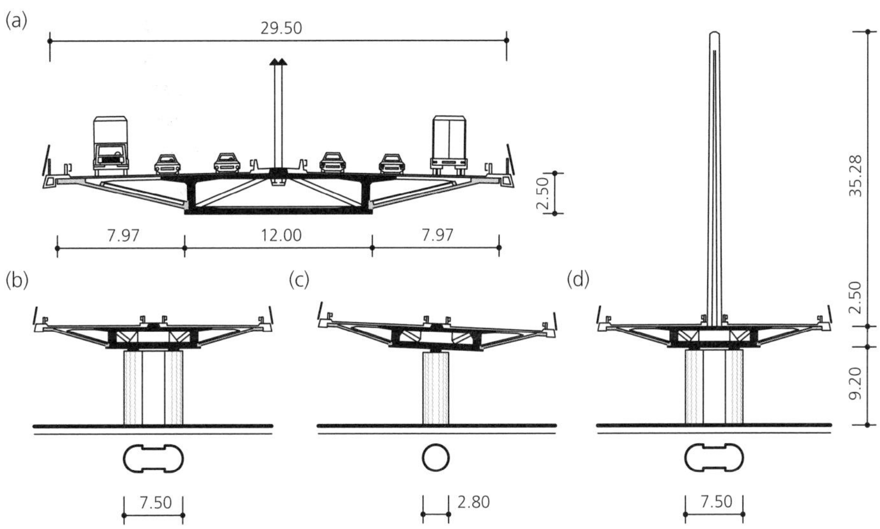

The bending moments originating in the deck close to the pylon of the Elbe Bridge are balanced by short, straight prestressing tendons situated at the top and bottom slabs of the box girder (Figure 7.24(a)). As the straight tendons do not balance the shear forces, it was necessary to widen the girder webs and increase the shear reinforcement of the top and bottom slabs.

As the deck of the Vršovice bridge is stressed by significant shear stresses, it was decided to balance not only the bending but also the shear stresses originating in the area close the pylon (Figure 7.24(b)). Therefore, short, draped external cables were added there (Figure 7.25). The cables are situated in the bridge axis, and function in a similar way to the stay cables. They are deviated and anchored in blisters, from which the vertical components of the cable forces are transferred by single, concrete-filled steel pipes into the intersections of the prestressed rods in the upper slab of the segments (see Figure 3.13(b)). In this way the dead load is transferred by the external cables directly into the pier diaphragm. Therefore, this arrangement secondarily reduces the transverse bending in the pier diaphragm. Similar external cables were situated close to the piers of all spans. These cables, together with continuity cables arranged in a similar way, significantly reduce the shear stresses originating along the whole length of the bridge. Therefore, it was possible to design very slender webs with a constant thickness of only 400 mm. The bridge, which has been heavily loaded since 1994, was again tested in 2019. The test showed that the structural function of the bridge is very good. No cracks were observed in the structure.

Figure 7.24 Dead-load balancing in non-suspended decks: (a) the bridge across the River Elbe, Czech Republic; (b) the bridge across the Vršovice railway switching yard, Prague, Czech Republic

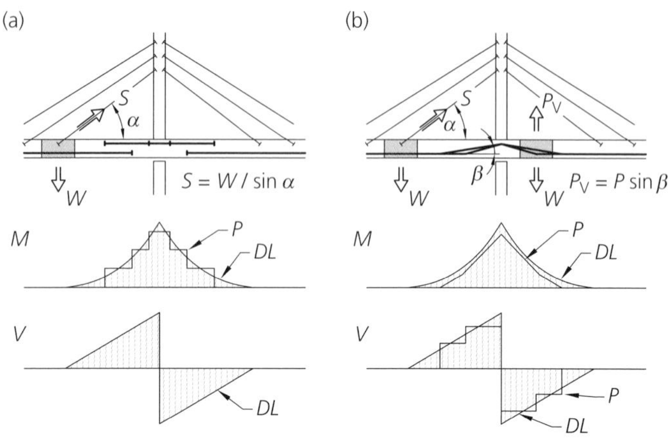

Figure 7.25 Layout of the prestressing tendons and external cables

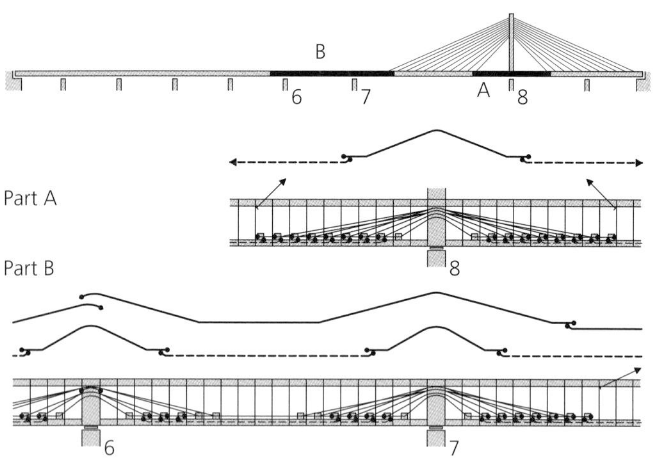

The deck of the cable-stayed bridge across the Danube Canal in Vienna, Austria (Pauser and Beschorner, 1976) (see Figure 4.78), is suspended on 2 × 2 × 4 parallel stay cables anchored in the span diaphragms situated at a distance of 44.40 m from the pylon (Figures 7.26 and 7.27). The deck was cast on the banks and subsequently rotated into the design position (see Figure 5.31). Before rotation, the deck was suspended on the stay cables and prestressed by internal prestressing tendons in the webs of the box girder. The prestressing tendons transfer the whole dead load into the pier and span diaphragms, from which the load is transferred to the pylons. After the rotation of the deck, the midspan closure was cast and prestressed by short-span tendons. The vertical components of the stay-cable forces, together with radial forces of the internal prestressing tendons, balance the dead load in the way shown in Figure 7.21(b).

Figure 7.26 Bridge across the Danube Canal, Vienna, Austria

Another example of the balancing of the dead load by a combination of stay cables and internal prestressing tendons is the design of the Delta Ponds Pedestrian Bridge in Eugene, Oregon, USA (Stráský, 2011). The bridge has a total length of 231.48 m and consists of a main cable-stayed structure crossing the highway and approaches. The main structure is 133.35 m long and is formed by a continuous deck that is suspended on a single pylon in spans 3, 4 and 5 (Figures 7.28 and 7.29). The main span crossing the highway is assembled from precast segments and a cast-in-place deck slab. The segments, which are formed by edge girders and the deck slab, are stiffened by diaphragms at the joints. The remaining spans are formed by a solid, cast-in-place slab of the same diameter. The main span was assembled after the back spans had been cast. The precast segments were erected in a cantilever, beginning at the pylon. The forces in the stay cables balance the dead load of the lighter main span. To keep symmetrical forces in the back-stay cables and in order not to load the pylon by horizontal forces, the deck in the back spans was prestressed by draped internal prestressing tendons, which balance the difference in the dead load between the main and the back spans (Figure 7.30).

7.8. Redistribution of stresses in the decks of progressively erected structures

The redistribution of stresses between members of different ages was discussed in Chapter 7 (see Section 7.3). It was shown that the compressive stresses are redistributed from the younger to the older concrete. Therefore, it is evident that this phenomenon has to be carefully checked in structures in which the decks are progressively erected from members of different ages. As it is difficult to design structures that involve significant stress difference, it is reasonable to try to find a way to mitigate these differences. A reasonable solution is to study the influence of the creep and shrinkage of the concrete and the level of prestressing of individual members in order to find the optimum solution.

It was shown in Section 7.7 that it is possible to determine the forces in the stay cables and prestressing tendons that balance the dead-load bending stresses. The structure is then centrically prestressed. In this way, the redistribution of the stresses can be studied at the

Figure 7.27 Bridge in Figure 7.26: (a) cross-section; (b) arrangement of the stay cables and prestressing tendons

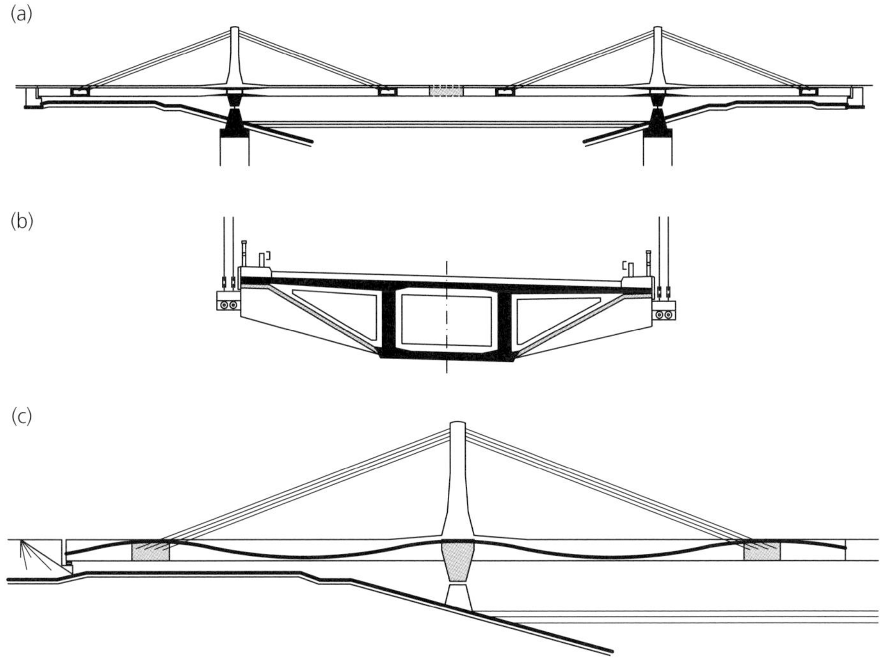

Figure 7.28 Delta Ponds Pedestrian Bridge in Eugene, Oregon, USA

level of the bridge cross-section. The calculation can be done by hand, and helps the designer understand the structural function. In the authors' opinion, this should always be done before a detailed analysis is undertaken.

The decks of the River Elbe and Vršovice bridges (discussed above) were progressively erected (Figure 7.31). First, the spine girder was assembled from precast segments, then the slab struts were suspended, and overhangs were cast in the formwork supported by the struts (Stráský, 2011). During the erection of the segments the structure was prestressed by the first group of prestressing tendons; then, after transverse prestressing of the overhangs, the deck was prestressed by the second group of prestressing tendons. Figure 7.32 shows the curves of the creep coefficients of the segments (φ_1^s and φ_2^s) and the overhangs (φ_1^o and φ_2^o) for the first and second prestressing, and the corresponding curves for the creep strain (ε_c) curves and the shrinkage strain (ε_{sh}).

The first group of tendons, together with the stay cables, create compression stresses f_1 = 10 MPa in the segments; the second group of tendons create compression stresses f_2 = 3 MPa in both the segments and the overhangs. This shows that if the age of the segments at the time of the first prestressing is 6 months, and the age of the overhangs at the time of casting is 7 months and at the time of the second prestressing 7.5 months, the strain curves composed of the shrinkage and creep strain curves of the segments and overhangs are similar. Therefore, there is no redistribution of the stresses between the segments and the overhangs.

The authors are aware that it is difficult to guarantee the construction of a bridge if the above assumptions are followed exactly. However, if the level of prestressing and the construction schedule is close to the values given, the degree of redistribution of the stresses is reasonable. The described approach was then incorporated into the design and more detailed analyses of the whole structure were performed.

Figure 7.29 Bridge in Figure 7.28: elevation of the main spans

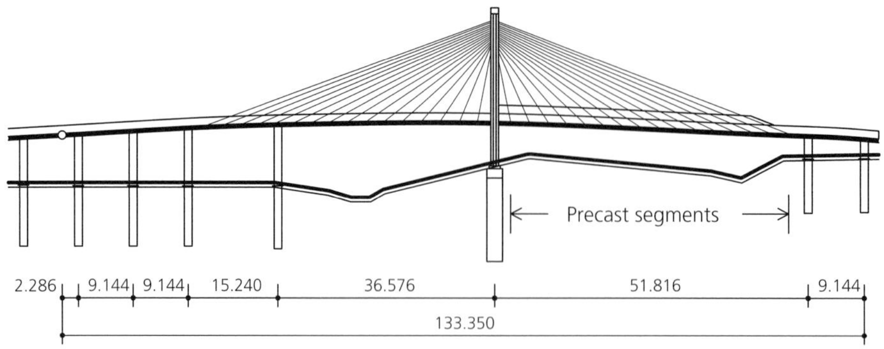

Figure 7.30 Bridge in Figure 7.28: dead-load balancing

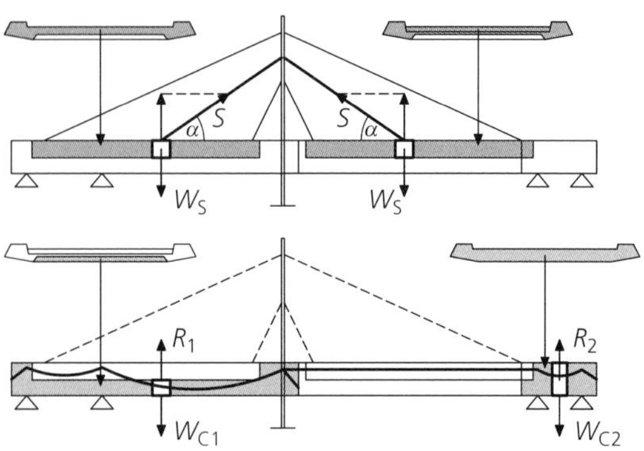

A similar progressive construction of the bridge deck has been used for several bridges in Slovakia (see Figures 5.89–5.91) (Stráský, 2015). This type of erection was first used in the construction of the 960 m long, 25.66 m wide motorway viaduct across the Hostovsky Creek Valley near Nitra, Slovakia (see Figures 4.60 and 4.61). The bridge deck has a typical span length of 69 m, and is formed by a spine and a one-cell box girder with large overhangs supported by precast struts. The spine girder has a variable depth, ranging from 4.0 m at the supports to 2.6 m at the midspan. The spine girder was progressively cast in formwork suspended on the overhead movable scaffolding system. During construction of the spine girder, the structure was longitudinally prestressed by internal bonded tendons situated within the basic cross-section. When the overhangs were cast and transversely prestressed, the structure was prestressed by external non-bonded tendons situated inside the central box (see Figures 10.25 and 10.26).

Figure 7.31 Progressive erection of the deck

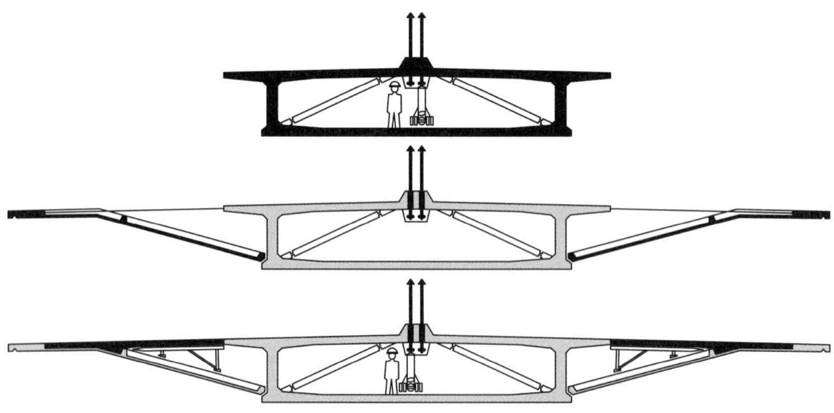

Figure 7.32 Strain in the deck due to the dead load, the forces in the stays and the prestressing: (a) box girder spine; (b) cast-in-place deck slab

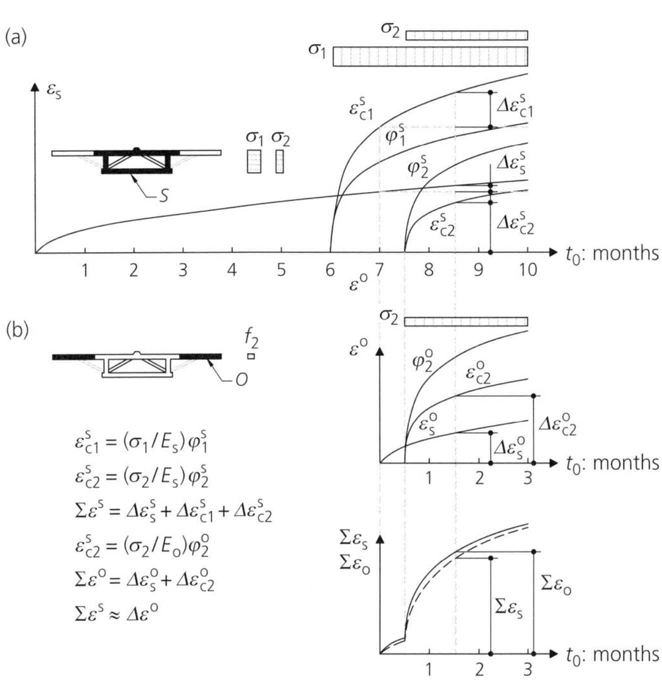

The viaduct was not only carefully analysed but has also been carefully monitored (during construction, in the loading tests and during service). To monitor the concrete stresses, strain gauges were placed in four sections. Two of these sections were in span 6 and two were in span 7. In both spans one section is situated at the midspan and one is situated at a distance of 0.5 m from the pier table. The gauges were connected to a Datataker data logger situated inside the bridge box. The concrete stress of the structural members in which the strain gauges had been placed was first measured after the members had been cast. After that, measurements were made before and after the post-tensioning of the bonded tendons, the stripping of the formwork, the erection of the struts, the casting of the overhangs, the post-tensioning of the external cables, the application of the additional dead load, the load tests and the bridge opening. During the service of the bridge, measurements have been made twice a year. Figure 7.33 shows the calculated and measured values, and it can be seen that there is good agreement between the results. Furthermore, the strain curves for the long-term deformations of the spine girder (A2 and A5) and the overhangs (A7) are nearly parallel, which confirms that the level of prestressing was correctly determined and the redistribution of the normal stresses between the spine girder and the overhangs is small.

Figure 7.33 Calculated and measured strains

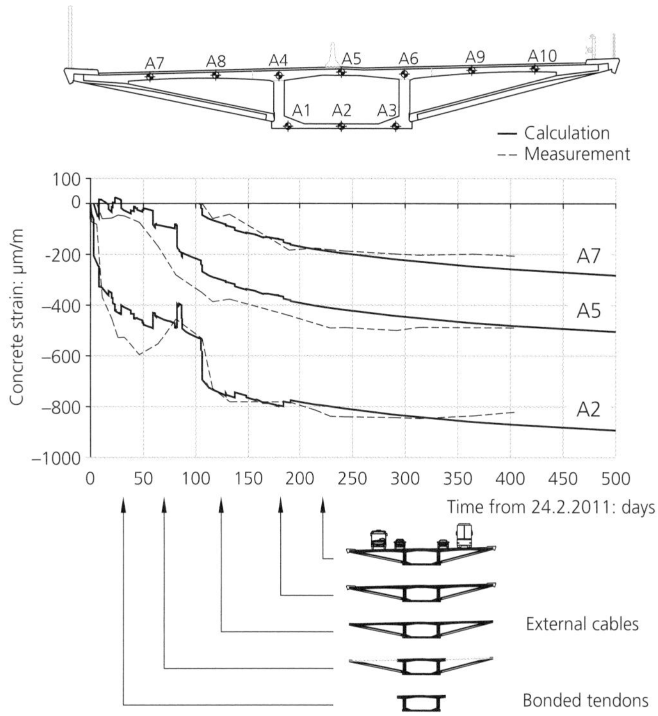

Figure 7.34 Bridge across the Ebro River, Zaragoza, Spain: (a) lifting of the central portion; (b) tensioning of the strands; (c) vertical deformation of the deck

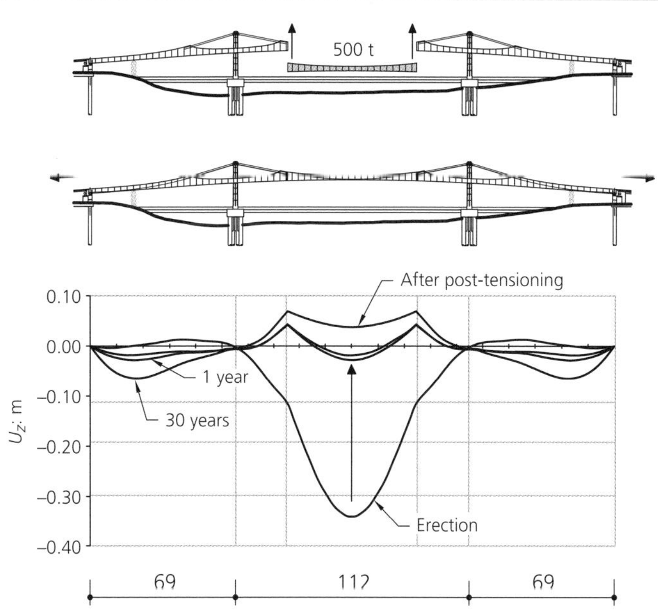

Figure 7.35 Bridge in Figure 3.34: redistribution of the stresses in the deck slab above the spine girder

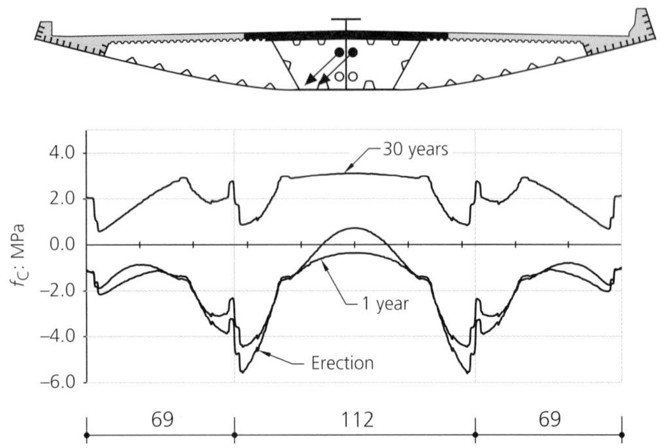

While in prestressed concrete structures the redistribution of stresses between the progressively cast structural members can be controlled by choosing the appropriate level of prestressing for the individual structural members, it is difficult to control the redistribution of stresses between the steel and concrete of composite structures. The bridge across the Ebro River at Zaragoza, Spain (see Figures 4.88 and 4.89) can serve as an example. The bridge forms a self-anchored suspension structure. A cable-stayed structure suspended on steel pipes was erected, and then, by tensioning the suspension cables passing through the pipes, the structure was modified into the self-anchored suspension structure (i.e. a structure prestressed by external cables) (Stráský *et al.*, 2014).

The side spans and the parts of the main span close to the pylon were erected first, and then suspended on the pylon via steep pipes (Figure 7.34(a)). After this, the central portion was lifted and connected to the already assembled structure, and the suspension cables were pulled and partially tensioned. In this way, the weight of the steel structure was economically resisted by the tension of the strands (Figure 7.34(b)). Subsequently, the deck slab was progressively cast, and the suspension cables were tensioned to the design values. Over time, the creep and shrinkage of the concrete have influenced the deformation of the structure and the redistribution of the compression stresses from the concrete to the steel. The degree of these changes is evident from Figures 7.34(c) and 7.35.

REFERENCES

FIP (1998) FIP Recommendations 1996. *Practical Design of Structural Concrete. FIP Congress,* Amsterdam, The Netherlands.

CEB (Comité Euro-International du Béton) (1986) *CEB-FIP Model Code 1978*. fib, Lausanne, Switzerland.

CEB (1993) *CEB-FIP Model Code 1990*. Thomas Telford, London, UK.

CEN (European Committee for Standardization) (2005) EN 1992-2:2005. Eurocode 2 – Design of concrete structures. Part 2: Concrete bridges: design and detailing rules. CEN, Brussels, Belgium.

Favre R and Markey I (1994) Generalization of the load balancing method. Prestressed Concrete in Switzerland 1990–1994. *Proceedings of the 12th Congress of the FIP, Washington, DC, USA.*

fib (International Federation for Structural Concrete) (2013) *fib Model Code for Concrete Structures 2010*. fib, Lausanne, Switzerland.

Leonhardt F (1964) *Prestressed Concrete. Design and Construction.* Ernst & Sons. Berlin, Germany.

Lin TY and Burns NH (1981) *Design of Prestressed Concrete Structures*. John Wiley, New York, NY, USA.

Navrátil J (2014) *Prestressed Concrete Structures*. VSB – Technical University of Ostrava, Czech Republic.

Pauser A and Beschorner K (1976) Betrachtungen über seilverspannte Massivbrüken, ausgehend vom Bau der Schrägseilbrücke über den Donaukanal in Wien. *Beton- und Stahlbetonbau* **71(11)**: 261–265.

Smerda Z and Kristek V (1988) *Creep and Shrinkage of Concrete Elements and Structures*. Elsevier, Amsterdam, The Netherlands.

Stráský J (1993) Design and construction of cable-stayed bridges in the Czech Republic. *PCI Journal* November-December: 24–43.

Stráský J (2011) *Stress Ribbon and Cable Supported Pedestrian Bridges*, 2nd edn. Thomas Telford, London, UK.

Stráský J (2015) Viaducts with progressively erected decks. In *Multi-Span Large Bridges* (Pacheco P and Magalhaes F (eds)). CRC Press, London, UK, pp. 27–36.

Stráský J, Cobo del Arco D, Novotny P and Raventós I (2014) Construction of "Lo Passador": a self-anchored suspension bridge at the Ebro River delta. *Structural Engineering International* **24(1)**: 114–117.

Stráský J, Navrátil J and Suský S (2001) Applications of time-dependent analysis of in the design of hybrid bridge structures. *PCI Journal* **46(4)**: 56–74.

ice Publishing

Stráský and Nečas
ISBN 978-0-7277-6385-3
https://doi.org/10.1680/dcpb.63853.131
ICE Publishing: All rights reserved

Chapter 8
Analysis of concrete bridges

Modern methods of static and dynamic analysis make it possible to express geometric and physical non-linearity of concrete structures, and to determine their failure in the limit state. However, the requirements for preparing input data, evaluating results and the capacity of computers are such that these calculations are performed only in special and/or study cases. An elastic analysis of a structure supplemented by checking its ultimate capacity using a strut-and-tie method is usually satisfactory.

8.1. Elastic modelling of concrete bridges

In the opinion of some engineers, recently developed software allows everything to be calculated. The availability of these programs and the improvement in of computer technology is such that a complex analysis of a complex structure can be done in an extremely short time. However, this does not mean that the calculations are done correctly, or that the models used really reflect the function of the structures. The main problems in static and dynamic structural analysis are to determine the appropriate computational model, to understand the method used to reach the solution, to correctly determine the input parameters and to correctly interpret the results.

When preparing input data, it is necessary to realise that common programs analyse isotropic structural members. Most bridge decks are designed as prestressed members in the longitudinal direction of the bridge and as reinforced concrete members in the transverse direction. Similarly, the substructure is mostly designed as reinforced concrete. Therefore, the characteristics of the structure or structural elements should be given with respect to the actual stiffness.

The majority of modern structural analysis is based on the finite-element method (Zienkiewicz and Taylor, 1994). The possibilities of structural modelling are illustrated by Figure 8.1 (Liebenberg, 1992), which shows a possible modelling of a two-cell box girder that is frame-connected to piers having a rectangular cross-section (Figure 8.1(a)). The bridge structure can be modelled as described below.

(*a*) The box girder and piers are modelled as a three-dimensional (3D) structure assembled from solid elements that are used for both the top and bottom slabs and the webs (Figure 8.1(b)). Two types of solid elements are usually used. The first type has six degrees of freedom in each node (three transitions and three rotations). However, the calculation is requires a lot of computing power and time, and so elements having only three transitions in each node are often used. In the latter case it is necessary to increase the density of the division of the structure into elements. If the three-transition elements are combined with elements that have six degrees of freedom in the nodes (shells, beams), it is necessary to form a connection that expresses a transfer of the rotation into the solid elements (e.g. to form a connection via three beam elements that connect three nodes of the solid element to the node of the beam element).

(*b*) The box girder is modelled as a 3D structure composed of shell elements that are stressed by both membrane and bending stresses (Figure 8.1(c)). The shell elements are used for both the top and bottom slabs, as well as for the webs. Each node of the shell elements has six degrees of freedom (three transitions and three rotations). The piers can also be modelled as shell elements.

(*c*) The box girder is modelled as a grillage structure consisting of three longitudinal girders that are mutually connected by transverse members expressing shear and bending stiffness of the transverse connection (Hambly, 1991) (Figure 8.1(d)). Piers can be modelled as grids or as beam elements. The grids are assembled from beam elements that have six degrees of freedom in each node (three transitions and three rotations).

(*d*) The box girder and piers are modelled as beam elements (Figure 8.1(c)). At their connection, infinitely stiff elements are used. All elements have six degrees of freedom in each node (three transitions and three rotations).

In addition to the methods described, the structure can be modelled as a space frame. The top and bottom slabs and webs are substituted by grids, which together create space frames. Software that uses the folded-plates or the finite-strips method has been developed. Although these methods are very effective for some problems, they are now used only for special cases. The beam-and-slab structures are sometimes analysed as slab structures. As most structures are also subjected to normal forces, and because the joining of the beams and slabs is mostly eccentric, it is better to always model the structure using shell elements that resist both the membrane and the bending stresses.

Figure 8.1 Modelling of two-cell box girder

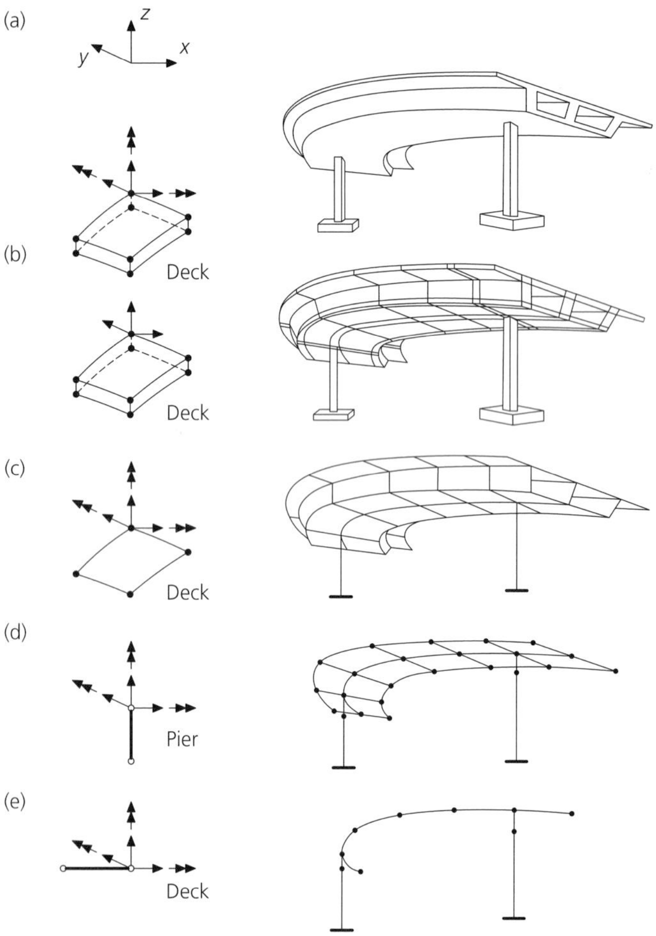

The indisputable advantage of using solid elements is that it is possible to describe the geometry of the structure without any simplification, and to correctly express the boundary conditions and the loading. Modelling structures as solid elements is particularly suitable for robust structures consisting of slabs, slab beams and slabs stiffened by edge beams (Figure 8.2(a)). The disadvantage of using solid elements is that most of the available programs do not allow for their time-dependent analysis. To calculate the ultimate load capacity, either a non-linear analysis, or the internal forces obtained from the integration of the stresses, must be used.

Modelling of structures as shell elements is suitable for thin-walled structures composed of planar elements (Figure 8.2(b)). The calculation makes it possible to express the space function of a structures, and describe the boundary conditions and the space action effect of the load. As with solid elements, the disadvantage is that some available programs do not allow their time-dependent analysis, and the ultimate load capacity must be calculated using a non-linear analysis or the internal forces obtained from the integration of the stresses.

The space behaviour of thin-walled structures can also be expressed by modelling them as 3D frames (Figure 8.2(c)). The frame consists of beam elements situated at the centre of gravity of the individual elements of the cross-section (cantilever, top and bottom slabs, webs), which are connected to each other by cross-beams.

The analysis of structures modelled as grids is suitable mainly for beam-and-slab structures, but it has also proven to be suitable for multi-cell box girders and for box girders connected at their top slab (Hambly, 1991). Because the structures are modelled as beams, the results of the analysis can be used directly for dimensioning the structural elements. The transverse connection of the longitudinal elements can be represented by substitute transverse elements. The method is described in greater detail in Section 8.4.

The decision about how to model a bridge structure depends on the stage of the project, the nature of the structure and the nature of the problem to be solved. The analysis should always be carried out in several stages, and the results of the solutions obtained should be compared. The choice of computational model and method of analysis is a highly skilled task that requires not only knowledge but also long-term experience. As there is no uniform guidance on how to model structures, any static analysis must be conducted under the guidance of experienced designers.

When modelling modern structures, the different types of members can be combined to express the function the whole structure and of the structural members that form the structure. An example of such

Figure 8.2 Modelling of the bridge deck: (a) solid elements, (b) shell elements, (c) beam elements

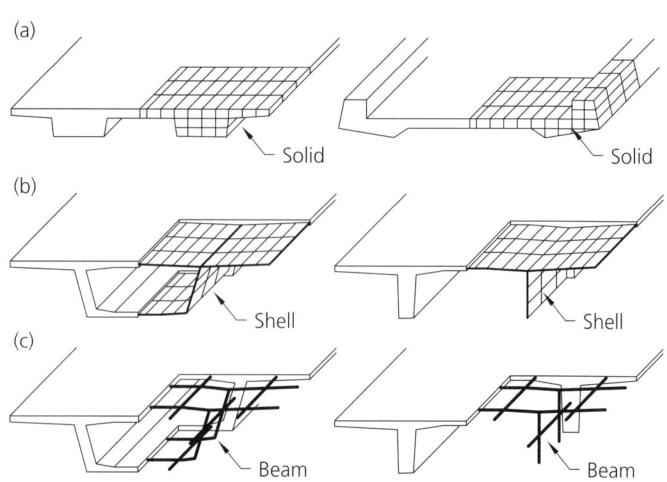

Figure 8.3 Structural arrangement

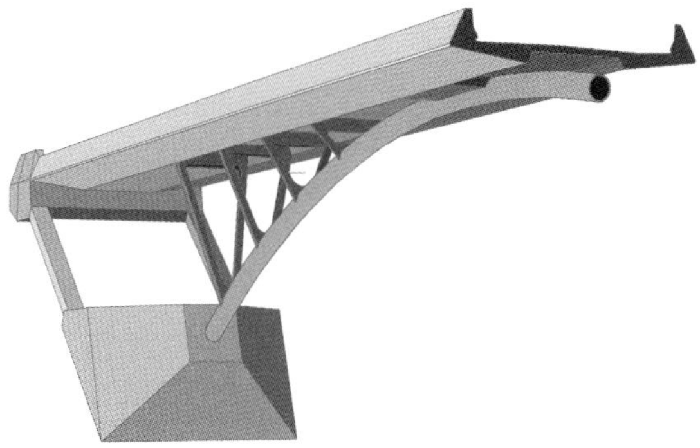

Figure 8.4 Modelling of the structure in Figure 8.3

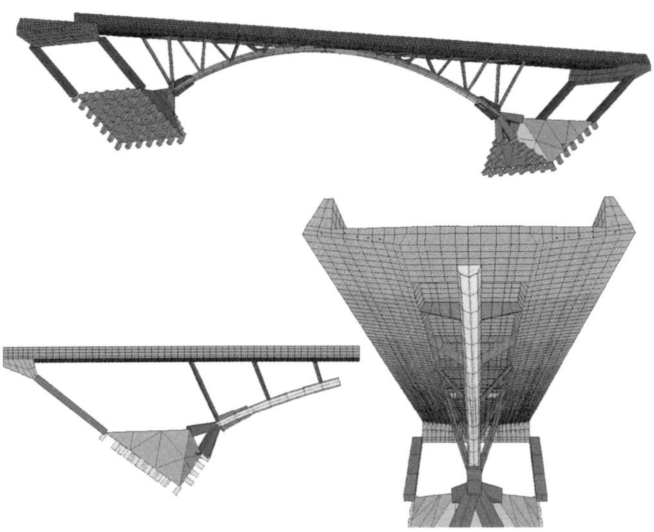

modelling is the calculation model used for the arch bridge near Rajhrad in the Czech Republic, described in Chapter 4 (see Figures 4.16 and 4.17). The bridge has a prestressed concrete deck with a channel cross-section that is supported via steel V-shaped struts by a steel arch (Figure 8.3) (Stráský and Husty, 1999). The arch is formed of 900 mm diameter pipe infilled with concrete. The struts, which are perpendicular to the arch, are formed of steel boxes filled with concrete. The concrete end deck diaphragms, which are supported by concrete struts fixed into the arch footings, transfer the load from the structure into the gravel soil.

The structure was analysed as a 3D frame structure assembled of beam elements, and its space behaviour was verified by finite-element analysis of the 3D structure assembled from solid, shell, tube and beam elements (Figure 8.4). Solid elements were used to model the concrete deck and footings, shell elements to model the steel struts, tube elements to model the steel arch and concrete fill, and beam elements to model the end concrete struts, and springs were used to express the function of the soil. The model served not only for the design of important structural details, but also for the correction of results obtained by the analysis of the structure modelled as a 3D frame (see Figure 8.34).

8.2. Analysis of two basic structural types

To aid the understanding of the static function and the modelling of structures, it is useful to examine the function of two basic types of bridge deck: a double-T beam and a one-cell box girder (Figure 8.5). A bridge deck composed of three spans (lengths 33 + 45 + 33 m) was modelled as a 3D structure assembled from shell elements that are stressed by both membrane and bending stresses (Figure 8.6). Before the results of the analysis are discussed, it is useful to remind the reader which internal forces and corresponding stresses originate in these elements (Timoshenko and Woinowsky-Krieger, 1959).

Figure 8.5 The studied bridge decks: (a) cross-section of the double-T beam; (b) cross-section of the box girder; (c) elevation

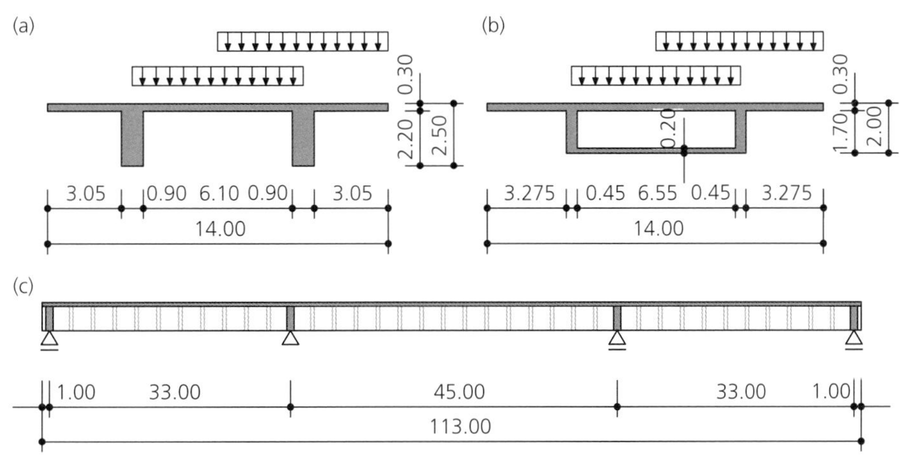

Figure 8.6 Modelling of the bridge decks in Figure 8.5: (a) box girder; (b) double-T beam stiffened by span diaphragms

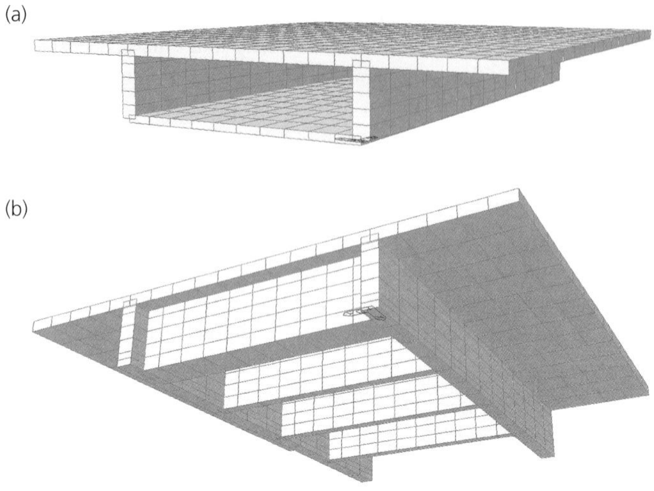

Figure 8.7 Internal forces and stresses: (a) shell elements; (b) plate elements

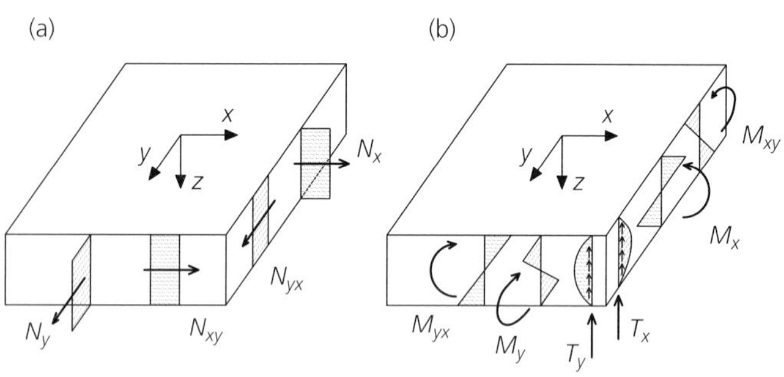

The elements combine the function of the shell elements, which are stressed by normal and shear forces acting in the plane of the element (Figure 8.7(a)), and the function of the plate (slab) elements, which are stressed by bending and torsional moments and by shear forces (Figure 8.7(b)).

The maximum normal stresses in the elements are given by the sum of the normal and the bending stresses.

$$\sigma_x = N_x/h \pm 6M_x/h^2 \quad (8.1)$$

$$\sigma_y = N_y/h \pm 6M_y/h^2 \quad (8.2)$$

The bridge decks are made from concrete with a characteristic strength of f_c = 50 MPa ($E_c(28)$ = 38.5 GPa), and have the dimensions given in Figure 8.5. They have three different arrangements

- without any diaphragms (see Figure 8.6(a))
- with 1.00 m wide diaphragms, which stiffen the deck at the supports
- with 1.00 m wide support diaphragms and 0.30 m wide span diaphragms, situated along the whole length of the deck at distances of 3 m (see Figure 8.6(b)).

To obtain symmetrical transverse deformations for a symmetrical loading, the deck was supported by multidirectional bearings supplemented by transversal soft springs, which allow deformation of the transverse deck. The structures were analysed for a uniform load of 10 kN/m^2 located at the main span on one longitudinal half of the deck, and between the webs.

The results of the analyses are presented in Figures 8.8 and 8.9, which show the deformations V of and the distribution of the normal stresses σ_x in the structures at the midspan and the intermediate support sections. It is evident from the figures that the influence of intermediate diaphragms guarantees the shape of the deck, and thus allows the possibility of using the pure torsion (St Venant) theory for analysis. It is interesting to note that, under the non-symmetrical load, the double-T deck stiffened by the span diaphragms has its maximum normal stresses at the deck edges at the support section. This is due to in-plane bending (warping) stresses.

The largest part of the bridges formed of a double-T deck is constructed without any diaphragms. As these structures are subjected to large deformations of the cross-section, it is evident that these decks cannot be modelled by one beam situated at the centre of gravity of the double-T deck. As will be shown below, these structures can be effectively analysed as structures modelled as a grid. Conversely, the decks formed of single-cell box girders can be analysed as a beam structure, although the influence of the deformation of the cross-section should be considered in a design of the thin-walled beams.

8.3. Beam analysis for the box girder

In a beam analysis, the space function of the bridge deck is transferred into the bridge axis. Although, at first glance, a beam analysis cannot express the space behaviour of the structure, procedures have been developed over the years to enable the design of complex structures such as the bridge over Nusle Valley, Prague, Czech Republic (see Figures 4.62–4.64).

At present the space function of the bridge deck can be more efficiently expressed by modelling it as a structure assembled from shell elements. However, it is advisable to recall the methods used in the

Figure 8.8 Deformations and courses of the normal stresses at the intermediate support: (a) double-T beam; (b) box girder

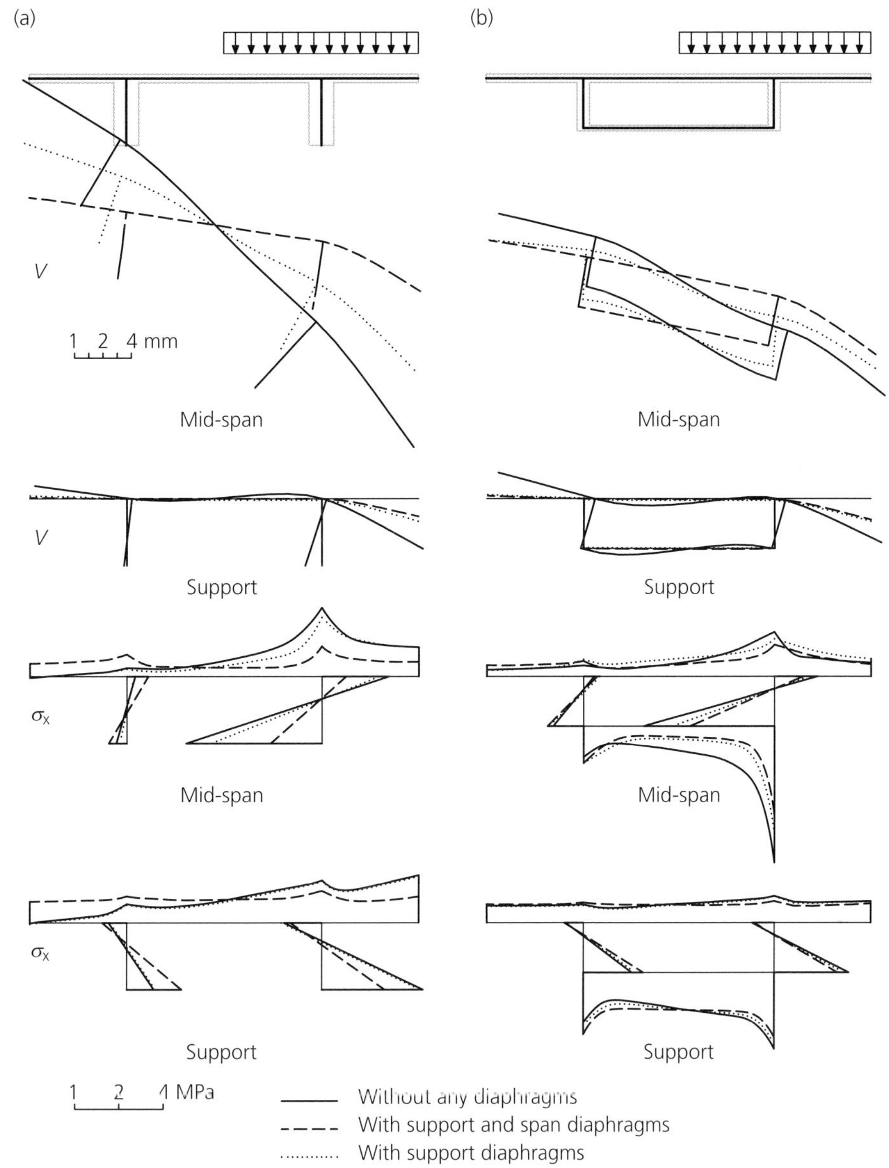

beam analysis, which will enable individual problems of the structural behaviour to be defined and explained.

8.3.1 Analysis of thin-walled beam

Figure 8.10 shows the deformation of a one-cell box girder loaded by a point load, as obtained from the analysis of the structure modelled as a 3D shell. It is clear that, due to its loading, the top slab is bent in the longitudinal and transverse directions, and consequently the whole structure deforms. In order to evaluate the individual effects and to understand the behaviour of the structure, it is advisable to follow the procedure used for the analysis of a box girder as a thin-walled beam. In this solution, the total deformation of the structure is determined in a step-by-step analysis of the local transverse bending, the longitudinal bending deformation, and the torsion and the dis-torsion of the cross-section.

In accordance with the principles of technical theory of elasticity, the effects of the external load on the girder can be decomposed into a separate analysis of the transverse deck strip (Figure 8.11) that is supported in the extension of the girder webs, and the analysis of the

Figure 8.9 Deformations and courses of the normal stresses at the midspan: (a) double-T beam; (b) box girder

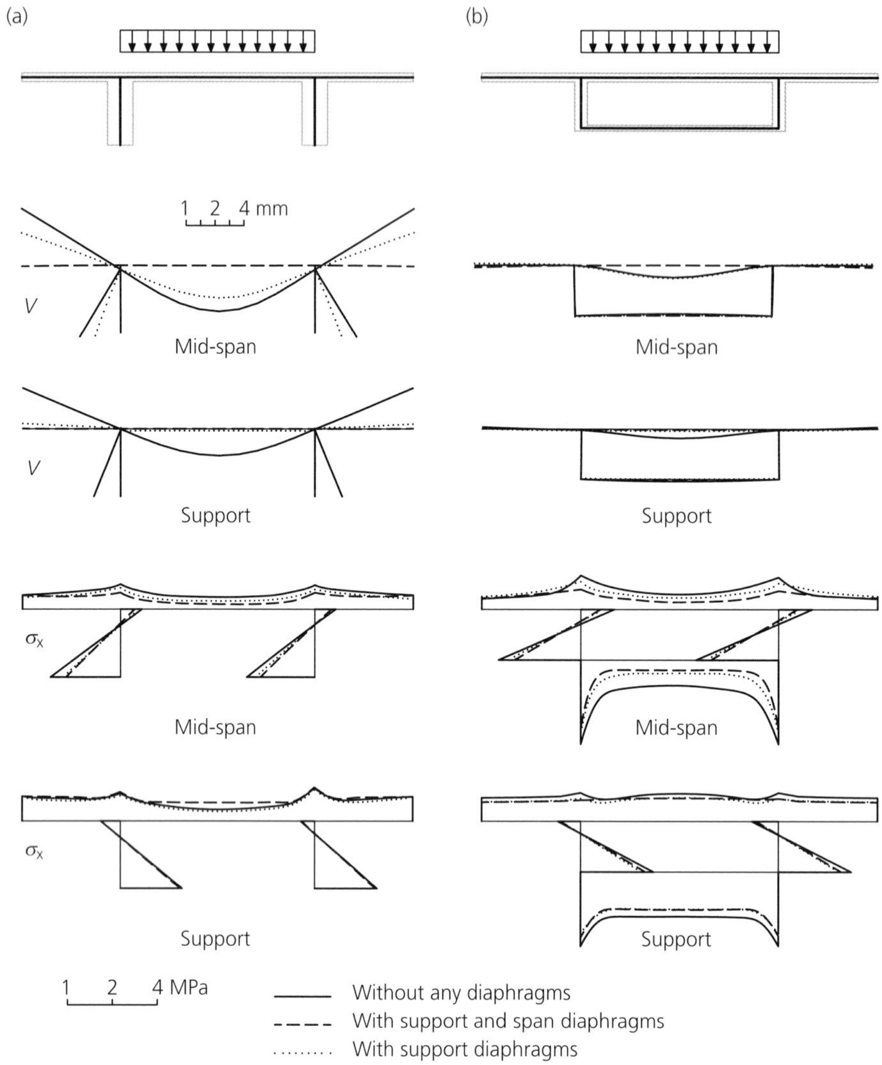

Figure 8.10 Deformation of the box girder

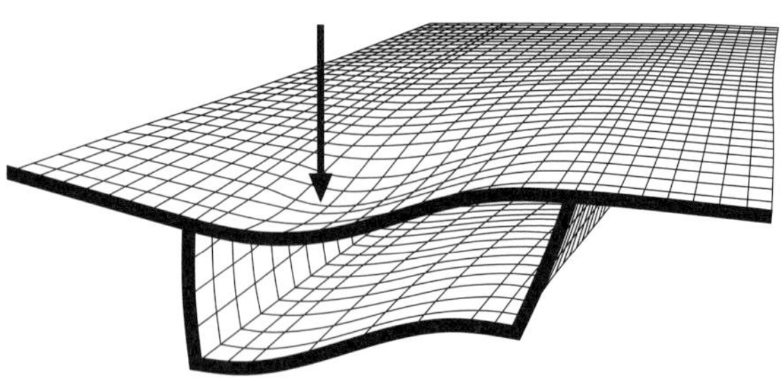

Figure 8.11 Analysed strip of the girder

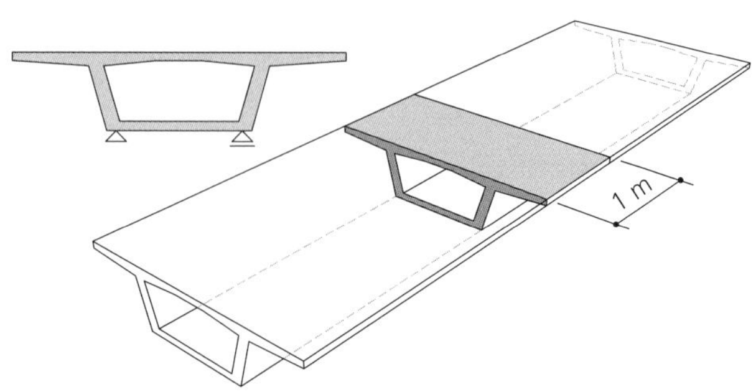

structure as a whole, loaded by the reverse reactions obtained from the analysis of the transverse strip. The step-by-step analysis is as follows.

(*a*) The 1.00 m wide transverse deck strip composed of a transverse frame and loaded by the point load is analysed at first. The analysis gives the transverse internal forces n_{P0} and $t_{P0,}$ and the moments m_{P0}, and the vertical reactions A and B at the supports (Figure 8.12(a)).
(*b*) The girder is loaded by the reactions A and B, which can be replaced by a pair of symmetrical forces $P_1 = B$ and a by a force $P_2 = A - B$ (Figure 8.12(b)). The symmetrical forces P_1 create normal and shear forces, and bending moments in the structure. The corresponding stresses are the shear stress τ_{01} and the normal stress σ_{01}. The force P_2 bends and torques the structure.
(*c*) The bending is separated from the torsion. The force P_2 is replaced by pairs of symmetrical and antimetric forces $P_2/2$ (Figure 8.12(c)). The symmetrical forces $P_2/2$ create normal and shear forces, and bending moments in the structure. The corresponding stresses are the shear stress τ_{02} and the normal stress σ_{02}. The antimetric forces $P_2/2$ stress the structure by torsion.
(*d*) The antimetric forces $P_2/2$ are divided into two pairs of forces $P/4$. These pairs are supplemented by horizontal forces $H = P/4(b/h)$ as shown Figures 8.12(d) and 8.12(e). The forces in Figure 8.12(d) stress the deck by a torsion, which can be divided into a pure (St Venant) torsion, which in the deck causes the shear stress τ_{KP}, and a warping torsion, which in the deck causes the shear stress τ_{KV} and the normal stresses σ_{KV}.
(*e*) The forces shown in Figure 8.12(e) cause a deformation of the cross-section of the deck (dis-torsion), and consequently create shear τ_{KD} and normal stresses σ_{KD}, transverse normal and shear forces n_{KD} and t_{KD}, and bending moments m_{KD}.

Figure 8.12 Analysis of thin-walled beam

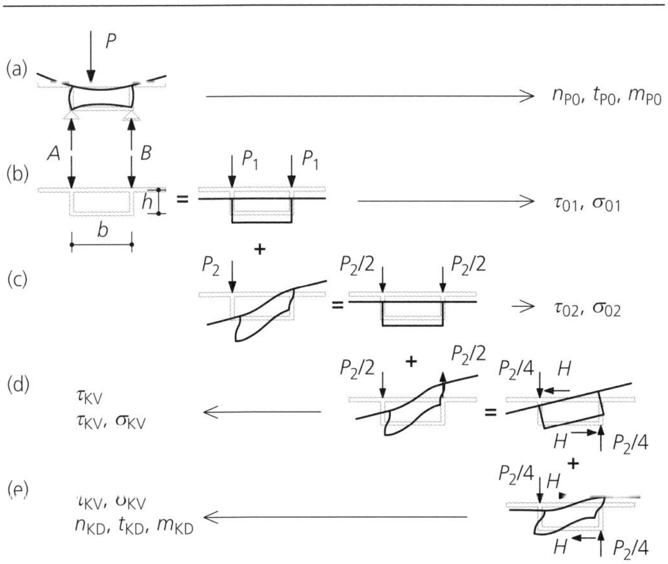

The resulting stresses in the structure are given by the sum of all the effects. In the simple torsion analysis, the shear stress is determined according to the classical (St Venant) torsion theory (see, for example, Timoshenko and Goodier (1951)). The static effects arising from the warping torsion are determined using the conventional procedure (see, for example, Vlasov (1961)). The static effects caused by the deformation of the cross-section are usually determined by the analysis of a beam on an elastic foundation (Winckler) in which the subsoil constant is substituted by a constant representing the stiffness of the transverse frame (Kristek, 1979; Schlaich and Scheef, 1982).

By using the equation for the analysis of a beam on an elastic foundation, the beneficial effects of the support diaphragms, which reduce the transverse stresses and ensure that the shear flow from the torsion is transmitted evenly through both webs and the top and bottom slabs, can be easily proved (Schlaich and Scheef, 1982). If no support diaphragms are designed, the deformation of the cross-section is greatly increased (see Figure 8.8(b)), and the torsional load is resisted only by the webs stressed by large shear stresses. The effects of warping torsion are relatively small in conventional concrete structures, and can be neglected.

Because modern concrete structures are commonly designed without intermediate diaphragms, the effect of cross-sectional deformation must be considered when designing thin-walled structures, multi-cell box girders and beam-and-slab decks.

8.3.2 Effective width

The uneven distribution of the normal stresses in the top and bottom slab shown in Figures 8.8 and 8.9 is caused by shear deformations of the slabs. This uneven distribution depends on the ratio of the bridge width to the span length, the position of structure of the section under consideration, and the type of load. Higher stress concentrations arise under a point load and in the continuous structures above the intermediate supports. To illustrate this point, Figure 8.13 shows the distribution of the normal stresses in the double-T beam deck of the bridges discussed in Chapter 10 (see Figure 10.56) for the four different loads. The three-span structure with spans of 25 + 36 + 25 m and an overall width of 19.50 m is made up of two 1.60 m high beams. The transverse spacing of the beams is 9.00 m.

In international specifications, the effect of the uneven distribution of normal stresses is taken into account by reducing the width of the slabs. This reduced width, called the 'effective width', is determined in such a way that in the studied section the stress determined according to the theory of elasticity is the same as the maximum stress determined by the 3D analysis of the structure assembled of solid or membrane elements (Figure 8.14). In the international recommendations, graphs are presented which, depending on the type of structure and the ratio of the span length to the span width, indicate the effective width b_s for the section above the support and b_m for the section at

Figure 8.13 Shear forces V and normal stresses σ_x above the support (p) and at the midspan (l) section of the double-T deck for: (a) self weight; (b) point load at mid-span; (c) uniform load at the main span; (d) deflection of the outer supports

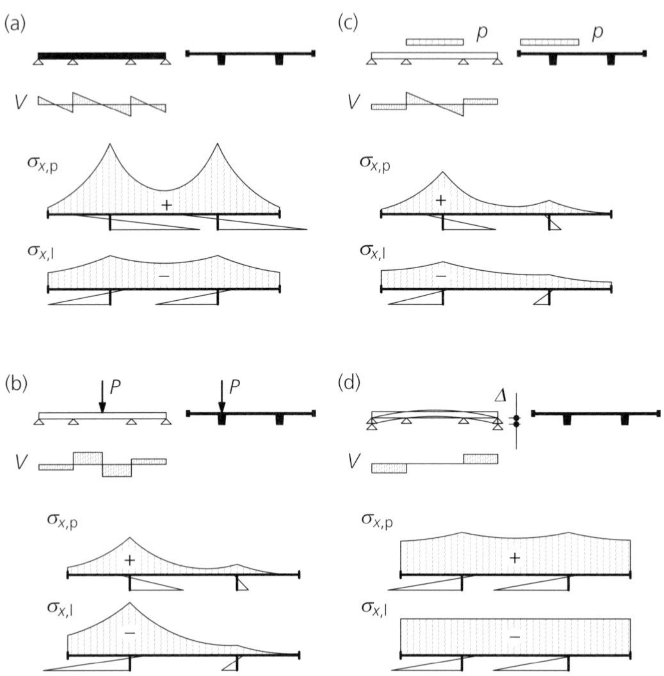

the midspan. Figure 8.15 shows the graph taken from AASHTO (2003).

According to the Eurocode (EN 1992-1 (CEN, 2005)), the effective width of the top and bottom flanges is given by

$$b_{\text{eff}} = \Sigma b_{\text{eff},i} + b_{\text{w}} \leq b \tag{8.3}$$

$$b_{\text{eff},i} = 0.2b_i + 0.1l_0 \leq b_i \text{ and } \leq 0.2l_0 \tag{8.4}$$

where: b is total flange width available for the particular web; b_i is the width available on one side of the web, measured from its face; l_0 is the distance between points of zero bending moments adjacent to the location being considered (l_0 is different for the span and the support parts).

When designing a prestressed concrete deck, it is necessary to realise that the compression stresses from prestressing are distributed across the whole section. Therefore, when preparing inputs for the computer analyses of structures, the whole area of the cross-section has to be entered. If the program calculates the section properties from the entered shape of the cross-section, it is necessary to transfer the eliminated parts to the level of the centre of gravity (see Figure 8.14).

Figure 8.14 Effective width

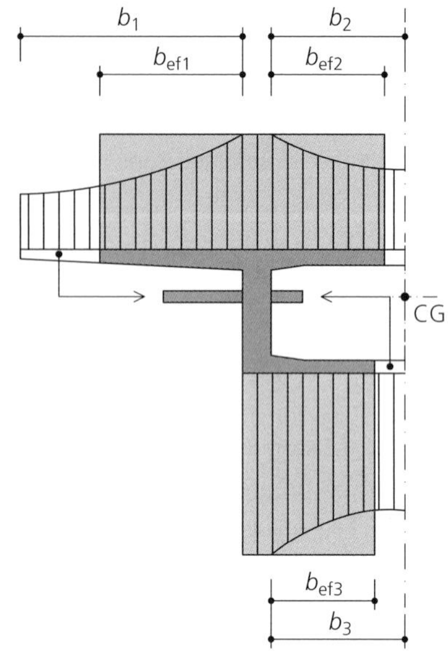

Figure 8.15 Determining the effective width

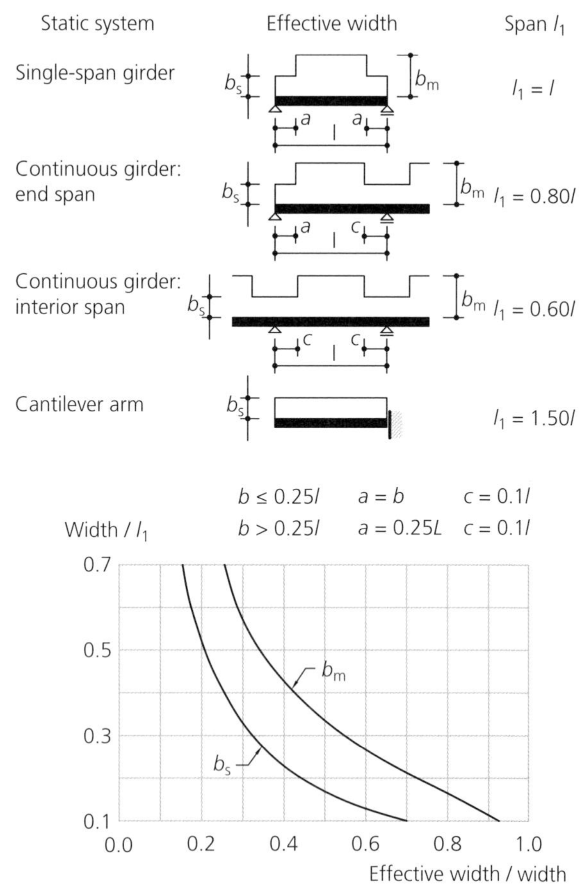

Figure 8.16 Cantilever deck: normal stresses due to post-tensioning of the straight tendons for *P* acting: (a) above the beam, (b) at the deck slab; (c) at slab edges

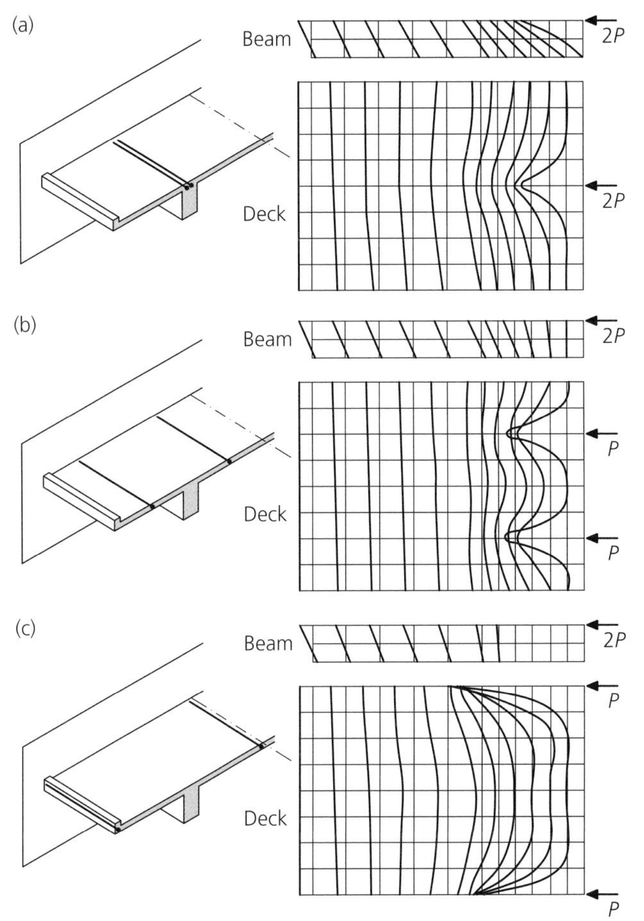

However, when using the effective width determined using the above procedure, the calculations for structures must be done very cautiously. To give an idea, Figure 8.16 shows the results of an analysis of a cantilever with a double-T cross-section modelled as shell elements. The 12 m long cantilever has a total width of 19.50 m and is composed of two 1.60 m deep beams at a transverse distance of 9.00 m. The cantilever is stressed by pairs of horizontal forces acting in the bridge deck. The horizontal forces used to model the action of the horizontal tendons located in the bridge deck were placed symmetrically with regard to the beam in three positions. It can be seen from the figure that for all loads the distribution of the normal stresses in the sections located at a distance equal to half the width of the structure is uniform. The explanation for this is very simple. As the cantilever is stressed by a constant bending moment, the shear forces are zero. Therefore, there is no shear deformation of the deck slab, and the normal stresses in the deck slab are uniform.

Figure 8.17 Three-span beam: normal stresses at the support and the midspan sections

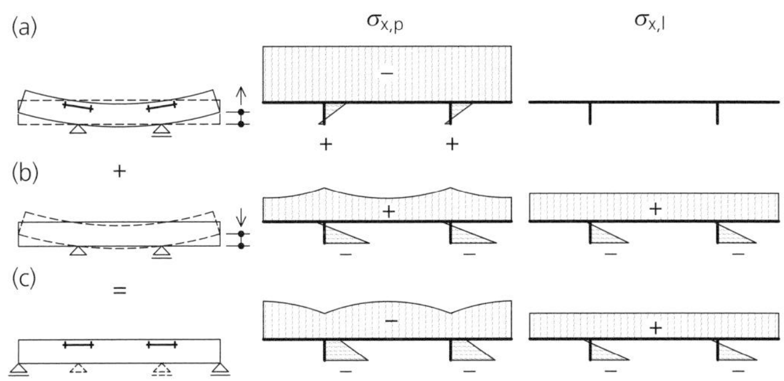

Figure 8.17 shows the results of the analysis of a three-span structure of the same cross-section as the example in Figure 8.16 prestressed by straight tendons situated above the internal supports. The tendon anchors are located at distances equal to the width of the structure. Therefore, the distribution of stresses at the support sections corresponds to the results in Figure 8.16.

In order to understand the effect of the tendons, the structure presented in Figure 8.13, modelled as a 3D structure assembled of shell elements, was studied. The structure was solved in two steps. In the first step, a simple beam with an overhanging side span was solved (Figure 8.17(a)). As the bending moments due to prestressing are constant, the shear forces are zero. Therefore, the distribution of the normal forces above the intermediate support is uniform. As the tendons are situated at the top slab, the structure is stressed by positive bending moments, and therefore the ends of the overhangs rise. In the second step, the structure was loaded by vertical deformations of the ends of the overhangs determined in the first step (Figure 8.17(b)). As shear forces occur only in the side spans, a non-uniform distribution of the normal stresses arises above the support. In the centre span, where there is zero shear force, the stresses are uniform. As there is a negative bending moment in the structure from the deformations of the overhangs, there are tension stresses in the deck slab. The resulting solution is given by the sum of both effects (Figure 8.17(c)). It is clear that in the cross-section above the support, there is no concentration of the resulting stresses, but there is a reduction in the normal stresses in the section above the beam. As there are no shear stresses at the section at the midspan, the distribution of the normal stresses in the deck slab is uniform.

It is clear from the above examples that the graphs for determining the effective width must be applied very carefully. They are suitable for the analysis of the static effects from the dead load and the uniform live load. When analysing the influence of prestressing, however, it is necessary to consider whether the equivalent effects from prestressing have a similar character as the permanent load. If the characters are similar, the effective width can be determined from the graphs; if it is different, the graphs must not be used.

8.3.3 Modelling of structures by beam elements

Most of the software available for the analysis of beam structures is based on structural analysis developed from the theory of elasticity. Therefore, the programs are suitable for the analysis of structures that do not have a significantly thin-walled character. During the analysis, the structure is modelled as beam elements of constant cross-section, which are characterised by the cross-section characteristics A_x, A_y, A_z, I_x, I_y and I_z, Young's modulus of elasticity E and the shear modulus G. When describing the geometry, it is necessary to determine the direction of the main central axes of the cross-section. The outputs of the structural analysis are the deformations u, v, w, φ_x, φ_y and φ_z and the internal forces N_x, T_y, T_z, M_x, M_y and M_z.

When preparing input data, it is necessary to realise that most bridge decks are designed as prestressed members, while the piers are usually designed as reinforced concrete members. Therefore, the characteristics of the structural elements should be given with respect to the actual stiffness. Typically, the cross-sectional characteristics of prestressed structures are entered as a value corresponding to the full cross-section, and those of reinforced concrete structures are entered as reduced cross-sectional characteristics (Figure 8.18). The bending stiffness of typical reinforced concrete members, which is given by the moment of inertia, is reduced from one to two-thirds of the gross stiffness. Usually, the structures are analysed using these values, after which, once the reinforcement has been determined, the area and moment of inertia of the cracked section is corrected and the analysis is repeated.

As a reduced stiffness of substructures also reduces the seismic forces, for which the structures have to designed, the reduction must be done carefully (for more on this see Priestly *et al.* (1996)).

It is also necessary to realise that if the structural member cracks in torsion, its torsional stiffness is dramatically reduced, to up to one-tenth of its original value (Figure 8.19) (Collins and Mitchell, 1987; Priestly *et al.*, 1996). This fact has to be considered for structures in which the bending moments in one structural member depend on the torsional stiffness of another one, such as the case of a bridge deck indirectly supported by outrigger cap beams. Therefore, it is reasonable to check the function of the deck for two boundary conditions: torsionally stiff and soft beam.

Figure 8.18 Substructure stiffness: (a) deformatin due to temperature changes; (b) effective section of the pier

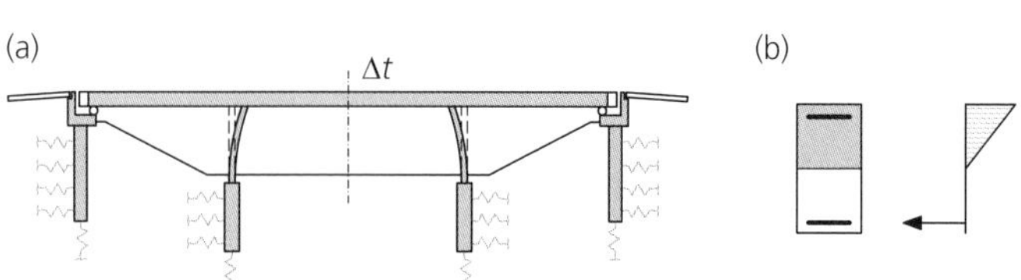

When developing an analytical model it is advisable to prepare a model that describes the real boundary conditions and the load action. As the theory of elasticity is based on the Bernoulli hypothesis of linear strain distribution, it is possible to assume that the individual sections remain planar and are perpendicular to the centreline after deformation. Therefore, it is possible to describe the positions of the supports and the force actions at points that are connected to the centreline by rigid elements (Figure 8.20).

Figure 8.21, which was developed from Figure 9.10 of Hambly (1991), shows an example of a plan curved structure. The deck is formed of a box girder and is frame connected, with single-column supports, and is supported by sliding bearings on the end abutments. The structure is modelled as a 3D frame structure composed of a series of straight elements mutually connected at nodes situated at the centre of gravity of the structure (Figure 8.21(a)). The box girder is connected to the supports by infinitely stiff elements. The end

Figure 8.19 Torsional stiffness: (a) torqued member; (b) torsional stiffness

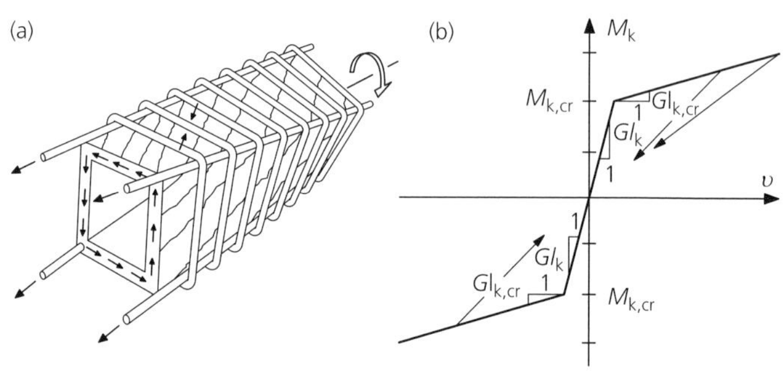

Figure 8.20 Deformation and modelling of the beam: (a) planar sections; (b) vertical load; (c) horizontal load; (d) torsional load

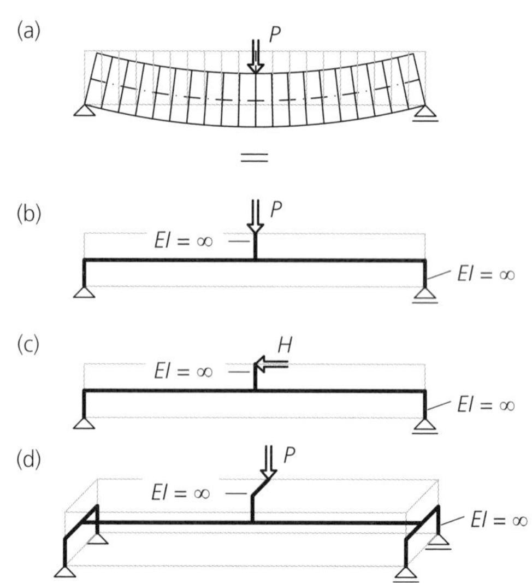

Figure 8.21 Modelling of the curved bridge structure

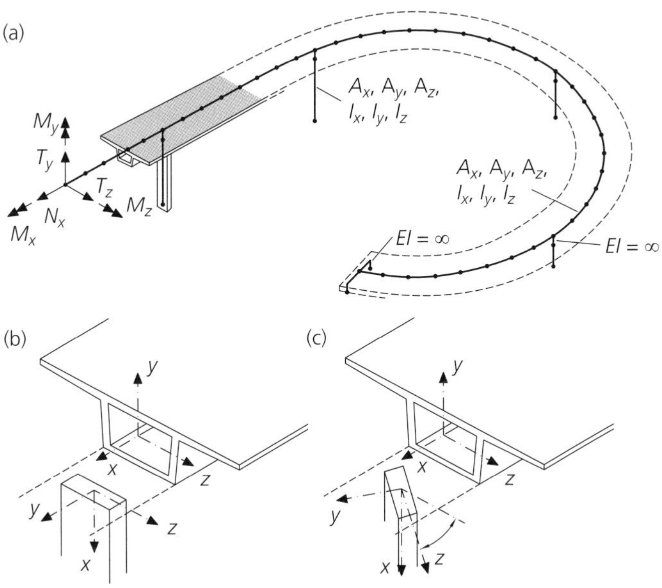

diaphragms and bearings express the supporting conditions at the abutments.

When modelling the structure, it is necessary to describe the direction of release of the bearings as well as the connection of the deck to the supports. The stress of the structure is fundamentally influenced by the relative position of the main central axes of the deck and piers. In a radial arrangement of the piers, the local axes of the bridge deck and the supports are parallel (Figure 8.21(b)), when the supports are in a skew arrangement, the angles between the axes has to be given (Figure 8.21(c)). Different computer programs describe this fact in different ways.

When applying the effects of prestressing, the direction of the radial forces has to be correctly determined. If the prestressing tendons are led in inclined webs, only the vertical components should be applied in a beam analysis. However, the influence on the transverse design of the deck should be considered (Figure 8.22).

Figure 8.22 Inclined webs of the box girder: (a) vertical load; (b) prestressing

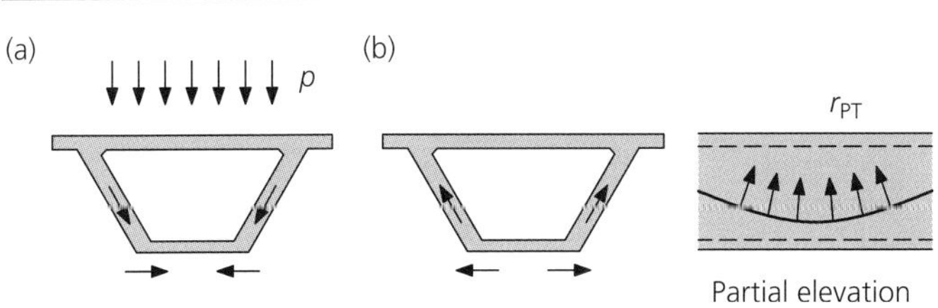

Figure 8.23 Variable girder depth: (a) elevation; (b) cross section

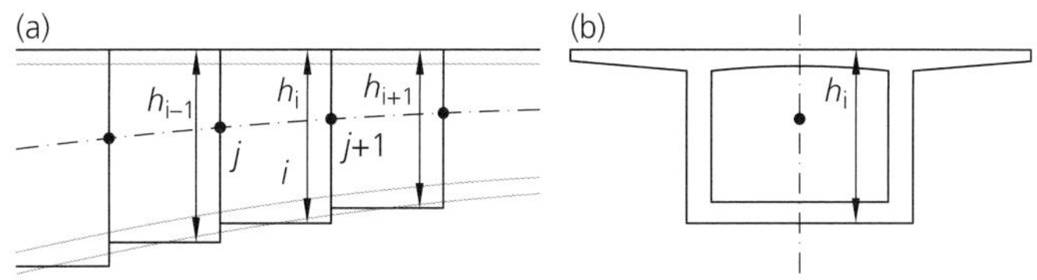

8.3.4 Variable girder depth

Originally decks of variable depth were designed as straight beams, where the influence of the haunches was expressed as the ratio of the moments of inertia. When these structures were loaded by normal forces (by prestressing), the influence of the eccentricity of the force was represented by adding additional transverse forces.

Figure 8.23 shows an example of the modelling of a box girder structure with a variable depth. The structure is composed of straight, mutually connected beam elements connected at nodes situated at the centre of gravity of each element. The cross-sectional characteristics of the element i between the nodes j and $j + 1$ are determined for the average height of the element h_i. As the junction of the nodes follows

Figure 8.24 Bottom slab of the box girders

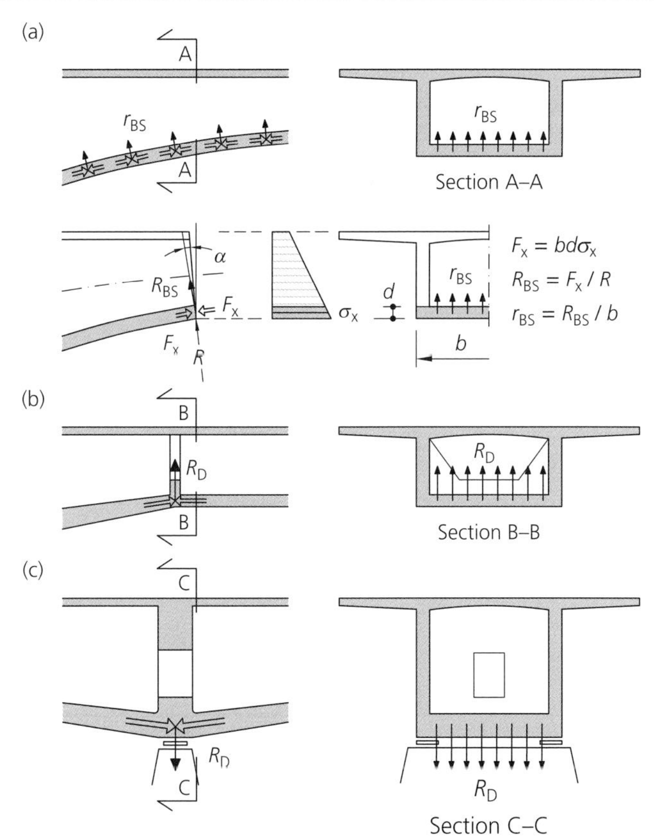

the curved centreline, the structure is designed directly for the internal forces obtained in the analysis.

Bridges with large spans are usually formed of one-cell box girders that have a variable depth. The bottom slabs then have a longitudinal parabolic or circular curvature (Leonhardt, 1979; Mathivat, 1983). Due to this curvature, the compression stresses originating in the slab create transverse radial stresses r_{BS} that stress the bottom slabs in the transverse direction (Figure 8.24(a)). As the curvature of the bottom slab is different from the girder centreline, these radial forces act against the shear stresses originating in the girders. Therefore, the sum of the radial stresses $R_{BS} = r_{BS}b$ transferred in the direction of the shear forces can be subtracted from them.

If the variable depth changes linearly (Figure 8.24(b)), the resulting forces originating in the line break create significant stresses in the slab. Therefore, it is usually necessary to stiffen the girder by a span diaphragm, which transfers the resulting stresses into the girder webs. The same is true if the deck is supported by bearings situated under the webs (Figure 8.24(c)).

8.4. Grillage analysis

The method in which the deck is modelled as a grillage mesh (Figures 8.25 and 8.26) has proved to be very efficient in the design of beam-and-slab decks and multicellular decks. The method is discussed in the excellent book by Hambly (1991). The analysis done by the first author (Stráský, 1979) is briefly presented here.

The method, which uses commonly available software for the analysis of 3D frames, can also express the influence of the deformation of the beam's cross-sections and the membrane function of the top and bottom slabs. In the analysis, the structure is divided into longitudinal beams and transverse links. The grillage analysis is usually carried out in four steps.

(*a*) First, the analytical model is determined. The superstructure is divided by longitudinal cuts into longitudinal beams for which the cross-sectional characteristics with respect to the effective width are determined. The continuous joints of the beams is replaced by singular transverse links (substituted transverse beams) (see Figures 8.25 and 8.26). The substituted beam 1–2 is formed by a transverse frame connecting adjacent longitudinal beams. It consists of the webs and top and bottom slabs, and it is supported at the centres of gravity of the longitudinal beams 1 and 2. If an analysis is done of two box girders connected by a top slab, the centre of gravity of the longitudinal girders can be connected to the lower corners of the box cross-section with infinitely rigid arms (Figure 8.26(d)). In this way the box girder is supported in the same way as the box girder shown in Figure 8.12(a). Instead of describing the function of the connection link by the cross-sectional characteristic, the substituted beam 1–2 can be described by a stiffness submatrix K_{22} (6, 6) of the transversal frame. This submatrix is obtained by analysing the transverse frame (the substituted beam 1–2) by loading the support 2 with unit deformations (Figure 8.27).

$$u = v = w = \varphi_x = \varphi_y = \varphi_z = 1 \tag{8.5}$$

After determining the characteristics of the longitudinal beams and the transverse substituted beams, the characteristics of the end diaphragms and supports are determined. The boundary conditions are then described.

Figure 8.25 Grillage mesh idealisation of the deck formed: (a) T-beams; (b) twin T-beam; (c) three-cell box girder; (d) two box girders connected by a deck slab

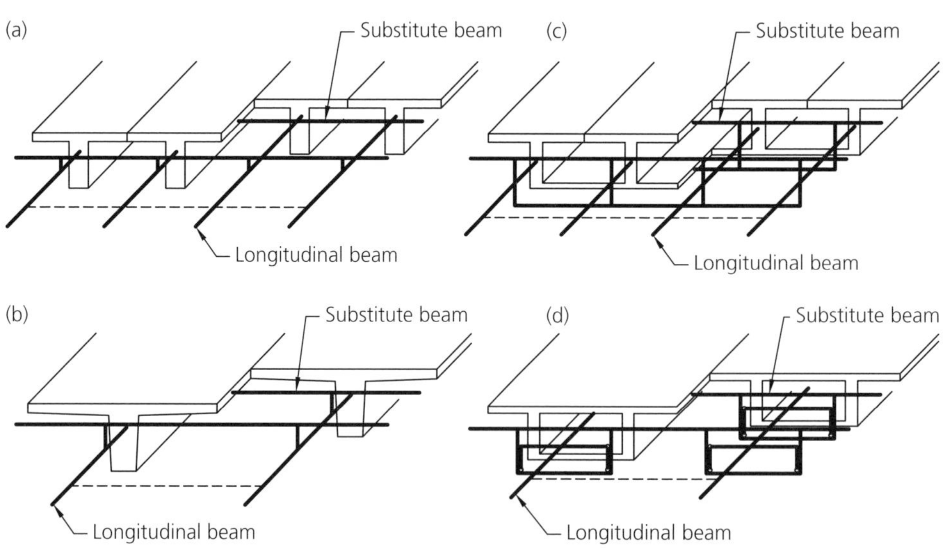

Figure 8.26 Longitudinal and transverse substituted beams of the deck formed: (a) T-beams; (b) twin T-beam; (c) three-cell box girder; (d) two box girders connected by a deck slab

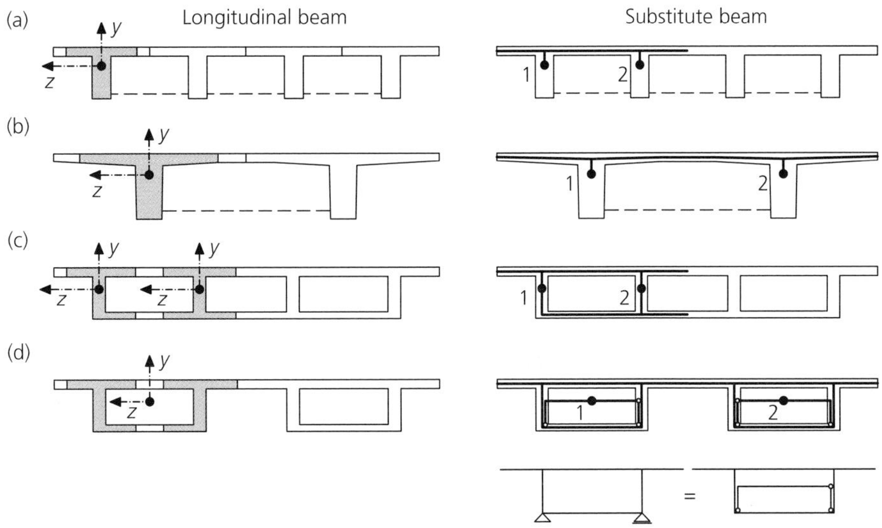

Figure 8.27 Deformations of the substituted beam of the deck formed by two box girders connected by a deck slab

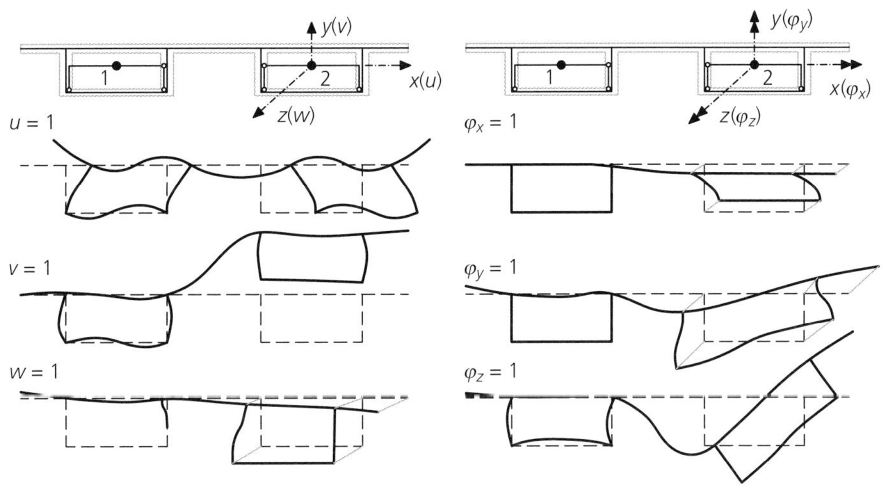

(*b*) The load is determined for this analytical model. This can be done in two ways. Either the longitudinal beams with the corresponding forces acting at their centre of gravity are loaded, or the nodes are loaded by the reactions that were determined in the analysis of the transverse substituted beams.

(*c*) By analysing the structure for the given load, the deformations of the nodes and the internal forces in the longitudinal members, cross-members and in the piers are obtained.

(*d*) Static effects in the transverse direction are obtained by solving the transverse frames–substituted beam, which are loaded by the corresponding loads and deformations of their supports (i.e. the deformations at the nodes of the grillage).

In the grillage analysis of structures, the continuous connection of beams is replaced by single links. The accuracy of the solution increases with the increasing number of the transverse links. However, parametric calculations have proved that sufficiently accurate results can be obtained even with a relatively coarse division. For conventional medium-span bridges, it is sufficient to determine the nodes in tenths of the span.

For the purposes of illustration, Figure 8.27 shows the deformation curves of the substituted beam 1–2 used to model the transverse connection of the structure formed of two box girders connected only by a deck slab. The deformation curves were obtained by solving the transverse frame loaded at support 2 by unit deformations. It can be seen from the figure that the stiffness matrix of the substituted beam 1–2 also describes the deformation of the cross-section and, due to the eccentricity of the top slab, the membrane action of the connection to the top slab.

The advantage of the grillage analysis is that it works with the internal forces for which the dimensioning of the structural members has been developed. For longitudinal beams and sections of the transverse frames it is possible to determine the influence lines of the distribution of the longitudinal internal forces between the longitudinal girders and the influence lines of the transverse forces acting in the connecting slab, and thus get an idea of the load positions for which the structures need to be designed. For illustration, Figure 8.28 shows the influence lines of a three-span structure formed by two box girders connected by the deck slab (Figure 8.29). The longitudinal beams are connected by the transverse substitute beams 1–2 described by their stiffness submatrix K_{22} (6, 6). As the girder support sections are stiffened by diaphragms, two different submatrices were determined.

The analytical model was analysed for several positions of the linear uniform load situated in the central span (see Figure 8.29(a)). For each position of this load, the internal forces in both the longitudinal and the transverse direction of the structure were determined. Some results of the analysis are presented in Figure 8.28. Figure 8.28(a) shows the portion of the total longitudinal moment M_z above the intermediate support that is resisted by girder 1; that is, it shows the influence line of the support moment M_{zP}. Figure 8.28(b) shows the value of the torsional moment M_{xP} which stressed the support section of girder 1.

Figures 8.28(c)–8.28(e) show the transverse bending moments m_z and shear force v_y that originate at sections **a** and **b** of the connecting slab at the midspan. Section **a** is at the connection of the deck slab with the box girder; section **b** is at the midspan of the deck slab. The transverse shear forces v^0 and the moments m^0 are the internal forces that would originate in the deck slab of the structure in which the transverse frame is firmly supported (Figure 8.26(d)).

The grillage analyses described were successfully used in the design of two types of progressively erected segmental structures, as discussed in Chapter 10 (see Figures 10.56–10.60). The possibilities of grillage modelling are demonstrated in Figure 8.30, which gives several results of the analyses of the bridge shown in Figure 10.57. The bridge was loaded by a 60 t three-axle truck at the middle of the central span in the bridge axis and near the crash barrier. The truck load was substituted by a uniform load acting on an area of 6.0 × 3.0 m.

Figure 8.28 Influence lines of a three-span structure formed by two box girders connected by the deck slab

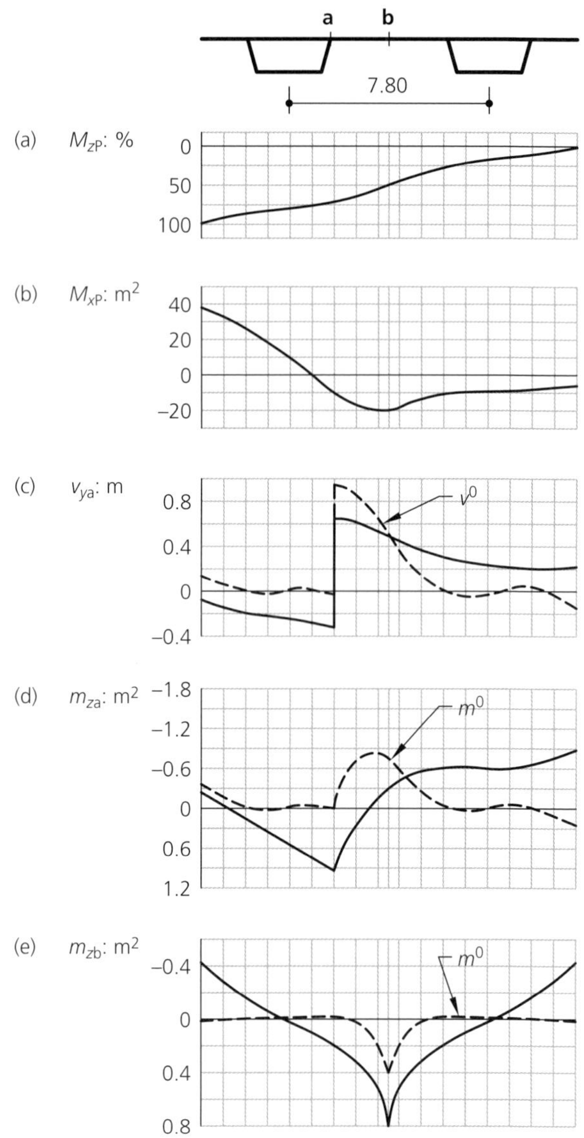

The figure shows curves for the vertical deformation v and rotation φ of the longitudinal girders 1 and 2, the deformation of the cross-section at the midspan and at the support, and the corresponding normal forces and bending moments. The analysis described allowed the girder deformations to be visualised and the bridge function to be understood.

The author's design firm currently uses the approach described here for the design of all bridges in which the deck is assembled of a precast beam and a composite slab, and for the design of double-T decks.

Figure 8.29 Analytical model for a three-span structure formed by two box girders connected by the deck slab: (a) cross section; (b) elevation; (c) calculation model; (d) grillage idealisation

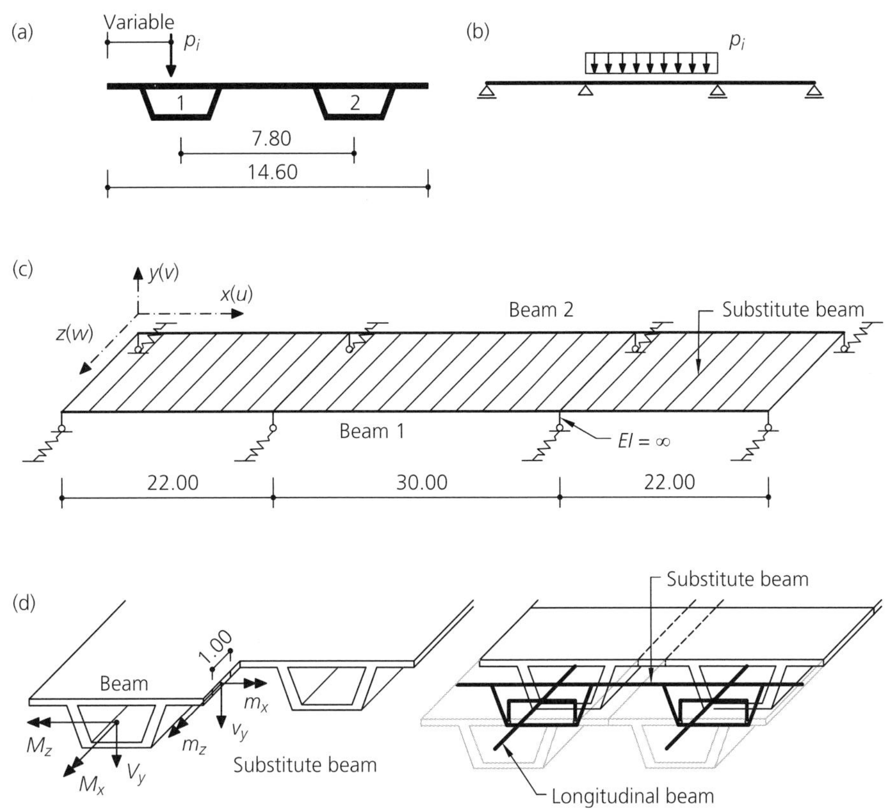

8.5. Analyses of bridge decks

Although the analysis of a structure modelled as a thin-walled beam makes it possible to capture both the effect of the warping torsion and the effect of the cross-section deformation, it is rarely used at present. However, as already mentioned, to understand the structural behaviour of modern bridges it is useful to follow the procedure shown in Figure 8.12. The results obtained in such an analysis are discussed further here.

As discussed earlier in this chapter, all structures can be modelled as a 3D structure assembled of solid elements or shell elements (see Figure 8.2). Although recent software allows very efficient modelling of a structure, it is not straightforward to obtain the required results. Therefore, this analysis is usually done only when necessary.

Box girder decks are usually analysed as beam structures, and the effects of the deformation on the distortion of the deck and/or detailed deck slab analyses, which used to be performed by using influence surfaces, is done on the individual parts (segments) of the decks. This enables the use of a very fine mesh of the elements. To express the boundary conditions of these parts, it is useful to connect them to the beam model, which is used for the analyses of the global function of the deck (Figure 8.31(a)). This approach has also been used in the design of wide bridge decks formed of spine girders with large overhangs supported by slab struts (Figure 8.32) (Novotny *et al.*, 2012).

Beam-and-slab decks and multicellular decks are usually modelled using grids. The longitudinal beams follow the gravity centrelines of the individual beams, and the transverse beams are situated at the deck slab (Figure 8.33). Beam-and-slab decks can also be modelled using shell elements that are connected to the grid in order to describe the global function of the deck (see Figure 8.31(b)).

To express the membrane function of the deck slab it is necessary to define its eccentricity. This is automatically done when the transverse connection to the longitudinal beams is described by a stiffness submatrix, as discussed in Section 8.4. As the transverse deforma-

Figure 8.30 Grillage analysis of the bridge: (a) plan; (b) vertical deformation *v* and rotation *φ* of the longitudinal beams 1 and 2; (c) deformation, normal forces *N* and bending moments *M* of the midspan section; (d) deformation of the intermediate support

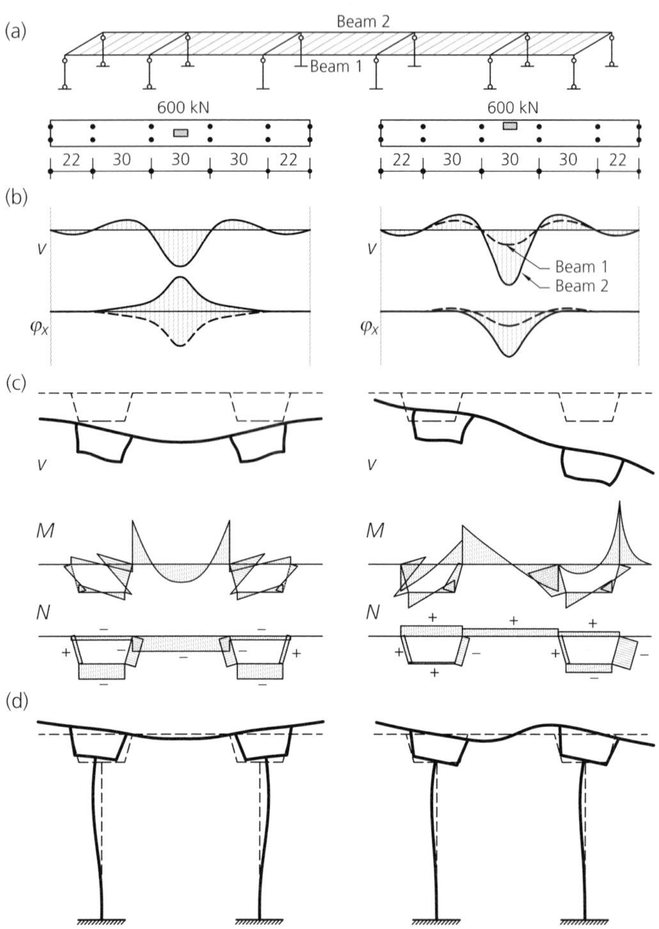

Figure 8.31 Modelling of the structures: (a) box girder; (b) double-T beam

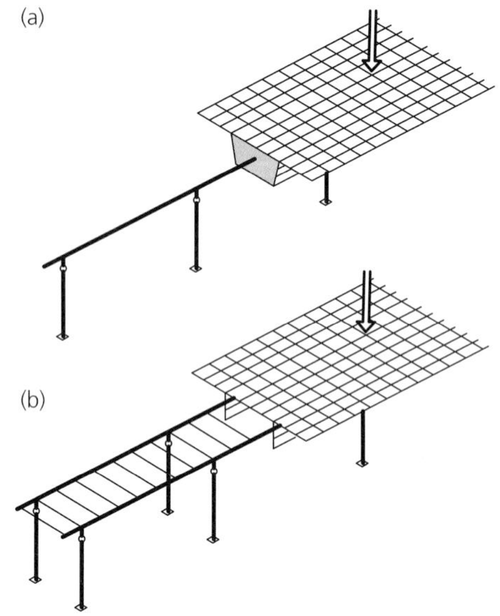

tions of the beams are negligible, it is also possible to express the eccentricity of the slab by connecting the transverse elements to the longitudinal beam by very stiff vertical members (see Figure 8.33). These members should also be used to describe the positions of the supports (see Figure 8.21).

Another option is to model the deck slab using shell elements that are eccentrically connected to longitudinal beams. In this case, it is necessary to accurately describe the positions of the supports.

8.5.1 Modelling of the deck connection to another structural member

The composite function of two structural members is usually expressed as a connection of the two members by rigid elements. Figure 8.34 shows the calculation model and dead-load bending moment diagram for the deck of the arch structure presented in Figures 8.3 and 8.4. The arch, which is formed by steel pipe infilled with concrete, supports the prestressed concrete deck of the channel section. The arch is monolithically connected to the beam in the middle of the bridge. In the computational model, this connection was captured by the connection of the arch to the deck beam by several rigid elements.

It is clear from the figure that this model accurately captured the global effect of the structure, but it gave the wrong results at the point of the connection of the arch to the deck beam. The moments and corresponding normal stresses change sign at the connection point to the short elements. There is a similar moment change in the arch. However, a detailed analysis of the structure assembled of solid elements showed a continuous course of stresses of the same sign (Figure 8.35). It is therefore clear that in the discontinuity regions the beam analysis must always be supplemented by a more detailed analyses.

8.6. Curved structures and members

The analyses of prestressed concrete bridge decks modelled as 3D structures assembled of solid or shell elements automatically determines the stresses originating in curved structures and/or in curved structural members. However, to obtain the correct results it is necessary not only to accurately describe the analysed structure, but also the position and function of the internal tendons and external cables. The problem is illustrated in Figure 8.36, which shows the midspan part of the plan curved cantilever structure prestressed by continuity tendons led in the girder web and with a span tendon situated at the bottom slab. In fact, there are more tendons, both in the web and in the bottom slab (see Figure 8.45).

Figure 8.32 Modelling of the box girder deck: (a) deck cross-section; (b) calculation model

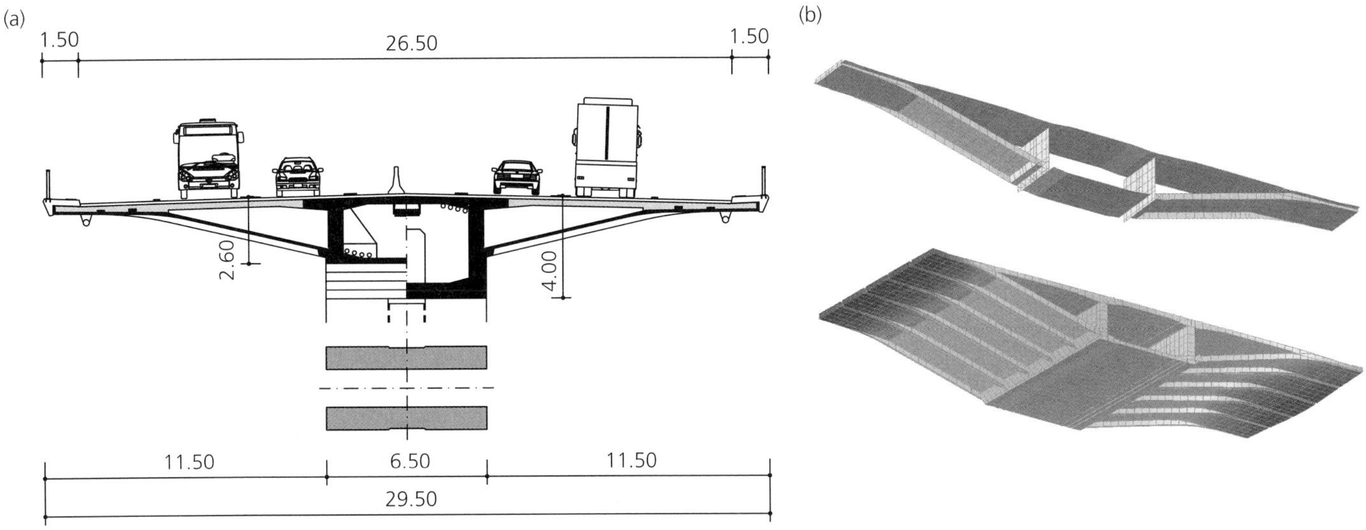

Figure 8.33 Grillage idealisations of the beam-and-slab decks: (a) double-T beam; (b) precast beams with a composite deck slab

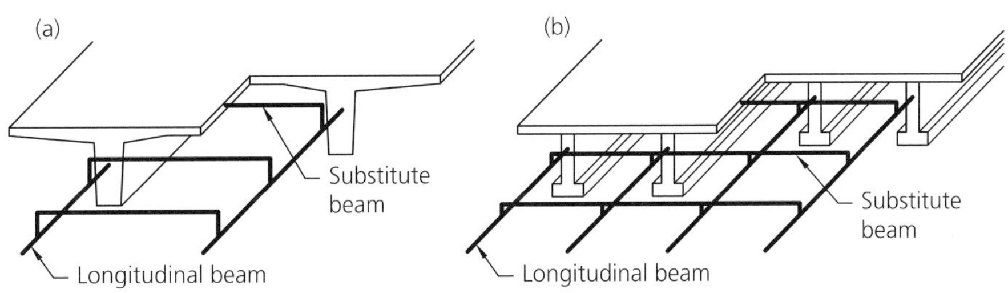

Figure 8.34 Bending moments in the deck of the arch structure

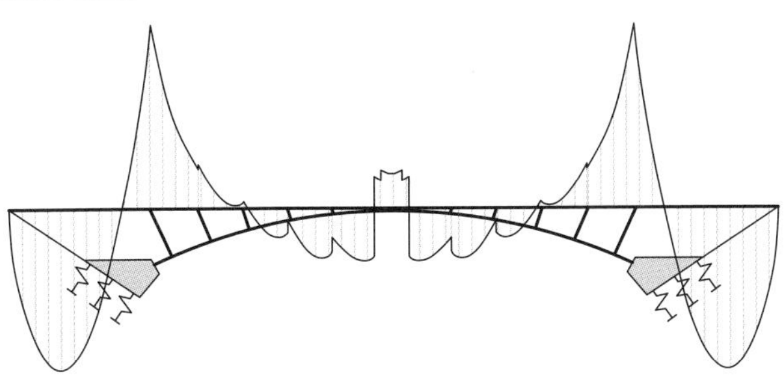

Figure 8.35 Normal stresses in the deck of the arch structure: (a) top view; (b) bottom view

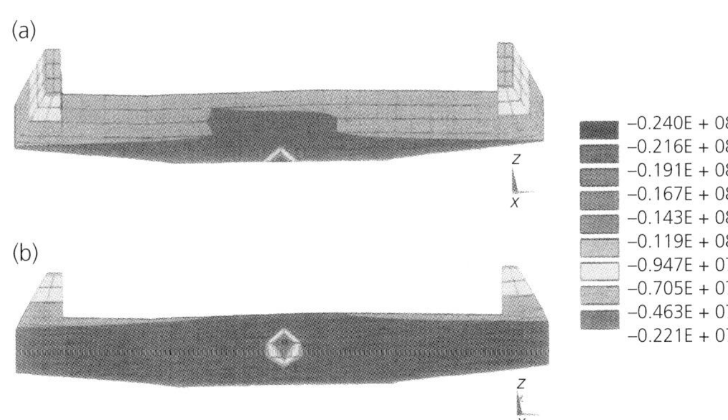

Figure 8.36 Continuity and span tendons at the midspan sections of a curved cantilevered deck. PT, prestressing

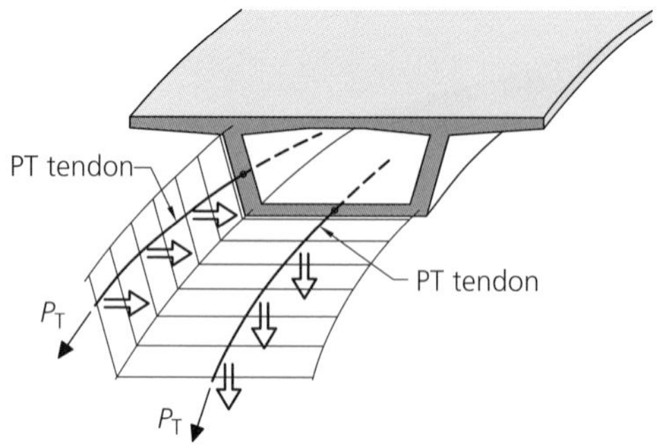

Figure 8.37 Static function of the arch ribs: (a) parabolic ; (b) circular

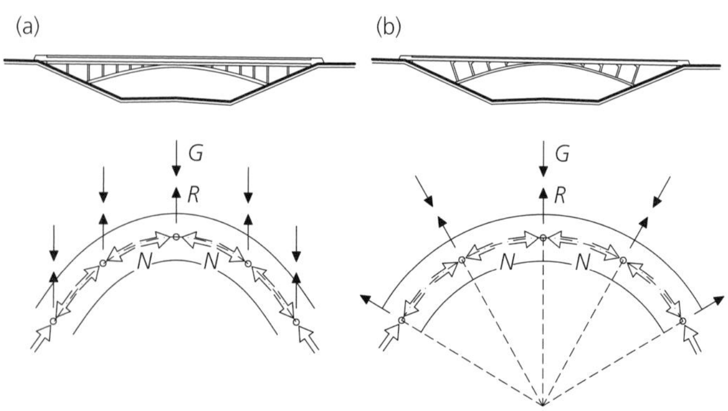

8.6.1 Arch ribs

In the beam analyses of 3D bridge structures, it is necessary to understand the function of the curved structural members and the prestressing. The curved structural members usually form the arch ribs of arch bridges. Due to the curvature, the compression stresses in each arch node have resultants acting in the direction towards the centre of the curvature. If the shape of the curvature of the member is a second-degree parabola, the resultant forces are uniform and parallel; if its shape is a circle, the resultant uniform forces are in the direction of the centre of the circle.

In correctly designed arch structures, which have funicular shapes, these stresses are balanced by the dead load, and the ribs are stressed only by the compression stresses (Figure 8.37(a)).

8.6.2 Transverse bending of the curved structures

In the beam or grillage analysis, the space function of the bridge deck is transferred into the beam axis. Due to the plan curvature, the individual structural members are subjected not only to longitudinal bending and torsion, but also to an additional transverse bending.

Figure 8.38 Transverse bending of the beams of the curved double-T deck: (a) elevation and cross section of the deck; (b) plan

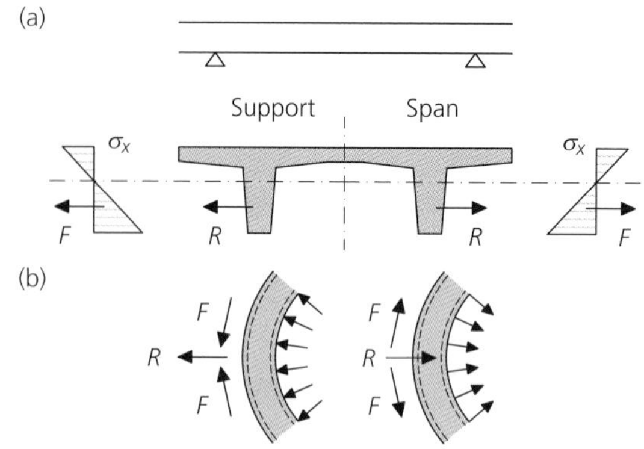

This phenomenon is demonstrated using the example of a bridge deck with a double-T cross-section.

The deck has a plane curvature with a radius of 250 m and the cross-section shown in Figure 8.38. It was analysed as a grid assembled of two longitudinally curved beams mutually connected by transverse members. Due to the plan curvature, the bending stresses originating in the webs create transverse resultant stresses. In the part of the beam below its centre of gravity, the sections close to the supports are stressed by stresses directing outward from the curvature; the span sections are stressed by stresses directing inward towards the curvature. Of course, the part of the girder above the centre of gravity behaves similarly.

It is evident that the girder webs are transversally bent.

Figure 8.39 Prestressing of the curved ribs: (a) rib and prestressing tendon; (b) radial forces in the tendon; (c) normal and radial forces in the rib and tendon

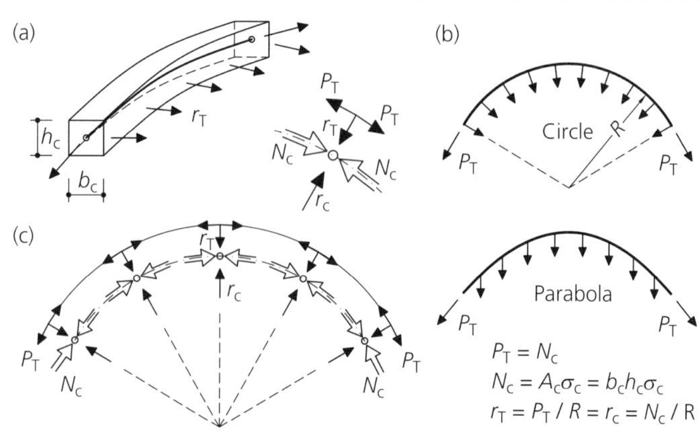

Figure 8.40 Prestressing of the curved plate. PT, prestressing

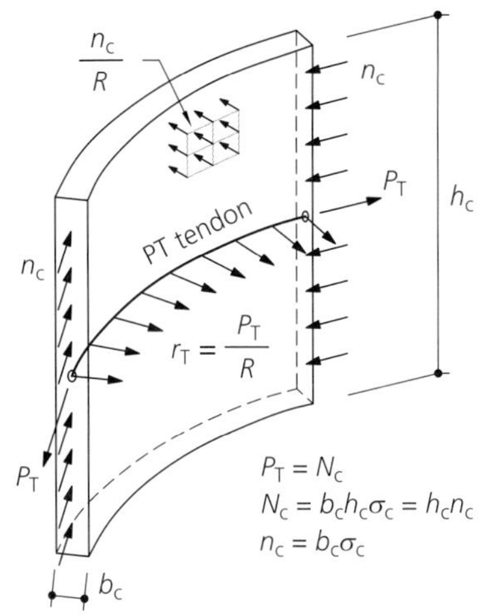

Figure 8.41 Loading of the webs of the prestressed curved box girder

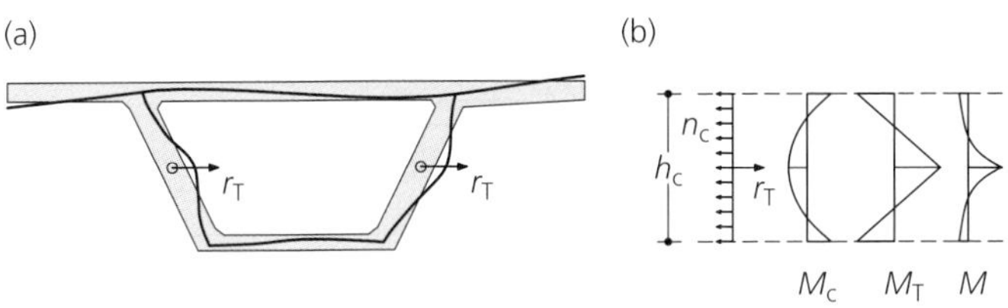

8.6.3 Webs of the prestressed curved box girders

Curved prestressed beams behave similarly to arch ribs. However, the resultant forces originating in each node are balanced by the radial forces of the tendons. For the condition of zero friction, the members are stressed by uniform compression (Figure 8.39). If the curve is a second-degree parabola, the resultant and radial forces of the tendons are constant and parallel; if the curvature of the member is a circle, the resultant and tendon radial forces are also constant, and are directed towards the circle centre.

Figure 8.39(a) shows the curved beam element prestressed by a tendon situated at the centre of gravity of the bar. If the same tendon is situated at the centre of gravity of a curved plate, it is evident that the transverse resultants of the compression stresses originating in the plate and radial forces are not locally balanced – the plate is transversally bent (Figure 8.40) (Corven, 2016).

The curved plate can function as a web of the box girder Figure 8.41(a). As the plate is transversally frame-connected to the top and bottom slabs, its free deformation is constrained. Therefore, the radial forces of the tendon are transferred into the whole of the girder section.

Figure 8.42 Loading of the webs

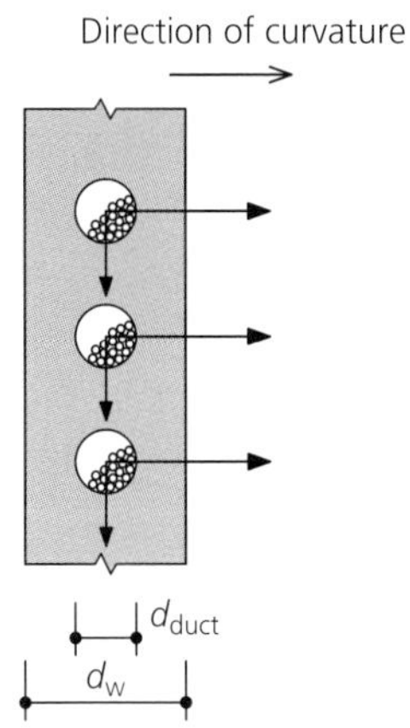

Figure 8.43 Failure of the prestressed curved webs

Figure 8.41(b) shows the transverse forces loading the webs of the one-cell box girder and the corresponding transverse bending moments. The resulting bending moment is a sum of the moments originating from the forces resulting from the longitudinal compression and the radial force caused by the tendon. While the prestressing force in Figure 8.40 is distributed into the whole plate cross-section, the prestressing force in the box girder is distributed into the whole section. Therefore, the resultants of the compression stresses originating in the webs are smaller.

Usually there are more tendons in the webs (Figure 8.42). If the radius of the plan curvature is small, the tendons are concentrated and the webs are very slender. The webs can fail (Figure 8.43). Therefore, it is necessary to carefully check that the stresses originating in the webs which are reduced by the non-grouted tendon ducts during post-tensioning. The maximum stresses usually originate in sections close to the quarter spans (Figure 8.44), where the

Figure 8.44 Prestressing of the curved box girder: (a) cross section; (b) elevation; (c) plan; (d) layout of the tendons in the webs; (e) tendons bent incrementally

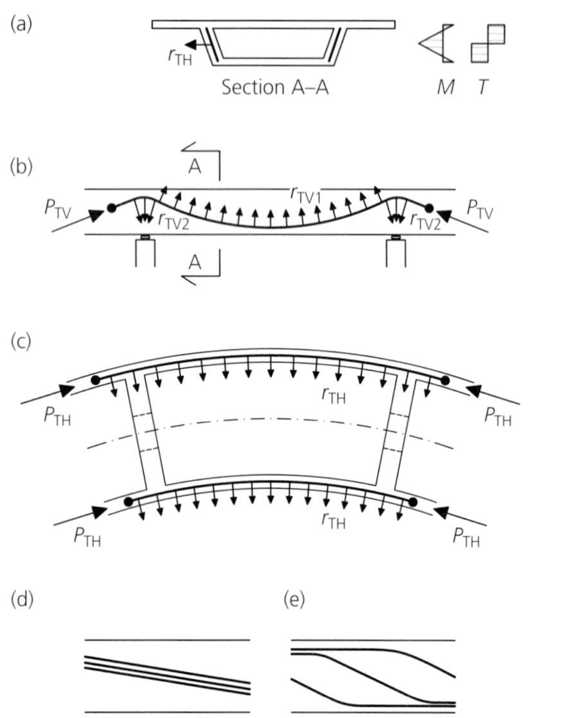

Figure 8.45 Static effects of the span tendons: (a) cross section; (b) elevation; (c) plan

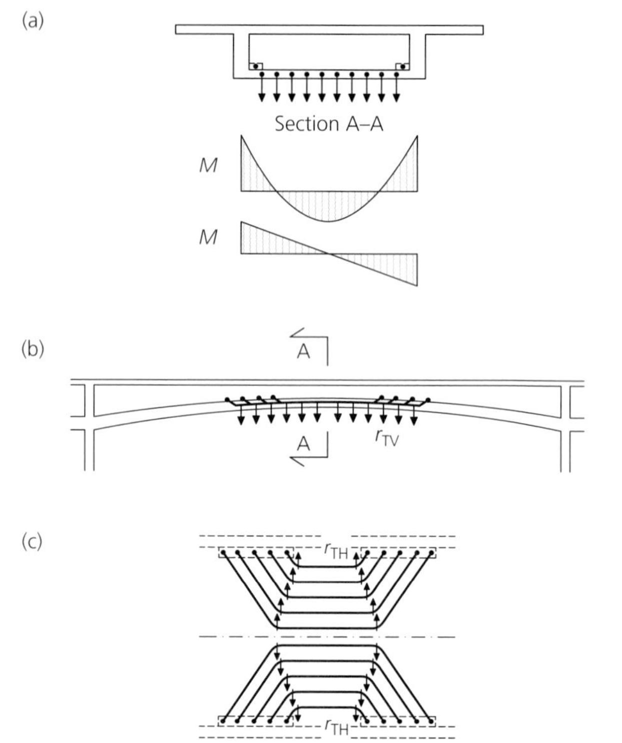

tendons move from the top to the bottom fibres of the box girder. To reduce the value of the radial forces, the tendons should be bent incrementally (Figure 8.44(e)).

8.6.4 Bottom slab of cantilever bridges

Cantilever bridges are usually prestressed by cantilever, continuity and span tendons. The span tendons are situated at the midspan sections. They are led in the bottom slab and are anchored at anchor blocks situated at the bottom corners of the box sections (Figure 8.45). The curved bottom slab functions in a similar way to the curved webs of the box girders (Figure 8.46).

While the bottom slabs of support sections are mainly compressed by the effects of the dead load, the sections close to the midspan are also compressed by the prestressing force. As the prestressing is distributed into the whole section, the bottom slab is compressed only by a portion of the prestressing force. The longitudinal compression stresses therefore create resulting stresses that are smaller than the radial forces created by prestressing tendons.

This is why the bottom slabs are stressed by significant radial forces, which cause their transverse bending and shear (Mathivat, 1983). Therefore, the bottom slab has to be sufficiently reinforced. Neglecting these effects caused failure of the bottom slab of a large cantilever bridge in Austria (Pauser, 1990).

In addition to the bending stresses, the bottom slabs are transversally stressed by normal compression forces originating in the tendon anchors, and by tension forces originating at their plan bends (Figure 8.45(c)).

8.6.5 Global effects of prestressing

Figures 2.21 and 2.22 show how the function of internal tendons and external cables can be modelled. They also show the vertical and horizontal components of the radial forces. It is evident that the transverse components can create significant torsional moments and distortion of the deck (Figure 8.47). This phenomenon is mainly

Figure 8.46 Prestressing of the curved bottom slab. PT, prestressing

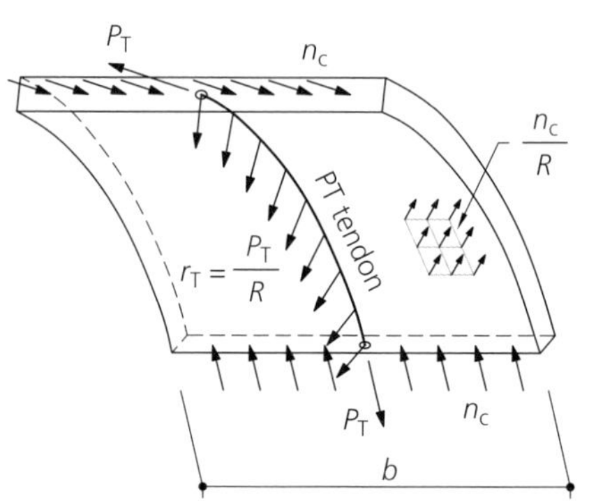

Figure 8.47 Prestressing of the curved bridge: (a) cross-section at the abutment; (b) cross-section above the intermediate support; (c) cross-section at the midspan; (d) elevation; (e) plan

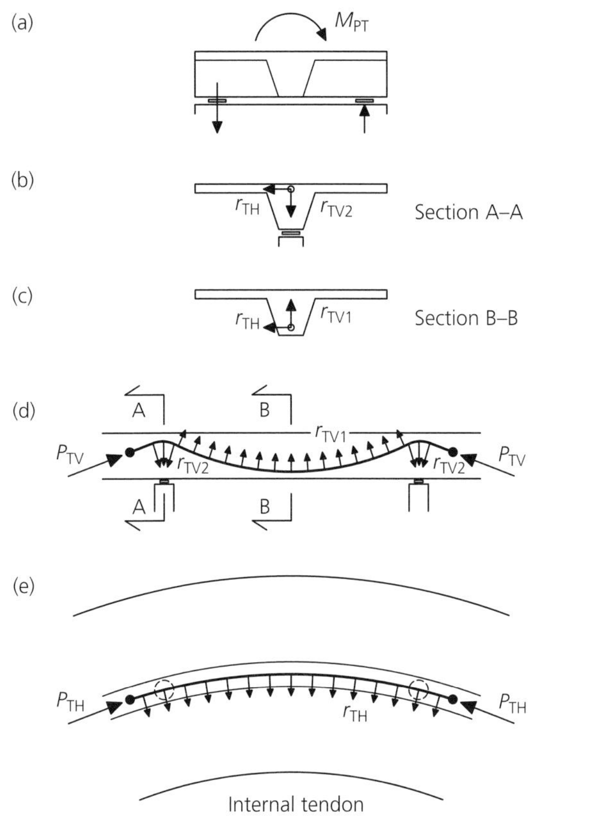

Figure 8.48 Curved bridge of the motorway junction at Rudná, near Ostrava, Czech Republic

Figure 8.49 Bridge in Caracas, Venezuela: (a) prestressing (PT) of the bridge deck; (b) axonometric view

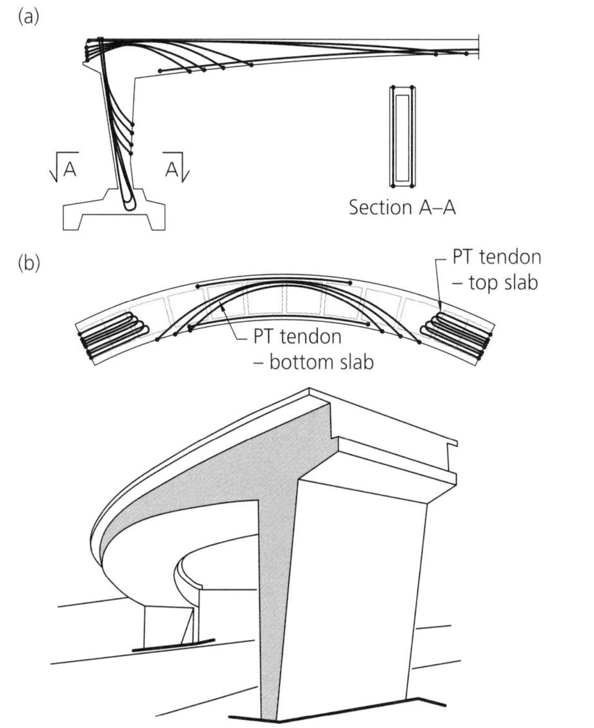

important in structures that are pin supported by a single column forming the intermediate piers (Figure 8.48). The curved decks are usually fixed into end diaphragms stressed by significant transverse moments, for which they have to be reinforced.

8.6.6 Effects of the horizontally draped tendons

The torsional stresses can be reduced by using horizontally draped tendons led in the top and bottom slabs (Leonhardt, 1979). These effects were utilised as early in 1962 and 1965 in the designs of the one-span curved frame structures built in Caracas, Venezuela (Figure 8.49), and Brno, Czech Republic (FIP, 1962, 1966). Both structures utilise horizontally draped tendons situated in the bottom slab of the box-section deck.

To understand the function of the horizontally draped tendons, a plan curved structure of three spans (lengths 33 + 45 + 33 m) was studied (Figure 8.50). The bridge deck, which is in a plan circle with a radius of 100 m, is formed of a box girder with the cross-section shown in Figure 8.5(b). Both, the top and bottom slabs were prestressed by draped tendons, as shown in Figure 8.50(a). The value of the prestressing force was P = 1000 kN. From Figure 8.50(c), which shows the courses of the torsional moments, it is evident that the torsional moments due to the dead load can be effectively reduced by horizontally draped tendons.

8.7. Cable-supported bridges

Cable-supported bridges form a hybrid structural system in which the deck distributes the load and transfers it into a cable system (see Figure 3.36). Due to the non-linear behaviour of the cables, global geometrically non-linear analyses of the bridge structures have to be performed (Stráský, 2011). The concrete decks are modelled as 3D bars or as shell elements that are connected to the cable elements describing the function of the suspension or stay cables (Figures 8.51 and 8.52). The analyses usually start from the initial stage, in which the bridge geometry and the forces in the cables are chosen. For concrete structures, this initial stage is strongly influenced by the effects of creep and shrinkage of the concrete. To minimise the redistribution of stresses, the initial forces in the cables usually balance the effects of the dead load (see Chapter 7).

Figure 8.50 Prestressing (PT) of the box girder slabs: (a) plan and cross-section of the deck; (b) developed elevation; (c) torsional moments

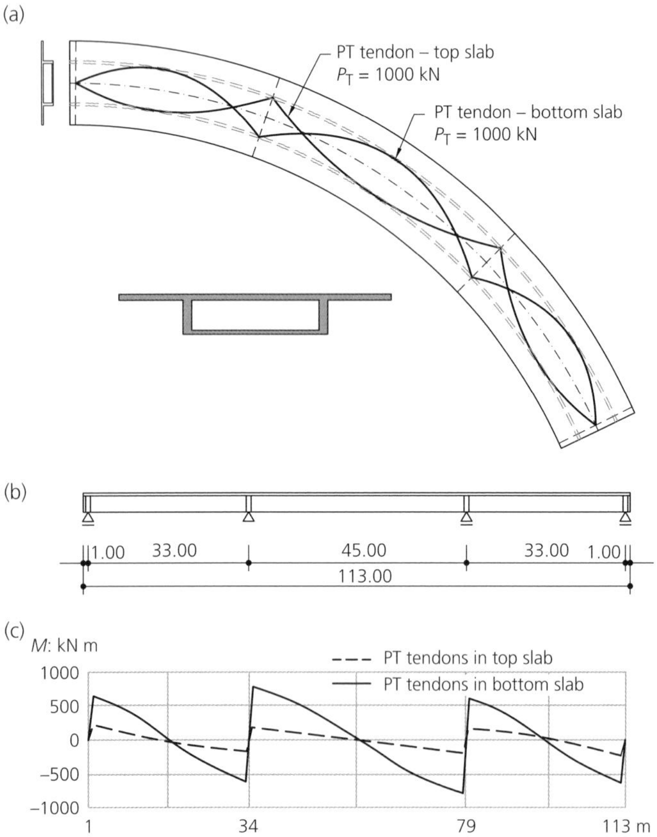

Figure 8.51 Suspension bridge – modelling of the deck as: (a) 3D bars; (b) shell elements

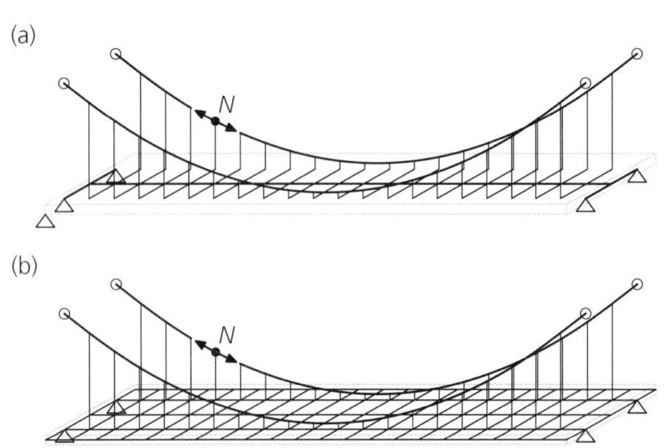

Figure 8.52 Cable-stayed bridge – modelling of the deck as: (a) 3D bars; (b) shell elements

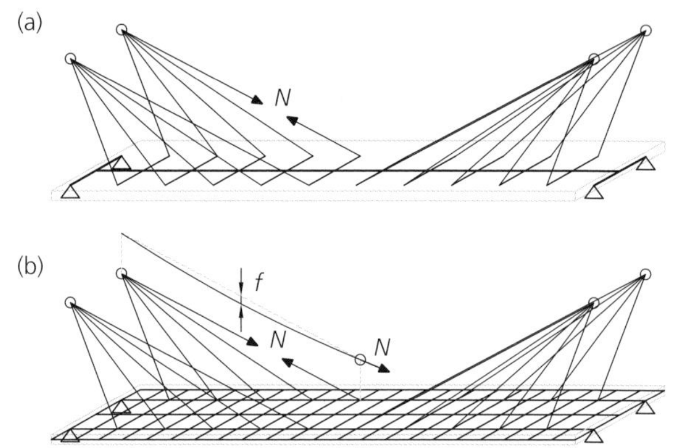

Figure 8.53 Bridge across the River Elbe, Czech Republic: function of the stay cables

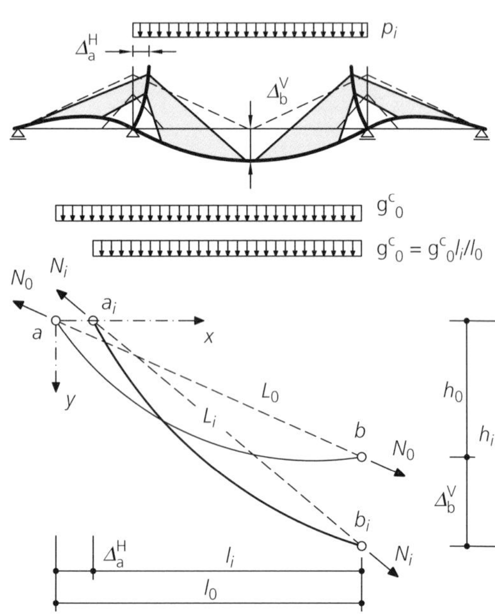

While suspension structures have to be always analysed as geometrically non-linear structures and static effects cannot be superimposed, cable-stayed structures can be (under certain conditions) be analysed linearly. Figure 8.53 shows the geometry of the longest cable of the bridge across the River Elbe in the Czech Republic (Figure 4.79), before and after loading of the main span. Due to the deformation of the deck and the pylon, the cable anchor points move and sag and tension force in the cable is changed. The original distance of the anchor points L_0 has changed to L_i, and the force at the anchor points N_0 has changed to N_i. The study presented in Stráský (2011) proved that, if the initial cable force N_0 is sufficiently high, the change in the forces $\Delta N = N_i - N_0$ is linearly proportional to the change in the length $\Delta L = L_i - L_0$.

In the case of the River Elbe bridge (Stráský, 1993), the initial stresses in the cables f_{s0} had to be larger than $f_{s0} > 0.1 f_{s,u}$ of ultimate stresses. This meant that, for these stresses, the stay cables could be modelled as pin-connected bars (Figure 8.54(a)). Therefore, common structures can be solved using linear programs. However, it is necessary to check that the stresses in the cables are always in the range that guarantees their linear behaviour. The same is true for structures in which the stay cables are situated under the deck.

For longer stay cables the initial stresses have to be higher. The geometrically non-linear behaviour is usually taken into account

Figure 8.54 Stresses in the stay cables

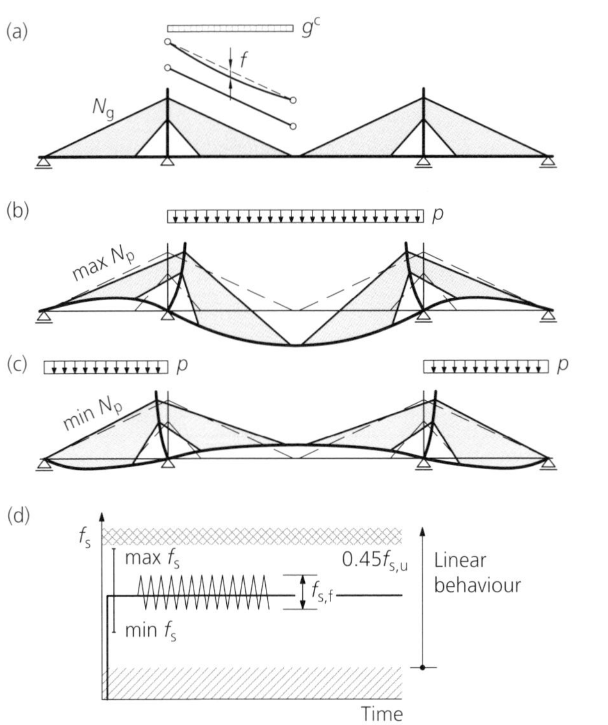

Figure 8.55 Ernst modulus

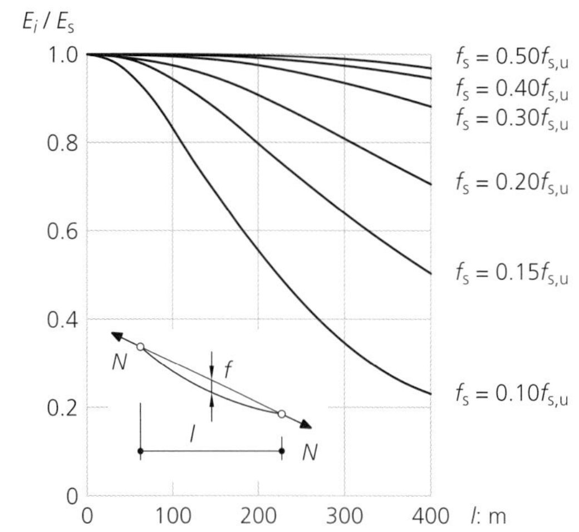

using the Ernst modulus E_i (Figure 8.55) (Ernst, 1965; Walther *et al.*, 1998).

$$E_i = \frac{E_s}{1 + 1 + (\gamma l)^2 E_s / 12 f_s^3} \tag{8.6}$$

where E_s is the modulus of elasticity of the steel, f_s is the stress in the cable, γ is the density of the cable, and l is the horizontal span of the cable.

The forces in the stay cables vary according to the position and intensity of the load. Figures 8.54(b) and 8.54(c) show the two positions of the live load that cause maximum stresses in the longest backstay cable of the Elbe River bridge. The linear analysis shows that the cable is stressed by tension due to the load situated in the main span, and is stressed by compression due to the load situated in the side spans. It is evident that the initial tension of the cable has to be designed in such a way that the maximum tension stresses are smaller than the permissible stresses, and the minimum tension is higher than the tension that guarantees the linear behaviour (Figure 8.54(d)).

8.8. Curved arch and cable-supported structures

The radial forces created by horizontally curved tendons led in the top or bottom slab can also be used in the design of horizontally curved arch or cable-supported structures. The internal tendons can be supplemented by external cables situated at the curbs, on the railings or at the bottom fibres of the deck.

The static and structural problems associated with the design of curved arch and cable-supported structures are discussed in Strasky's book (Stráský, 2011). Therefore, only basic information is given here. In Chapter 7 it was shown that the forces in the suspenders of

Figure 8.56 Ruck-a-Chucky Bridge, Auburn, California, USA

arch and suspension structures and the vertical components of the cable-stay forces should balance the effects of the dead load (see Figures 7.16 and 7.18). The same is true for curved structures.

The decks of the curved structure can be suspended on both edges or on only one edge (inner or outer). The first type of structure requires a space arrangement of the suspenders or stay cables that balances the torsion of the deck. An excellent example of this arrangement is the design for the Ruck-a-Chucky Bridge at Auburn, California, USA

Figure 8.57 Bridge in Figure 8.56 – balancing of forces: (a) vertical; (b) horizontal

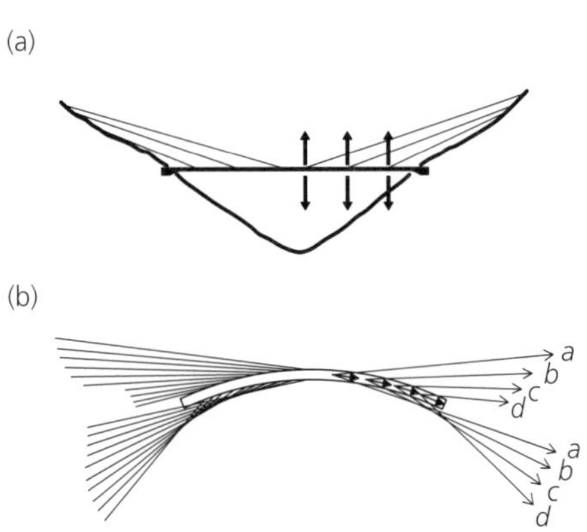

Figure 8.58 Bridge in Figure 8.56 – balancing of forces for suspension on the inner edge: (a) cross-section; (b) plan

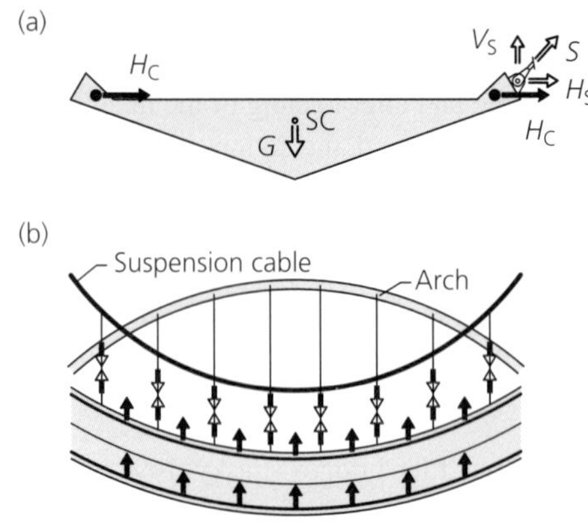

(Figure 8.56), by T. Y. Lin International (Lin and Burns, 1981). Although this bridge has not been built, its design clearly shows how all the internal forces can be balanced by external prestressing through the arrangement of the stay cables.

The planned bridge has a span of 396.24 m and crosses the reservoir with a plan curvature of 628.00 m. The deck is suspended on stay cables arranged in a hyperbolic paraboloid formation to create an array of tensile forces, which produce pure axial compression in the curved deck. The components of the vertical force of the cables balance the weight of the deck (Figure 8.57(a)). The resultants of the horizontal components act in the direction of the curved axis, and are designed to reduce the horizontal bending moments at critical points to zero (Figure 8.57(b)).

If the structure is suspended on its inner edge (Figure 8.58), the deck is torqued by a couple of vertical forces formed by the deck weight G

Figure 8.59 Bridge in Figure 8.56 – balancing of forces for suspension on the outer edge: (a) cross-section; (b) plan

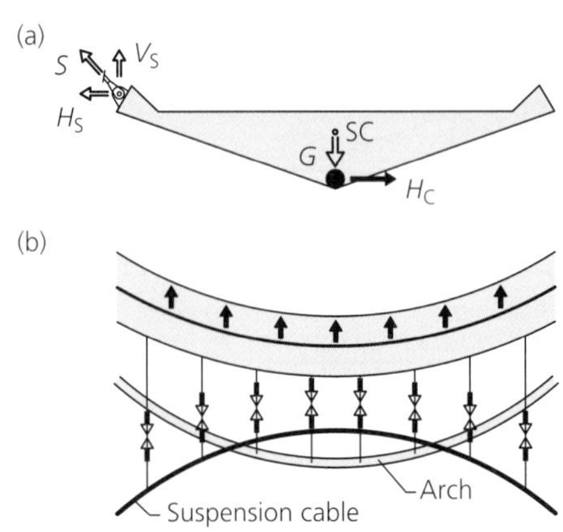

Figure 8.60 Model of the curved structures

and the vertical component of the suspender force V_S. This moment can be balanced by a moment created by the horizontal component of the suspender force H_s, together with radial forces originating in the curved cables H_c situated above the shear centre SC.

If the structure is suspended on its outer edge (Figure 8.59), the deck is torqued by a couple of vertical forces formed by the deck weight G and the vertical component of the suspender force V_S. This moment can be balanced by a moment created by a horizontal component of the suspender force H_s, together with radial forces originating in the curved cables H_c situated at bottom fibres of the deck.

The problems associated with the design of curved suspension and cable-stayed structures have been studied at the Brno University of Technology, Brno, Czech Republic. The analytical results were checked using model measurements (Figure 8.60) (Kolacek *et al.*, 2011).

Both the cable-stayed and the suspension structure studied had a curved composite deck, which were suspended on their inner edges

Figure 8.61 Balancing of the forces

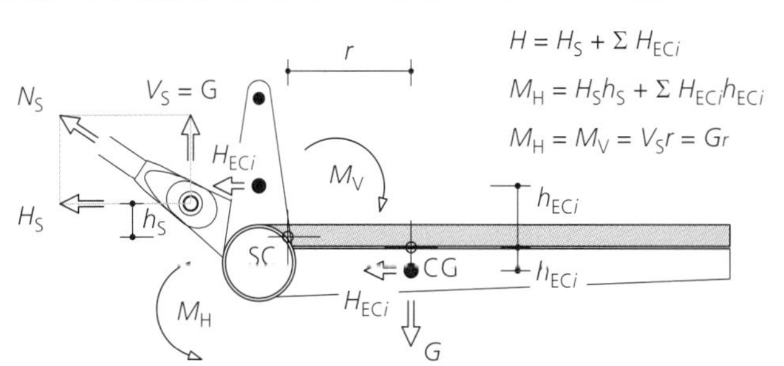

Figure 8.62 Analytical model

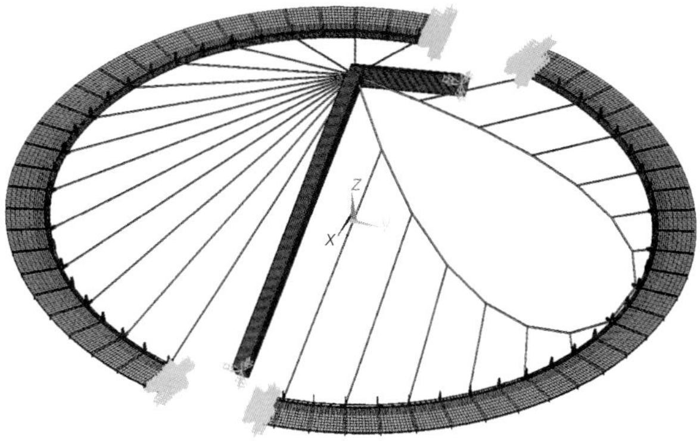

by a common, A-shaped pylon. The 18.40 m tall pylon was supported by two concrete blocks, in which both decks were anchored. The semicircular decks were in a plan curvature with a radius to the inner edge of 30 m. The deck of the cable-stayed structure was suspended on stay cables in a radial arrangement, while the suspension structure was suspended on radial suspenders attached to suspension cables arranged in a funicular shape.

The 4.00 m wide deck of both structures was formed by 625 mm diameter steel pipe and a 150 mm thick concrete deck slab (Figure 8.61). The deck was stiffened by cross-beams located at distances of 3.13 m, and the railing posts and the anchor plate of the suspenders or stay cables were situated at the same locations. The deck was prestressed by radial external cables passing by the railing posts and cross-beams. The prestressing of the external cables, together with the horizontal components of the stay cable or suspender forces,

Figure 8.63 Ultimate load

guaranteed the shape of the structure, and eliminated the torsion of the deck due to the dead load.

The structure was carefully analysed for all effects of the dead load, the prestressing and several positions of the live load (Figure 8.62). Then, using the theory of model similarity, a 1:10 scale model of the structure was constructed and checked for the same loadings. Finally, the function of the model at the ultimate load was checked (Figure 8.63). The results of the tests confirmed the correctness of the structural modelling and the good function of both bridges.

The results of the studies and tests were used in the design of several pedestrian bridges. Two examples are presented here.

The first example is the Harbor Drive pedestrian bridge in San Diego, California, USA, built in 2010 (Stráský, 2011). The bridge has a plan curvature with a radius of 176.80 m, and is a self-anchored suspension structure that is suspended by hangers located on the inside of the curve (Figure 8.64). The suspended span is 107.60 m long, and is monolithically connected to stairs at both ends. The stairs of lengths 13.54 and 21.97 m form part of the structural system that transfers the stresses into the abutments supported on the piles. The 39.80 m tall pylon, which supports the main cable, is founded on the convex side of the deck, leans over the deck, and supports the main cable on the inside of the curve. It is stabilised by two back-stays and internal post-tensioning.

The main cable stretches from the abutment, to a deviator at the top of the stairs, to the anchorage at the top of the pylon. It is made of prestressing strands encased in stainless steel pipe. The hangers are attached to the steel pipe of the main cable and to the handrail on the bottom. The top of the handrail also carries a large post-tensioning cable, which is anchored at the deviators at the top of the stairs. This cable is overlapped by the internal cables that prestressed the stairs.

Figure 8.64 Harbor Drive pedestrian bridge, San Diego, CA, USA: structural arrangement

Figure 8.65 Bridge in Figure 8.64: deck

Figure 8.66 Bridge in Figure 8.64 – balancing of forces: (a) cross-section; (b) vertical; (c) horizontal

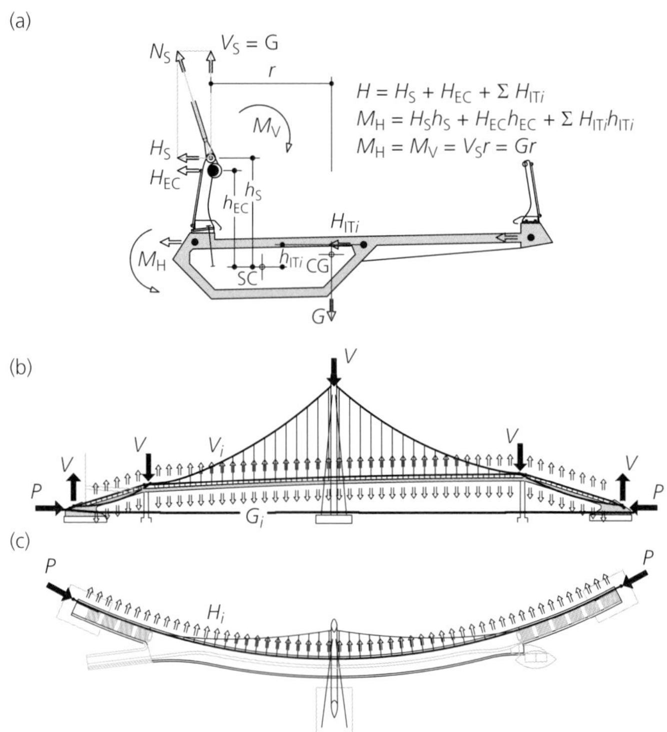

The suspended deck is formed of a non-symmetrical box girder with overhangs on one side that are supported by the ribs (Figure 8.65). The girder is prestressed not only by the internal tendons situated at the top slab, but also by the horizontal components of the hanger

Figure 8.67 Bridge over the River Olse, connecting Cieszyn in Poland with Český Těšín in the Czech Republic

Figure 8.68 Bridge in Figure 8.67 – structural arrangement

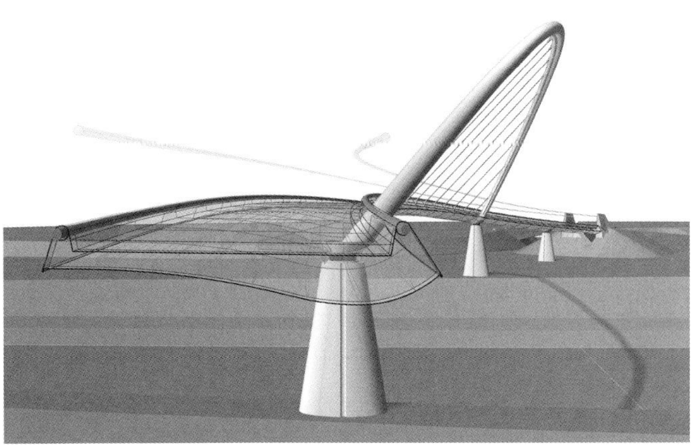

forces and by the external cables. Therefore, the inner railing, in which the hangers are anchored, is a part of the structural system. The geometry of the deck, the position of the anchoring of the hangers and the position of the external cables and internal tendons

Figure 8.69 Bridge in Figure 8.67 – radial forces

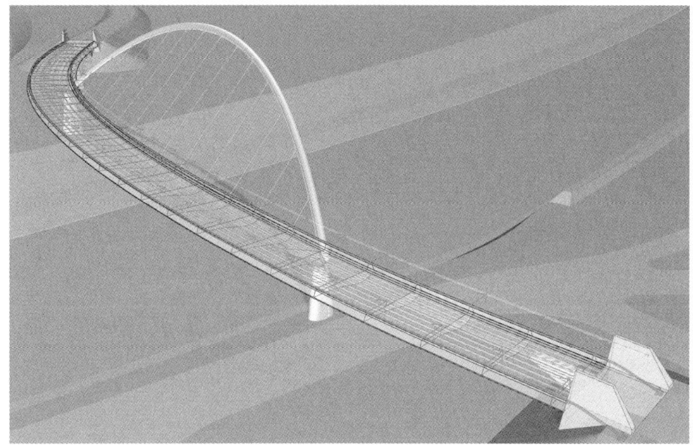

were determined in such a way that the horizontal forces balance the moment created by the eccentricity of the suspension ribs (Figure 8.66).

The second example is the arch bridge built in 2011 (Stráský *et al.*, 2013) over the River Olse, connecting Cieszyn in Poland with Český Těšín in the Czech Republic (Figure 8.67). The bridge has a total length of 95.40 m, a plan curvature with a radius of 100 m and a crest elevation. The bridge has four spans (lengths from 13–45 m). The deck is formed of a slender box girder of a non-symmetrical streamline cross-section that is stiffened on one side by an inclined arch in the main span (Figure 8.68). The deck is fixed into the end abutments and is supported by elastomer pads on intermediate piers. To balance the torsional moment due to the dead load, the deck is prestressed by radial cables situated at the edge curbs (Figure 8.69). Both the girder and the arch are made of steel and concrete composites.

8.9. Strut-and-tie analyses of selected problems

As mentioned in the previous sections, when designing bridge structures it is necessary to consider the space function of the structures not only for the service load but also for the ultimate load. During the design it is necessary to study the course of the forces, properly design the reinforcement and check the ultimate capacity of the structure. There is no standard procedure, each structure is different; the structure must first be understood, and then properly designed. The examples below illustrate the possible problems and suggest a way to solve them.

The ultimate capacity of a structure is usually determined by analysing the space strut-and-tie models that express the function of both the B and the D regions of the structure. B regions are

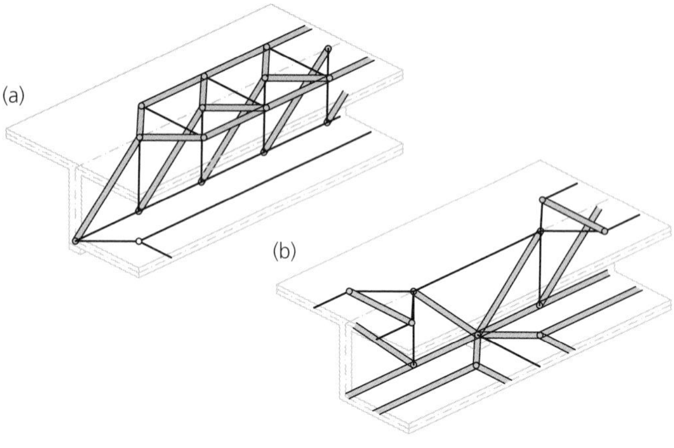

Figure 8.70 Direct support of the box girder

structural parts, in which the Bernoulli hypothesis of plane strain distributions is assumed to be valid. D (discontinuity) regions are the parts of the structures where the strain distribution is significantly non-linear (e.g. near concentrated loads, supports, openings and other discontinuities) (Schlaich *et al.*, 1987). Strut-and-tie modelling is accepted in all the current design specifications and codes.

8.9.1 Direct support of the box girders

Figure 8.70(a) shows the strut-and-tie model of the part of a structure above a simple support of the side span of a one-cell box girder, and Figure 8.70(b) shows the strut-and-tie model of the same structure above the intermediate support.

It is evident that the reactions from the bearings must be transferred not only into the girder webs but also into the top and bottom slabs of the bridge deck. In the part of the structure above the side support, the compressive force originating in the web must be distributed into the upper slab, where the transverse tensile force has to be resisted by the transverse reinforcement of the slab. In order to get the tensile force from the bottom slab into the webs above the bearing, it is also necessary to design a transverse tension reinforcement in the bottom slab.

Conversely, for the inner support, it is necessary to check the top slab for transverse compression forces. It is also necessary to realise that the transverse reinforcement of the bottom slab must be situated in the strip between the bearings.

8.9.2 Support diaphragms

Figure 8.8(b) shows the deformation of the cross-section of a box girder structure at the midspan and support sections for following options: the box girder is not stiffened by any diaphragms, the girder is stiffened by support diaphragms, and the girder is stiffened by span and support diaphragms. The box girder without any diaphragms significantly deforms transversally, indicating that the shear stresses are not resisted by the top slab. This means that the structure does not function as a torsionally stiff box girder.

To utilise the torsional resistance of box girders, the structural arrangement has to guarantee the transfer of the diagonal shear stresses, which have transverse and longitudinal components, into the supports. It is useful to realise that a box girder fixed into a stiff anchor block and a box girder supported by two bearings behave differently. In the first case, the shear stresses are resisted directly by the anchor block; in the second case, the box girder transfers all the stresses into two bearings. To ensure a similar function, the box girder has to be stiffened by support diaphragms that ensure the shape of the box girder and transfer the shear forces into the bearings. The end diaphragms must also ensure that the anchoring of the longitudinal reinforcement resists the longitudinal component of the shear stresses.

Figure 8.71 shows a two-span continuous box girder that is supported by a pair of bearings situated under the webs. This means that the girder is fixed in the torsion at all supports. The box girder has a cross-section of width b and depth h, and is loaded by a uniform line load $\pm p$ acting above the girder web. This load loads the structure with a uniform torque $m_K = pb$. To the left of the inner support (2), the beam is subjected to the torsional moment $^{L}M_{K,b} = -m_K(L/2)$; to the right of this support the beam is subjected to the torsional moment $^{P}M_{K,b} = m_K(L/2)$.

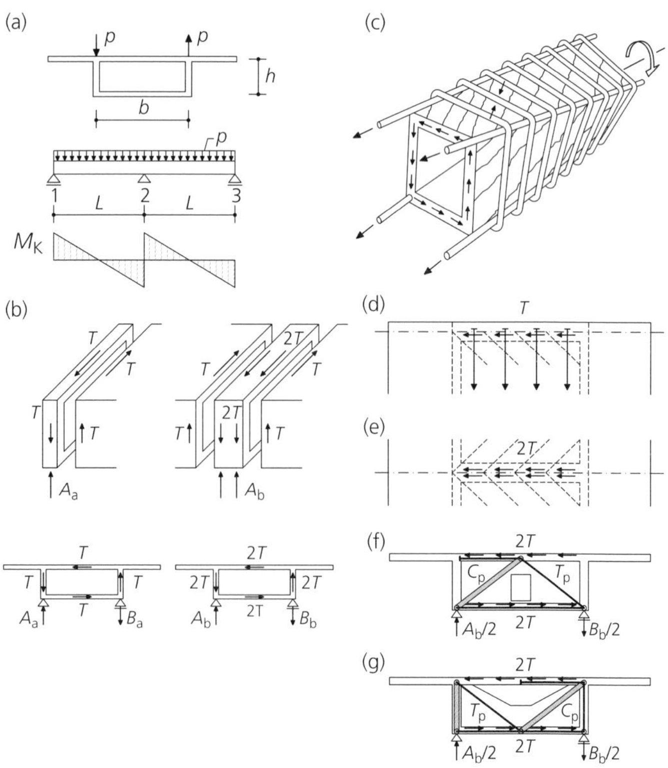

Figure 8.71 Static function of the pier diaphragms

The inner support (2) must resist the torsional moment $M_{K,b} = -^{L}M_{K,b} + {}^{P}M_{K,b} = m_K(L)$. Thus, the vertical reactions $A_b = -B_b = M_{K,b}/b = m_K(L)/b = p(L)$ originate at the inner support (2).

The shear flow $T = -^{L}M_{K,b}/2A$, where $A = bh$, originates in the girder box section to the left of the inner support (2) due to the torsion. Then, $T = -^{L}M_{K,b}/2(bh)$. The part to the right of the support in the girder box section is subjected to a shear flow $T = -^{P}M_{K,b}/2(bh)$. The strip above support 2 is therefore stressed by the shear flow $2T$ (Figure 8.71(b)). Similarly, the strip above the end support (1) is stressed by the shear flow T.

The resultant of the shear flow in the webs T_W in the section above support 2 is given by

$$T_W = 2T(h) = A_b/2 = -B_b/2 \tag{8.7}$$

and is resisted directly by the supports. The resultant of the shear flow in the bottom slab T_{BS} is given by

$$T_{BS} = 2T(b) \tag{8.8}$$

and is resisted by the horizontally fixed bearing. However, the shear flow has to be transferred by the reinforcement into the fixed bearing.

The resultant of the shear flow in the top slab T_{TS} is given by

$$T_{TS} = 2T(b) \tag{8.9}$$

This resultant tries to deform the girder cross-section, and is transferred by the support diaphragm to the supports.

Figure 8.71(c), which is taken from Collins and Mitchell (1987), shows the calculation model for the torqued box girder at the ultimate limit state. The torsion is resisted by a system of compressed concrete struts and tensioned transverse ties (stirrups and longitudinal reinforcement).

Figure 8.71(d) shows the top view of the box girder's top slab above the end support, which is stressed by the torsional ultimate moment. It is evident that the transverse components of the pressure exerted by the struts must be transferred to the supports by the support diaphragm, and the longitudinal component has to be resisted by the longitudinal reinforcement.

Similarly, Figure 8.71(e) shows the top view of the box girder's top slab above the inner support (2). It is evident that the transverse component of the pressure exerted by the struts must be transferred to the supports by the support diaphragm. The method of transmission depends on the diaphragm arrangement.

Figure 8.71(f) shows the transfer of the horizontal force of the top slab into the bearings via a classic support diaphragm, with a manhole opening located at the bottom plate. The resultant of the shear flow $T_{TS} = 2T(b)$ is transmitted from the centre of the top plate into the supports by a compressed strut and a tie. In order to get the shear flow into the node of the top slab, a transverse reinforcement resisting force $T_{TP}/2$ must be provided there. At the bottom corners of the box girder, the vertical components of the compressed strut force C_P and the tie force T_P are resisted directly by the bearings stressed by the vertical reactions $A_b/2 = -B_b/2$. The horizontal components of the force T_P are resisted by the compression capacity of the bottom slab.

When designing the deck slabs in the area close to the pier diaphragms, it is necessary to realise that any trucks situated there will create not only local transverse moments but also longitudinal bending moments $^{L}M_X$. The corresponding stresses have to be added

Figure 8.72 Global and local bending of the deck slab

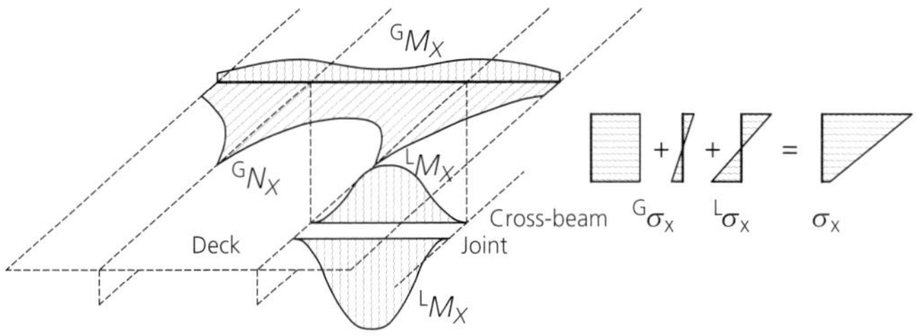

Figure 8.73 Transfer of forces in deep diaphragms: (a) cross section; (b) torsional moment; (c) force F_L; (d) force F_R

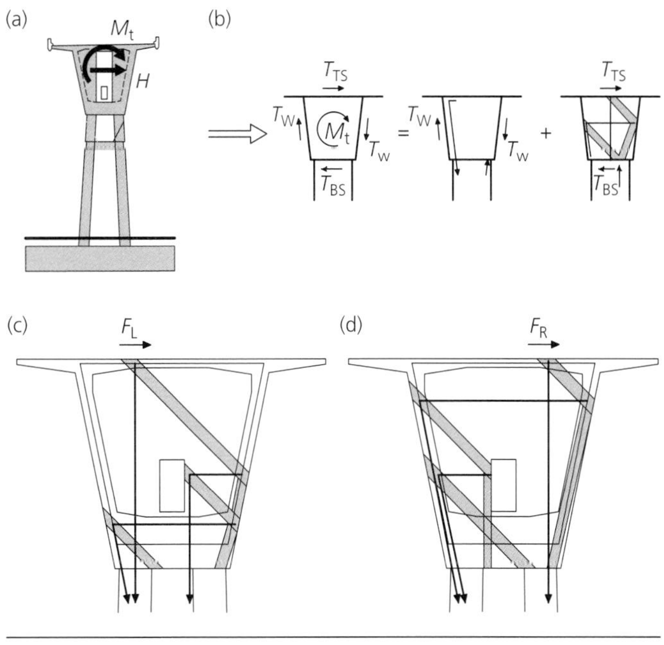

to the stresses originating in the deck from the global bending (^GN_X and ^GM_X) (Figure 8.72).

It is for this reason that in some structures the diaphragms are not connected to the top slabs (Figure 8.66(f)). The shear flow originating in the top slab is then transferred from the top corners of the box. Half of the shear forces $T_T/2$ is transferred to the opposite corner by the transverse tensile reinforcement. The horizontal force is transferred from the upper corners to the bearings by the strut-and-tie system shown in Figure 8.71(g). The horizontal components of the forces C_P and T_P are transferred into the bearings by the compression capacity of the bottom slab.

Figure 8.74 Indirect supports of a double-T deck

The pier diaphragms of cantilever bridges have a large depth, and therefore it is necessary to carefully analyse the transfer not only of the torsional moment but also the horizontal forces originating from wind and seismic actions. Figure 8.73 shows the strut-and-tie model used by the authors in their analysis of the pier diaphragms of the cantilever bridges built in 1944 on Freeway 1 in Taiwan. The bridges, which have spans of up to 165 m, are formed of 12.20 m wide box girders that have depths ranging from 3.50 m to 8.00 m at the piers (see Figures 11.56 and 11.58).

The dimensions of the decks allowed the bridges to be modelled as 3D beam structures, which were analysed for the effects of the dead and live loads, wind and seismic actions. Significant horizontal forces H and torsional moments M_T, which originate mainly from the seismic load, are transferred from the deck via the pier diaphragms into the piers. The horizontal force H acting at the centre of gravity of the girder was proportionally distributed into the top (H_T) and bottom (H_B) slabs. The shear forces acting in the top slab T_{TS}, bottom slab T_{BS} and the webs T_W were determined from the torsional moments acting to the left and the right of the diaphragms.

As the decks of all the bridges are frame connected to piers, the forces acting in the webs and the bottom slab were transferred directly into

Figure 8.75 Indirect support: tested structure

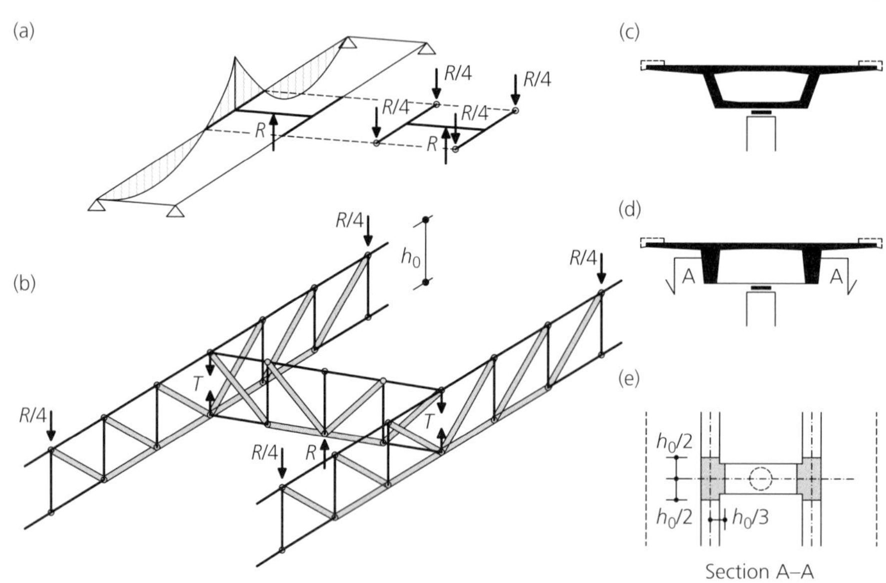

Figure 8.76 Indirect supports

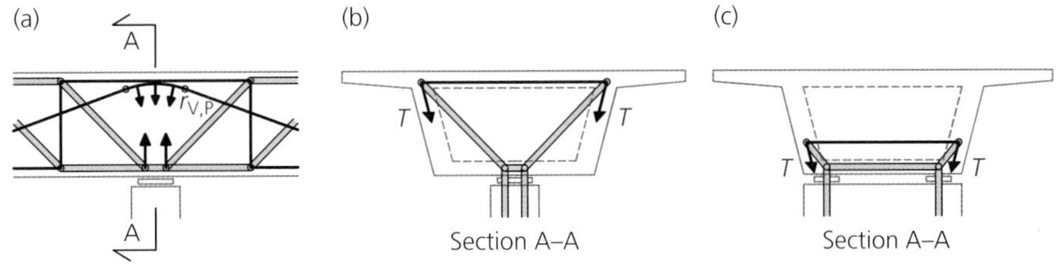

the piers. The horizontal force acting in the top slab $F = H_T + T_{TS}$ is transferred into the connection between the bottom slab and the piers, and this was modelled by the strut-and-tie model shown in Figures 8.73(c) and 8.73(d). Due to the opening, the force F was divided into two forces F_L and F_R, and two models of the transfer of forces were used.

8.9.3 Indirect supports

The webs of double-T and box girder bridges are often indirectly supported by transverse beams (diaphragms) that are supported by slender piers situated in the bridge axis (Figures 8.74. 8.75(c) and 8.75(d)). During the service of several structures, horizontal cracks have originated in the webs at the support area. These cracks indicate that no so-called 'hanging-up' reinforcement has been designed in the girder webs.

The need for reinforcement in this area, which results from a truss analogy (strut-and-tie method), was confirmed by experiments conducted by Professor Leonhardt in Germany (Leonhardt *et al.*, 1971). Figure 8.75(a) shows a two-span structure formed by two beams indirectly supported by a single bearing situated on an intermediate support. The reaction R from the bearing is transferred into the longitudinal beams by a transverse beam. A part of the structure situated between so-called zero points – between the points where the moment from the self-weight is zero – was removed from this structure. This part was analysed in detail and tested. The design of the test elements was designed with and without hanging-up reinforcement.

Figure 8.77 Diaphragm with a manhole: (a) elevation; (b) cross-section

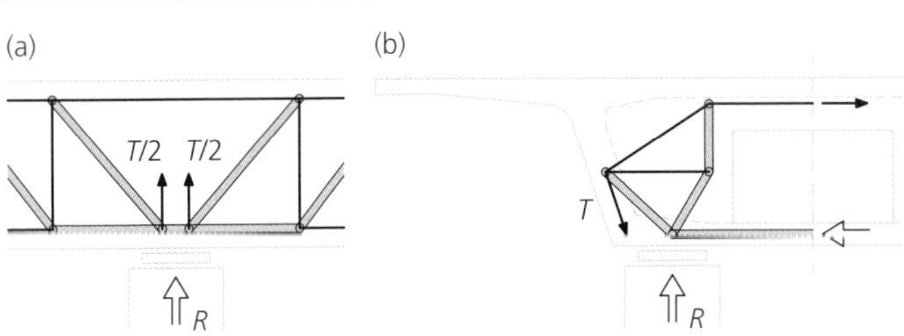

Figure 8.75(b) shows the model used for the calculations on the tested structure. It can be seen that the bearing force is transferred by the transverse beam through a system of struts and ties into the upper nodes connecting the top chords of the transverse and longitudinal beams. In the directly supported beams, the struts start from the bearing. As the transverse beam substitutes for the bearings, the struts of the truss model must start from the same place, that is, from the node, where the bottom chords of the longitudinal and transverse beams are connected. It is clear from the model that the top and bottom nodes must be connected by tensile elements, that is, by ties that transfer (hang up) the force $T = R/2$. Model tests have clearly confirmed the necessity of the reinforcement determined by the strut and tie model.

According to Professor Leonhardt's recommendation, it is necessary to design the hanging-up reinforcement for the vertical force T. The reinforcement must be placed in the girder the area equal to the effective height h_0 and in the transverse beam the length of $h_0/3$ from the girder axis (see Figure 8.75(e)). It is recommended that, with respect to the crack width, 50% of the reinforcement is formed by vertical or skew prestressing. As the radial forces from the supports due to the prestressing r_{vp}, act at the top fibres, they do not need to be 'hung up' (Figure 8.76(a)), and therefore they can be subtracted.

Figure 8.78 End cross-beam: (a) cross-section; (b) strut-and-tie model

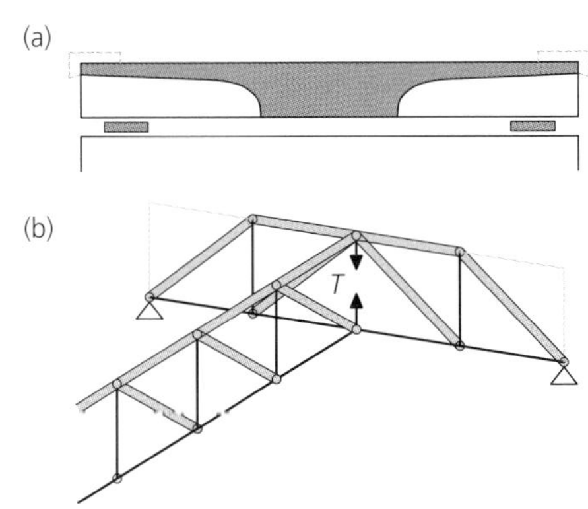

Figure 8.79 Indirect supports of an end cross-beam

Figure 8.80 Strut-and-tie model of a hinge

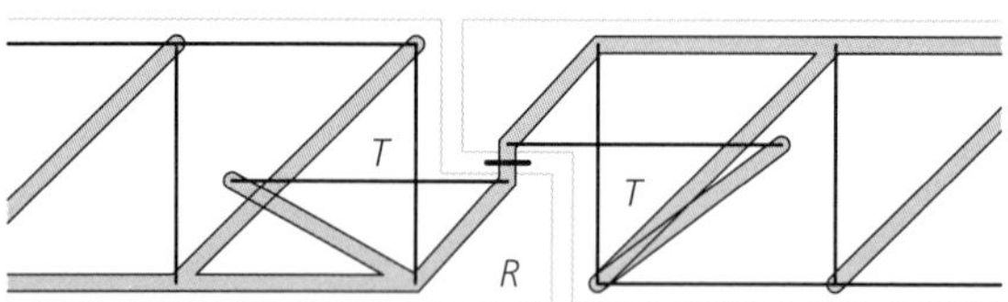

It is necessary to realise that the hanging-up reinforcement is required not only for structures supported by single supports (Figure 8.76(b)) but also for all structures in which the girder webs are not directly supported by bearings. Figure 8.76(c) shows the calculation model for the structure supported by two bearings. If the diaphragm has a manhole, the calculation model is more complicated. Figure 8.77 shows the strut-and-tie model that the authors used to checking the function of a precast segmental structure.

Hanging-up reinforcement has to be designed whenever the structural member under consideration is not directly supported. Figure 8.78 shows the end cross-beam of the structure shown in Figure 8.79. The deck of this bridge is formed by a solid spine girder with large overhangs, which is supported in the bridge axis on intermediate supports. The transverse stability of the bridge is ensured by pairs of bearings supporting the end beams. Hanging-up reinforcement has to be designed at the connections of the longitudinal and transverse beams.

8.9.4 Joints of the Gerber's beams

Figure 8.80 shows the calculation model for a part of a structure at the location of a joint. It is clear that the girder webs must be reinforced on both sides of the joint by hanging-up reinforcement that transfers the whole reaction $T = R$. It is also clear that the horizontal reinforcement of the short brackets must be properly anchored in the compressed concrete.

REFERENCES

AASHTO (2003) American Association of State Highway and Transportation Officials) *Guide Specifications for Design and Construction of Segmental Concrete Bridges*, 2nd edn. AASHTO, Washington, DC, USA.

CEN (2005) EN 1992-2:2005. Eurocode 2 - Design of concrete structures. Part 2: Concrete bridges: Design and detailing rules. CEN, Brussels, Belgium.

Collins MP and Mitchell D (1987) *Prestressed Concrete Basics*. Canadian Prestressed Concrete Institute, Ottawa, Canada.

Corven J (2016) *Post-Tensioned Box Girder Design Manual*. Federal Highway Administration, Department of Transportation, Washington, DC.

Ernst HJ (1965) Der E-Modul von Seilen unter Berücksichtigung des Durchhanges. *Der Bauingenieur* **40(2)**.

FIP (1962) Pont Helicoïde de Caracas. Réalisations Françaises. *4th FIP Congress, Rome, Italy*.

FIP (1966) Most v Rybarske ulici v Brne. Prestressed concrete in Czechoslovakia. In *Proceedings of the Fifth Congress of the Federation Internationale de la Precontrainte (FIP), Paris, France*.

Hambly EC (1991) *Bridge Deck Behaviour*. E&FN Spon, London, UK.

Kolacek J, Necas R and Stráský J (2011) Curved cable-stayed and suspension pedestrian bridges. *Footbridge 2011: 4th International Conference, Wroclaw, Poland*. Dolnośląskie Wydawnictwo Edukacyjne, Wroclaw, Poland.

Kristek V (1979) *Theory of Box Girders*. John Wiley, New York, NY, USA.

Leonhardt F (1979) *Vorlesungen über Massivbau. Grundlagen des Massivbrückenbaues*. Springer, Berlin, Germany.

Leonhardt F, Koch R and Rostásy F (1971) Aufhängebewerung bei indirekter Lasteintragung von Spannbetonträgern, Versuchsbericht und Empfehlungen. *Beton und Stahlbetonbau* **66(10)**: 233–240.

Liebenberg A (1992) *Concrete Bridges – Design and Construction*. John Wiley, New York, NY, USA.

Lin TY and Burns NH (1981) *Design of Prestressed Concrete Structures*. John Wiley, New York, NY, USA.

Mathivat J (1983) *The Cantilever Construction of Prestressed Concrete Bridges*. John Wiley, New York, NY, USA.

Novotny P, Juchelkova P, Jurik M and Pawelczak M (2012) Bridges with progressively erected cross sections. *Design of Concrete Structures using Eurocodes. Third International Workshop, Vienna, Austria*.

Pauser A (1990) *Talübergang Schottwien – Urhsachen und Behebung eines Schadens*. Austrian Concrete Association, Vienna, Austria.

Priestly JN, Seible F and Calvi GM (1996) *Seismic Design and Retrofit of Bridges*. John Wiley, New York, NY, USA.

Schlaich J and Scheef H (1982) *Beton-Holkastenbrücken*. IABSE, Zurich, Switzerland.

Schlaich J, Schäfer K and Jennewein M (1987) Toward a consistent design of structural concrete. *PCI Journal* **32(3)**: 74–150.

Stráský J (1979) *Analysis of Concrete Bridge Structures by Grillage Idealization Method*. (In Czech.) Inzenyrske stavby, 8/79, Czechoslovakia.

Stráský J (1993) Design and construction of cable-stayed bridges in the Czech Republic. *PCI Journal* November–December: 24–43.

Stráský J (2011) *Stress Ribbon and Cable Supported Pedestrian Bridges*, 2nd edn. Thomas Telford, London, UK.

Stráský J and Husty I (1999) Arch bridge crossing the Brno–Vienna Expressway. *Proceedings of the Institution of Civil Engineers – Civil Engineering* **132(4)**: 156–165.

Stráský J, Romportl T, Kocourek P and Rayor G (2013) Integral arch bridges. *ARCH 2013*. 7th International Conference on Arch Bridges, Split, Croatia.

Timoshenko SP and Goodier JN (1951) *Theory of Elasticity*. McGraw-Hill, London, UK.

Timoshenko SP and Woinowsky-Krieger S (1959) *Theory of Plates and Shells*. McGraw-Hill, London, UK.

Vlasov VZ (1961) *Thin-Walled Elastic Beams*. US Department of Commerce, Washington, DC, USA.

Walther R, Houriet B, Walmar I and Moïa P (1998) *Cable Stayed Bridges*. Thomas Telford, London, UK.

Zienkiewicz OC and Taylor RL (1994) *The Finite Element Method*. McGraw-Hill, London, UK.

ICE Publishing

Strásky and Nečas
ISBN 978-0-7277-6385-3
https://doi.org/10.1680/dcpb.63853.165
ICE Publishing: All rights reserved

Chapter 9
Structures assembled of precast beams

Bridge decks assembled of precast beams have been designed since prestressed concrete was first used, when Eugène Freyssinet built the first precast bridge in Algeria in 1936. The bridge deck was assembled of four 19 m long I-girders (Troyano, 2003).

9.1. Precast girder decks

Over the years, different structural types of precast beam structures have been developed. The bridge decks are usually assembled from precast longitudinal members (beams) and composite deck slabs (fib, 2001). The precast members can be formed by square beams, I-beams, T-beams, inverted T-beams, U-beams and V-beams (Figure 9.1). For very short spans, precast beams with an inverted-T section that are totally infilled with cast-in-place concrete are widely used (Figure 9.2).

As for short spans the static effects created by heavy vehicles (i.e. point loads) control the design, decks assembled of box girders are structurally very efficient (Figure 9.3). Detailed static analyses have confirmed that box girders mutually connected by longitudinal hinges distribute point loads well. Therefore, in many cases the decks of short-span structures have been assembled of box girders mutually connected by additionally cast longitudinal joints reinforced by overlapping reinforcing bars. In several countries these structures have been built without composite slabs. Although these decks are structurally very efficient, their use is disqualified due to the impossibility of checking the function of the internal cells. If water gets into insufficiently drained cells it can freeze in winter and damage the deck. Therefore, such decks are not permitted in some countries at present. Box girders cast with short cantilevers mutually connected by additionally cast longitudinal joints (see Figure 9.40) have also been successfully used in the design of prefabricated overpasses.

For longer spans, the design is controlled by the uniform load, and beam structures with T or I cross-sections are suitable. Although

Figure 9.1 Typical sections of precast beams: (a) square beam; (b, c) I-beams; (d) T-beam; (e, f) inverted T-beams; (g) U-beam; (h) V-beam

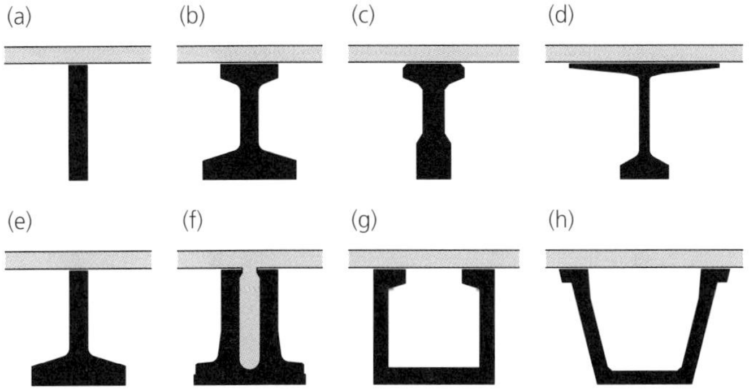

Figure 9.2 Short-span bridge deck with infilled precast beams: (a) I beams, (b) inverted T-beams

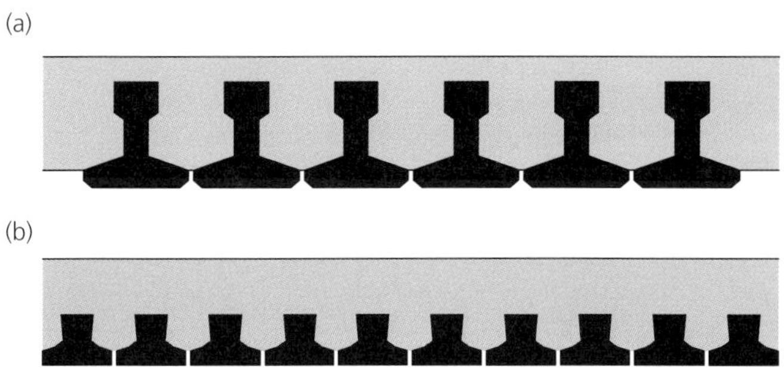

Figure 9.3 Short-span bridge decks assembled of box girders connected by: (a) joints at the centreline of the girder; (b) joints at the top slab

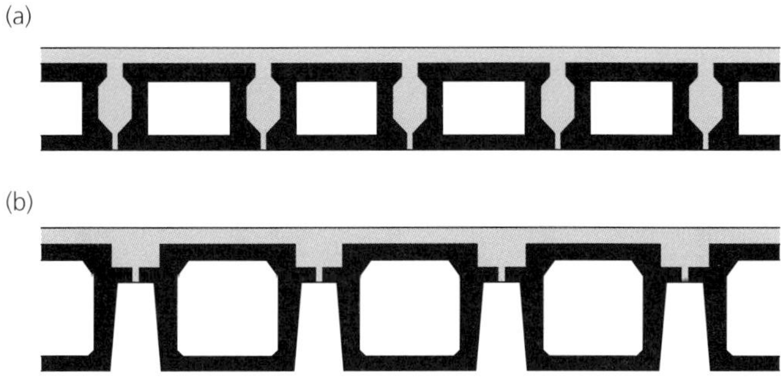

pre-tensioned precast beams made from lightweight concrete and with lengths of up to 68 m have been recently erected in the USA (West, 2019), pre-tensioned girders are only commonly used for spans up to 45 m. As it is rather complicated to transport beams with spans longer than 30 m, the longer girders are usually assembled from two or three precast segments that are post-tensioned together (Figure 9.4). The segments are usually match-cast and the joints are filled with epoxy resin.

In the past, the transverse cooperation of the beam structures was achieved through the use of several additionally cast span diaphragms. However, detailed static analyses proved that composite deck slabs properly connected with precast beams can guarantee sufficient cooperation of the girders and the integrity of the bridge deck (Figure 9.5). Therefore, at present, diaphragms are usually cast only at the supports. In some countries, girders without pier diaphragms are used. To simplify the formwork for the deck slab, precast slab members are usually inserted between precast girders (Figure 9.6).

Figure 9.4 Precast beams of length 40 m assembled from segments of lengths 13 + 14 + 13 m

Figure 9.5 Cross-section of a typical beam bridge

In North America, where a bridge deck slab without a waterproof membrane or asphalt overlay is used, the bridge structure should allow replacement of the composite deck slab. To guarantee the stability of during the erection of the beams and the replacement of the deck slab, the beams are mutually connected by steel diaphragms that are erected before the deck slab is cast.

In common precast structures, each beam is supported by bearings (Figure 9.7(a)). As, when it is necessary to replace the bearings, it is necessary to lift the whole span, the beams have to be supported by transverse steel girders situated close to the supports. The steel girders must have sufficient stiffness to guarantee the geometry of the deck. Therefore, it is suitable to reduce the number of bearings and to support the deck by only two bearings (Figure 9.7(b)). In this case, the support diaphragms have to be designed for the transfer of the load from individual beams into the bearings (see Figures 9.25 and 9.27).

Figure 9.6 Formwork for and reinforcement of a composite slab

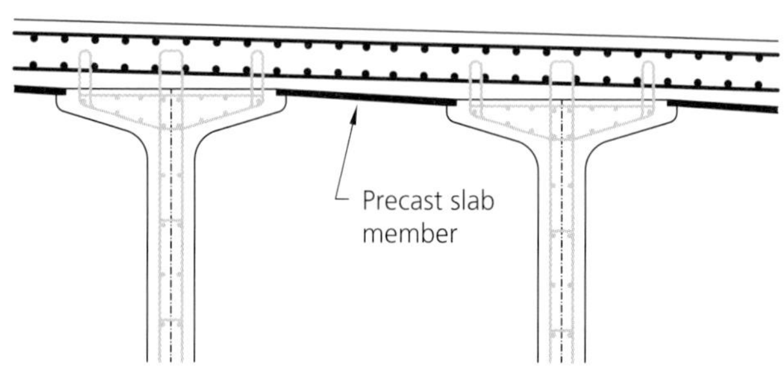

Figure 9.7 Supporting a precast deck: (a) direct support of each beam; (b) indirect support via a support diaphragm

(a)

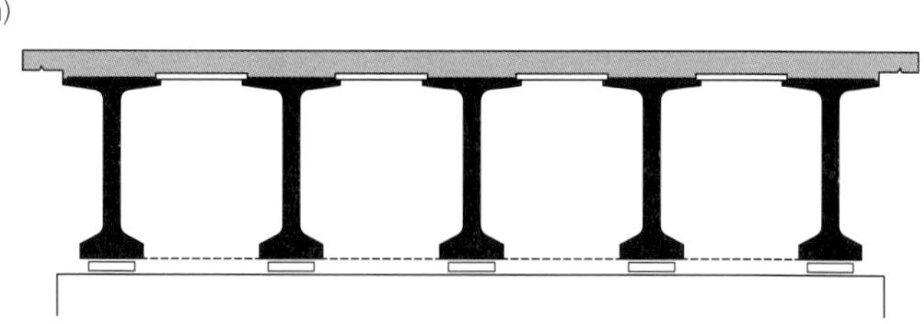

(b)

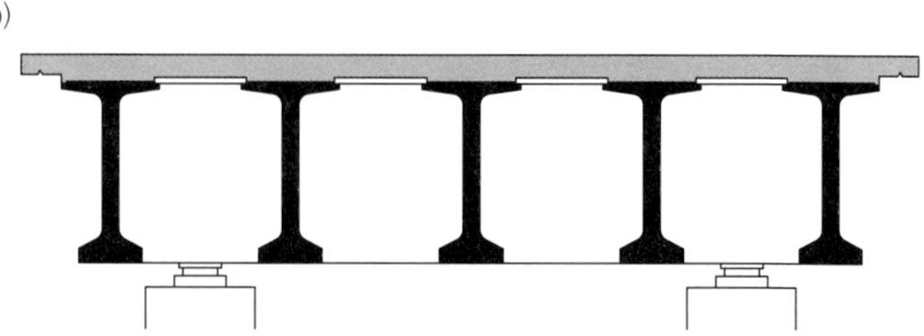

9.2. Continuous structures

In the past, multi-span girder structures were assembled of precast girders supported by pier caps (Figure 9.8(a)). The girder decks formed simple spans with expansion joints at each pier. As the bearings are situated eccentrically to the axes of the piers, the piers are stressed by significant bending moments caused by a live load situated at one span only. In addition, the pier caps are stressed by torsional moments. Therefore, the piers and pier caps are relatively heavy.

9.2.1 Structural arrangement

To avoid the well-known problems associated with the construction and maintenance of expansion joints, the girders should form a continuous structure. The girders can be longitudinally connected by continuous composite slabs cast above the piers (see Figure 9.8(b)). This arrangement eliminates the expansion joints; however, the bridge structures still function as a chain of simple beams, and the heavy pier caps remain.

By casting the joints between the girders, by keeping a continuous composite slab and by additional prestressing of the support area by tendons, a continuous structure for the effects of the additional dead and live loads can be created (see Figure 9.8(c)). As the bending moments due to the self-weight of the deck are zero at the support, classical prestressing creates tension at the bottom fibres of the deck at the supports (Figure 9.9). To avoid this tension, the resulting prestressing force has to be uneconomically situated within the core of a section. That is the reason why prestressing is seldom used, and deck continuity is achieved by reinforcing bars situated in the composite slab and by casting the joint between the girders (see Figure 9.8(d)).

To improve the aesthetics of precast bridges, it is possible to combine the function of the piers, caps and diaphragms. In this way, continuous structures supported by slender piers can be created (Figure 9.10). If the geotechnical conditions allow, integral or semi-integral structures with monolithic connections between the piers and the diaphragms can be designed. The simplest and most aesthetically pleasing solution is a design in which the girders are supported by temporary supports during construction (Figures 9.10(a) and 9.11). Then the pier diaphragms are cast in the formwork that is supported by these supports. For example, this approach was used in the construction in 2002 of the motorway viaduct near Olomouc in the Czech Republic. The twin bridge has a total length of 205 m, in seven spans (lengths 24 + (5 × 31) + 24 m). The decks of both bridges are assembled of six lines of precast T-beams, cast-in-place pier diaphragms and a composite deck slab. The decks are supported by pairs of slender columns (Figures 9.12 and 9.13). The continuity of the bridge structure is given by a progressively cast deck slab and pier diaphragms. The girders and composite deck slab are reinforced to achieve the effects determined in the strut-and-tie analyses (described later in this chapter).

To speed up construction and to eliminate the need for temporary supports, transverse beams supported by bearings situated on

Figure 9.8 Supporting precast beam decks

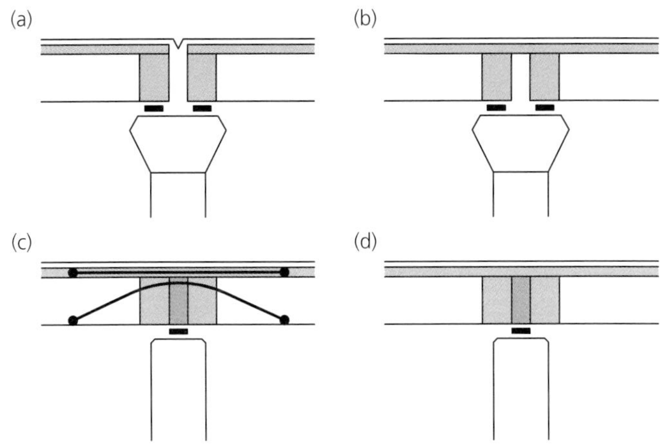

Figure 9.9 Bending moments and stresses: (a) layout of the prestressing tendons, (b) bending moment due to prestressing, (c) bending moment due to self weight

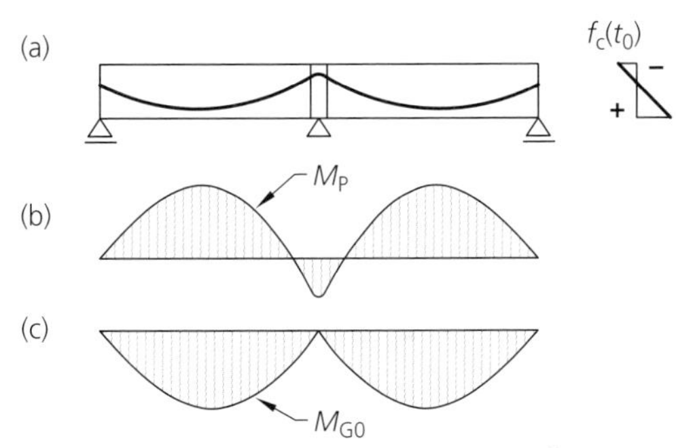

Figure 9.10 Supporting precast beam decks

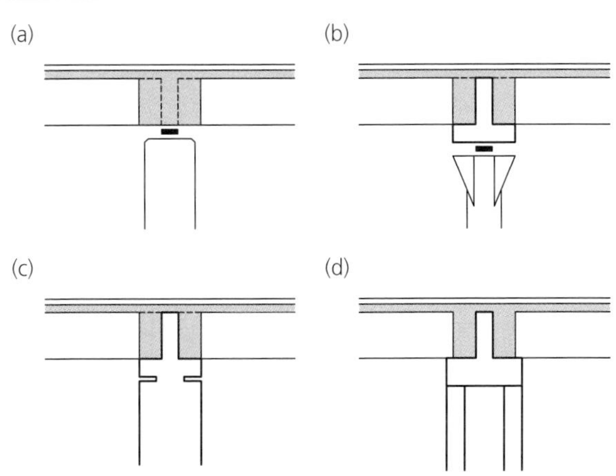

Figure 9.11 Precast beams on temporary supports

Figure 9.12 Motorway bridge near Olomouc, Czech Republic: completed bridge

slender columns are also used (see Figures 9.10(b)–9.10(d)). In this case, the pier diaphragms are erected in two steps. First, a transverse beam with an inverted-T shape is erected. The beam can be cast in situ or erected as a precast member. Second, once the transverse beam has been temporarily fixed to the piers (Figure 9.14(b)), the longitudinal girders are erected, and the remaining portion of the diaphragm and a composite deck slab are cast. When the concrete has gained sufficient strength, the temporary fixing is removed. Bridges having this type of arrangement are widely used in Slovakia.

If semi-integral structures are designed, the transverse beams are connected to the piers by a frame or hinge (Figure 9.10(c)). In the case of a hinge connection, the transverse beams have to be temporarily fixed into the columns. This fixing can be guaranteed by pairs of forces originating at the connection (Figure 9.15). The tension force is resisted by the tension capacity of the reinforcing bars in the joint, and the compression force is resisted by the compression capacity of the plywood boards inserted into the piers. The boards are cut out after casting the composite slab (Figure 9.16). This solution has been used

Figure 9.14 Precast diaphragms supported by bearings: (a) partial elevation; (b) erection stiffening

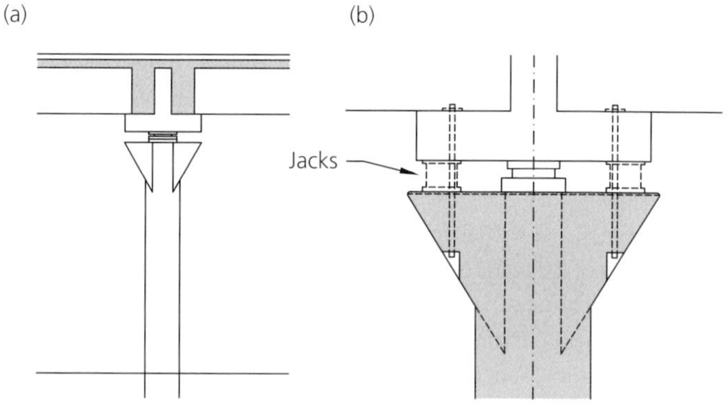

Figure 9.13 Bridge in Figure 9.12: (a) cross-section; (b) partial elevation

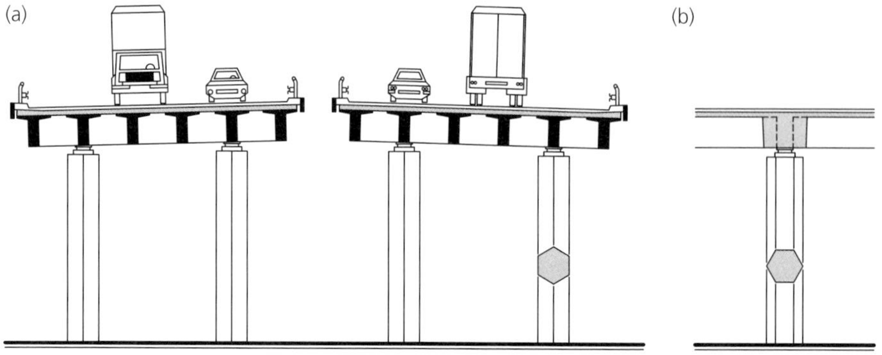

Figure 9.15 Integral pier: (a) bridge cross-section; (b) partial elevation

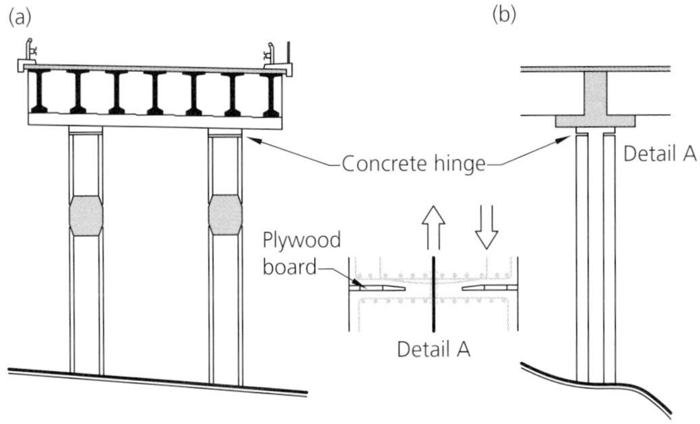

Figure 9.16 Construction of a semi-integral bridge: (a) before erection of the beams; (b) after casting of the composite deck slab

(a)

(b)

successfully in the construction of several semi-integral structures recently built on the D1 and D3 motorways near Žilina, Slovakia.

To extend the length of semi-integral structural systems and to eliminate the need for erection stiffening of the connection between the pier diaphragms and the columns, an alternative structural system has been recently developed (see Figure 9.10(d)). This system was used in the construction of the 960.30 m long viaduct on the D1 motorway near Višňová, Slovakia. The viaduct is formed of two bridges of different widths (14.20 and 15.30 m), which are assembled of eight and nine lines of precast girders. The girders are seated on the transverse beams supported by piers situated in the bridge axis. The viaduct is divided longitudinally into three sections (lengths of 280.15 + 400.00 + 280.15 m) (Figure 9.17).

The piers are formed of two slender walls that are mutually connected by a longitudinal wall member at their bottom parts. The length of the connection was determined by two opposing requirements: the piers have to be sufficiently flexible to allow longitudinal movement of the deck due to changes in the concrete volume, but they must be sufficiently rigid to ensure the integrity and stiffness of the structure. All the walls are hinge connected to the pier diaphragms. The expansion piers are also formed by two walls that support the divided pier diaphragms (Figure 9.18).

The girders were erected in the direction from the end abutments to the middle of the bridge. The arrangement described allows the erection of individual girders without the need for any temporary pier stiffening (Figures 9.19 and 9.20).

To reduce the tension stresses in the composite slab in the sections above the piers, the deck slab was cast span by span between the points where the bending moments due to the dead load are zero. The casting was performed in the direction from the free end to the previously cast section (Figure 9.21).

9.2.2 Static analysis

The static analysis of continuous bridge structures has to consider the construction sequences and take into account the fact that the bridge deck is assembled from members of different ages.

The influence of the redistribution of the internal forces in a continuous structure composed of simple beams is discussed in Chapter 6. The self-weight of the precast beams means that they function as simple beams, but once they have been connected all the other loads mean they act as continuous beams (Figure 9.22). Over time, the monolithic connection of the beams prevents the free rotation of the girders due to concrete creep, and therefore a secondary (additional) bending moment $M_{G0}(t_\infty)$ originates at the supports. The value of this secondary moment depends on the age of the girder and the time of its connection, and on the creep function used in the analysis. In a typical structure, the value of the secondary moment is up to 80% of

Figure 9.17 Bridge on the D1 motorway near Višňová, Slovakia: (a) cross-section; (b) elevation

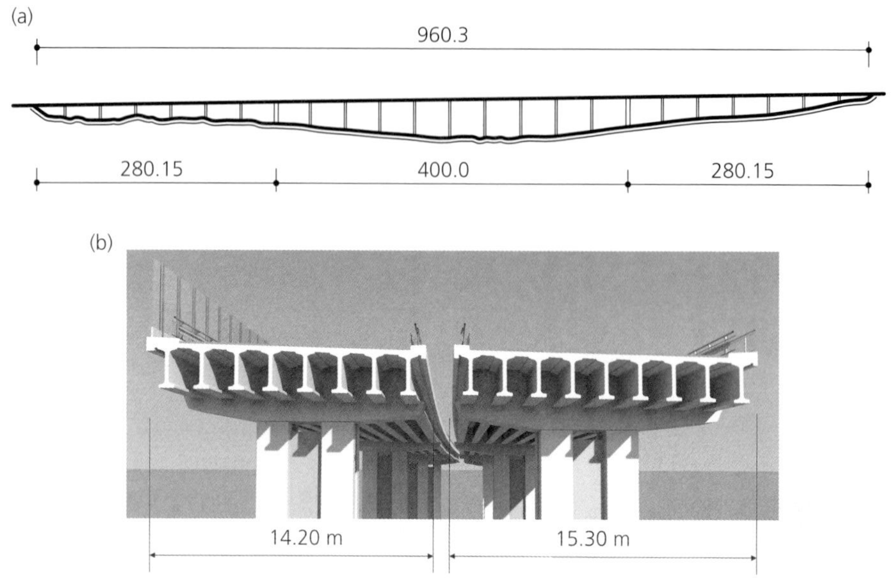

Figure 9.18 Bridge in Figure 9.17: partial elevation

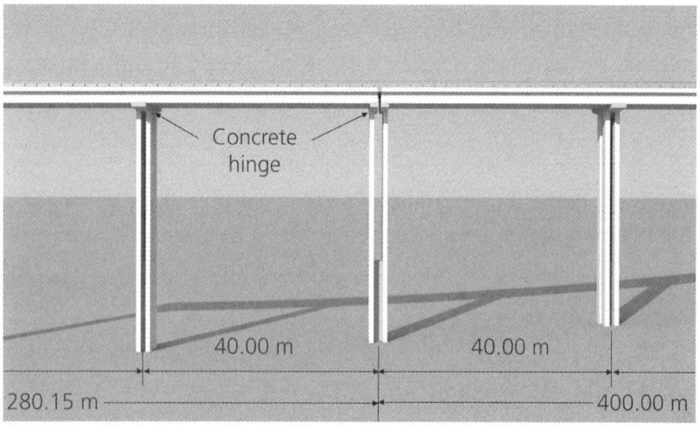

Figure 9.19 Bridge in Figure 9.17: erection of precast beams

the value of the moment that originates at the supports of the continuous structure $M_{G0}(c_b)$. Therefore, over time, tension stresses originate at the top fibres and compression stresses at the bottom fibres at the support.

Figure 9.20 Bridge in Figure 9.17: progressive bridge construction

The function of the prestressed beams is rather more complex. If the radial forces due to the prestressing are smaller than the self-weight of the girders (curve 1 in Figure 9.23), a negative bending moment originates at the support, and if the radial forces due to the prestressing are larger than the self-weight of the girders (curve 3 in Figure 9.23), a positive bending moment originates at support. Corresponding tension stresses have to be resisted by reinforcing bars. If the radial forces have the same value as the self-weight of the girders (curve 2 in Figure 9.23), the structure keeps its shape and there is no redistribution of bending moments – the bending moments and corresponding stresses are zero.

Figure 9.21 Bridge in Figure 9.17: progressive casting of the deck slab

Figure 9.22 Redistribution of the bending moments: pre-cast beams

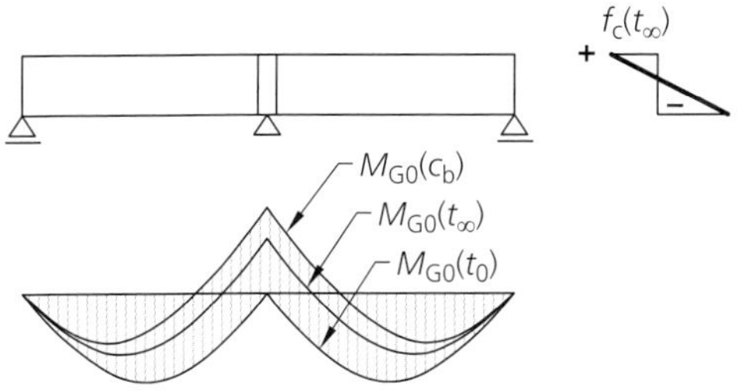

Figure 9.23 Redistribution of the bending moments: prestressed beams

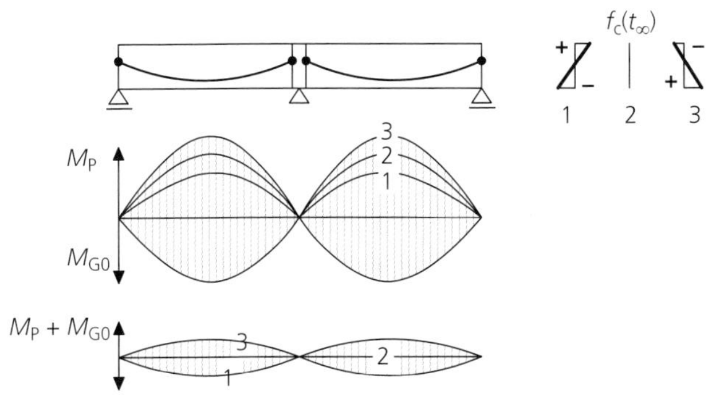

In a composite structure the situation is somewhat more complex due to the shrinkage and subsequent creep of the concrete. Figure 9.24 shows the change in the static function of a two-span bridge structure in which the continuity is secured by a composite deck slab and a pier diaphragm (see Figures 9.11–9.13). The figure also shows the global function of the structure, as well as a strut-and-tie model that describes the function of the structure at the support area.

Figure 9.24(a) shows the bending moments before the erection support is released. Prefabricated girders transfer the self-weight and weight of the composite deck slab to the erection supports that are loaded by the reactions R_G. After releasing the erection supports, the structure is loaded by the reactions R_G. The structure is stressed by the bending moment from the released reactions (Figure 9.24(b)). As these reactions act at the bottom surface of the girders, they have to be transferred by vertical rebars to the composite slab – to the top nodes of the strut-and-tie model. From these nodes the stress is transferred to the bearing by compressed struts that are connected horizontally by a tension tie situated at the composite slab. It is evident that the girders have to be sufficiently reinforced by vertical bars protruding into the slab, and together with horizontal bars situated in the top slab these have to create correctly constructed nodes. Due to the effects of the additional dead and live loads, the bridge deck functions as a typical two-span beam structure (Figure 9.24(c)).

Figure 9.24 shows the static function of a structure in which the bearings are situated directly under the girders. If the girders are indirectly supported (Figure 9.25), the transverse diaphragms have to transfer the load into the bearings. It is necessary to realise that the vertical reactions R_1 act at the bottom fibres of the diaphragm, and have to be transferred by vertical rebars into the top nodes. Then the forces are transferred into the bearings by a strut-and-tie model of the classical arrangement.

As the prefabricated girders prevent free shrinkage of the monolithic concrete of the composite slab, additional bending stresses originate in the bridge deck. In addition to the bending moments originating in the simple beams, continuity moments originate over time (see Figure 9.24(c)). At the supports, corresponding stresses are combined with tension stresses caused by the live load and by temperature changes.

If these stresses are smaller than the tension capacity of the concrete f_{ct}, the structure should be designed for these stresses. However, if these tension stresses are larger than f_{ct}, cracks originate in the slab, and the sections above the supports act as reinforced-concrete members. As a result of the cracking, the stiffness of the support area is reduced, and so the bending moments above the supports are reduced and redistributed into the span sections.

As Eurocode 2 Part 2, which covers the design of concrete structures (CEN, 2005) does not cover this phenomenon, the designer should follow the procedure described in Eurocode 3 Part 2, which covers

Figure 9.24 Static function of a two-span structure

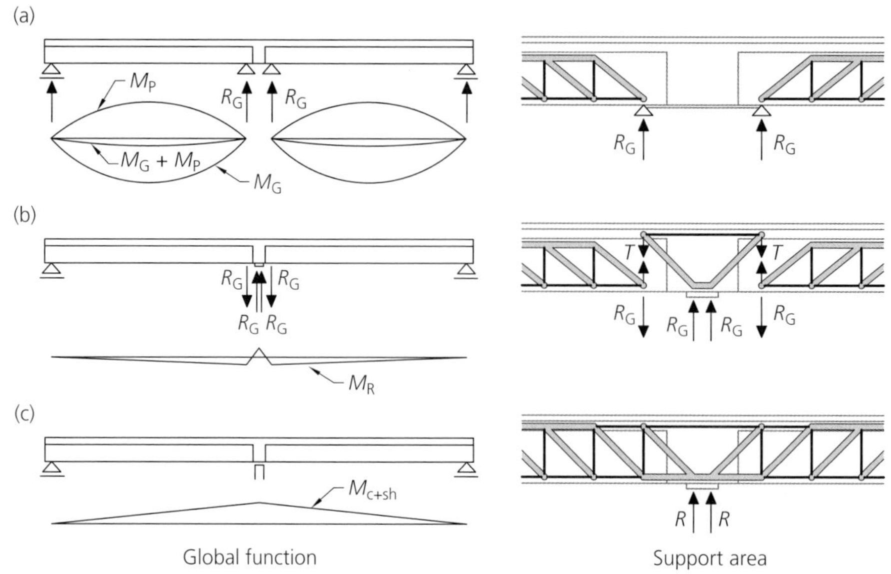

Figure 9.25 Static function of the pier diaphragm

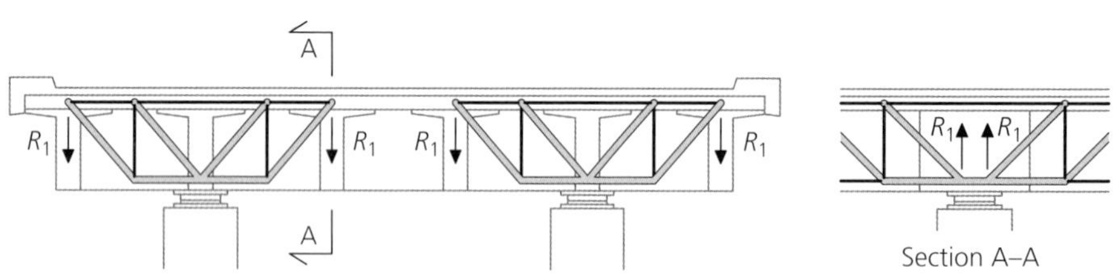

the design of steel structures (CEN, 2006). According to the Eurocode, if the tension in a composite slab is larger than f_{ct} the length of the composite slab should be 15% of the span adjacent to the eliminated intermediate support, and only rebars should be considered in the analysis.

However, detailed time-dependent analyses of typical concrete composite structures have proved that the tension in composite slabs is usually smaller than f_{ct}, and therefore it is not necessary to consider the elimination of the concrete slabs. The older concrete of precast girders also creeps, and therefore the tension stresses in the deck slab are significantly smaller than for steel–concrete composite structures.

Usually the prestressing radial forces are larger than the self-weight of the girder (see curve 3 in Figure 9.23). Therefore, due to creep and shrinkage of the concrete, over time positive bending moments originate at the sections near the diaphragms. Corresponding tension stresses originating at the bottom fibres of the beams have to be resisted by rebars protruding from the beams and overlapping the rebars of the neighbouring girder (Figure 9.26).

Figure 9.26 Reinforcement of the support area

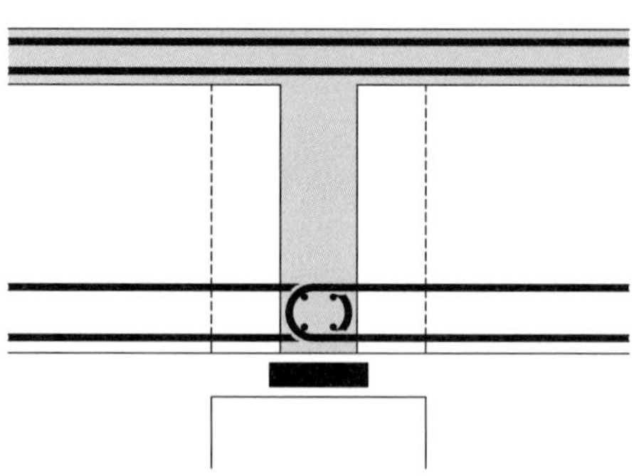

Structures supported by transverse beams (see Figures 9.10(b), 9.10(c) and 9.14–9.16) function similarly. However, the transfer of the forces from the girders into the beam in both the longitudinal and transverse directions has to be checked carefully. During construction, the bottom flange of the beam must transfer the vertical reactions from the girders into the bottom nodes, and from there *via* tension rebars into the top

Figure 9.27 Static function of the pier diaphragm: (a) erection; (b) service

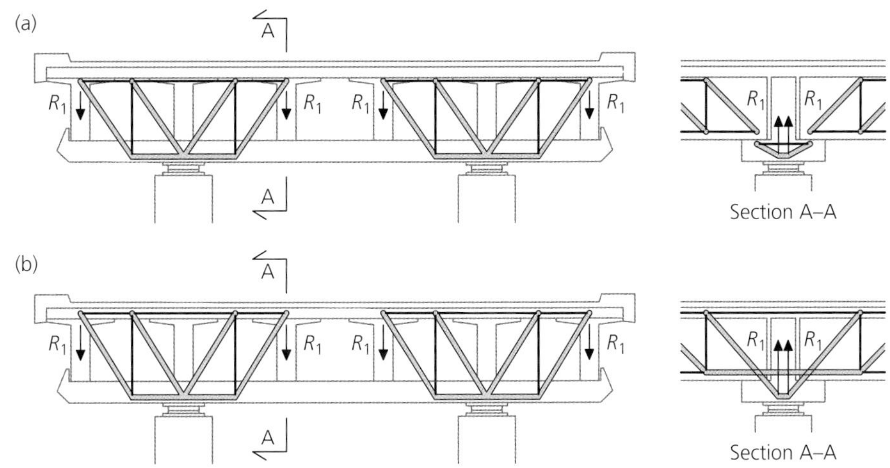

nodes (Figure 9.27(a)). The strut-and-tie model for the additional dead and live loads is shown in Figure 9.27(b). To fully employ the composite slab, the beam's stirrups have to protrude into the composite slab, where they must be correctly anchored to create nodes with the transverse rebars.

9.3. Long-span bridges

In the past, precast beams were also used in structures having longer spans (Troyano, 2003). They were usually inserted between pier tables that were formed of frame structures supported by inclined columns in various arrangements (Figure 9.28(a)). Precast beams also formed central parts of the first cable-stayed structures suspended on A-shaped pylons (Figure 9.28(b)). An interesting solution is arch structures in which the spandrel walls were omitted and precast girders were supported by central parts of the arch structures (Figure 9.28(c)).

Figure 9.28 Long-span structures with a precast beam deck

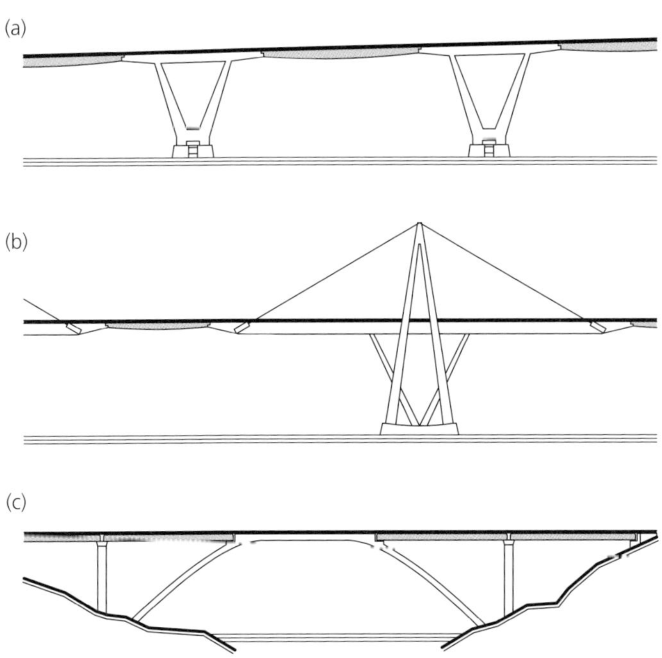

As the girder decks function as simply supported structures, expansion joints are situated between the supporting members and the girders. To avoid problems associated with the performance and maintenance of these joints, other solutions have been developed.

9.3.1 Splice beams – long-span beam structures

In the USA, precast girders have been used for spans up to 100 m (Figure 9.29). They are assembled of precast members (segments) that are spliced together by post-tensioning of continuity tendons. A typical three-span structure is assembled of three pre-tensioned span sections and two pier sections made from reinforced concrete. For lengths of the central span up to 70 m, the pier segments have

Figure 9.29 Typical elevations

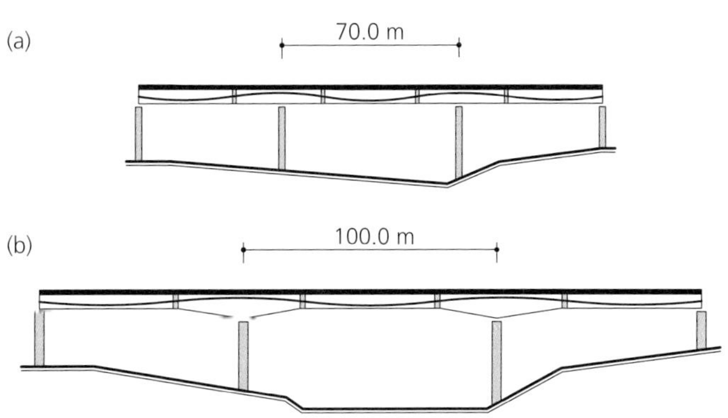

constant depth (Figure 9.29(a)); for longer spans the pier segments are designed with symmetrical haunches (Figure 9.29(b)).

Straight bridges usually have decks assembled of T-girders and a composite deck slab (Figure 9.30(a)), while curved bridges are assembled of U-girders and a composite deck slab (Figure 9.30(b)).

Span segments are pre-tensioned by strands situated in the bottom flanges (Figure 9.31(a)). Ducts for continuity post-tensioning are situated in the webs (Figure 9.31(b)). The bottom flanges of pier segments with haunches have a variable depth. For example, the main span of the Interstate 10 bridge across Escambia Bay in Florida, USA (Rudie *et al.*, 2008), has a main span of length of 76.2 m, and the depth of the span segments is 1.98 m (Figure 9.31(a)). The bridge is stiffened transversely by diaphragms situated at the piers and the splices.

Figure 9.30 Typical cross-sections

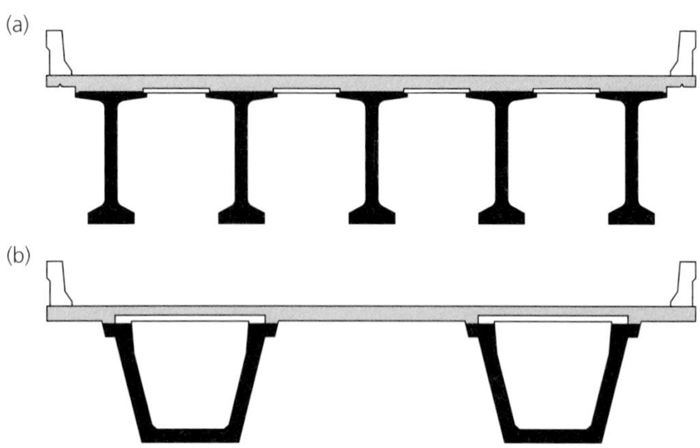

Figure 9.31 Cross-section: (a) at mid-span; (b) above the pier

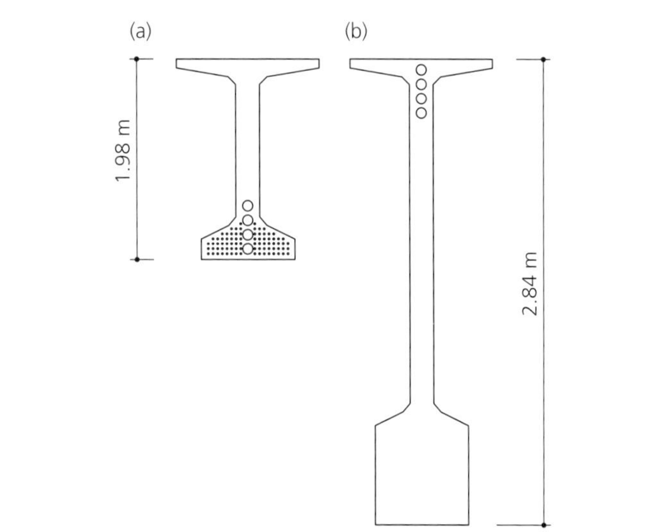

The construction process of this bridge is illustrated in Figure 9.32. After construction of the substructure and the erection of temporary towers in the side spans, the back span (Figure 9.32(a)) and pier segments (Figure 9.32(b)) were erected. Then the first splices and pier diaphragms were cast, and the drop-in span sections were set (Figure 9.32(c)). The central sections were provided with 'strong backs' (Figures 9.33 and 9.34) that allow safe placing of the drop-in span without the need to use temporary towers. Then the second splices and diaphragms were cast and two continuity tendons were post-tensioned (Figure 9.32(d)). When the composite slab had been cast, the remaining two tendons were post-tensioned (Figure 9.32(e)).

Figure 9.32 Construction sequences

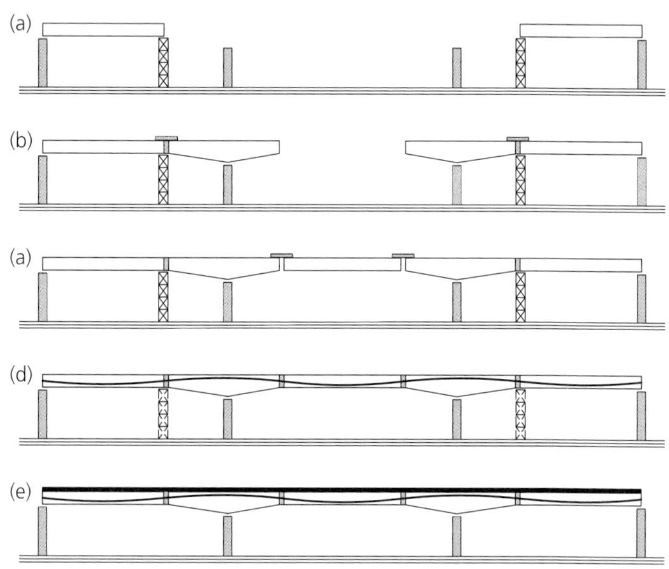

Figure 9.33 Interstate 10 bridge across Escambia Bay, Florida, USA: beam erection

Figure 9.34 Bridge in Figure 9.33: strong backs

Figure 9.35 Bridge in Figure 9.33: erection of the precast beam segments

Figure 9.36 Composite pier slab

Figure 9.37 Double composite beam bridge: elevation

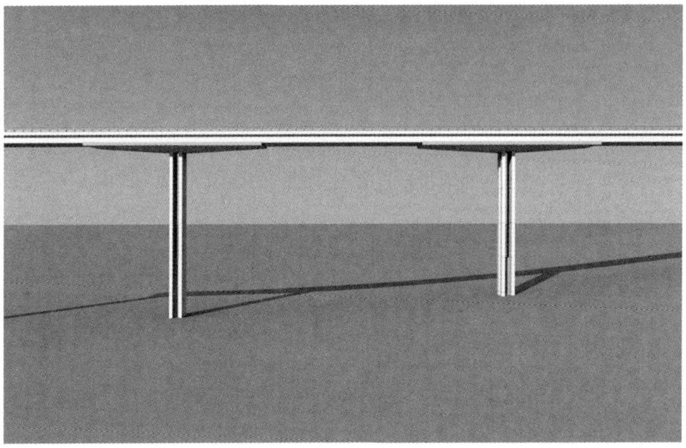

U-girders (see Figure 9.30), which together with a composite slab create a torsionally stiff deck, are mainly used for curved bridges (Stelmack *et al.*, 2016). The U-sections are pre-tensioned by strands situated at the bottom slabs. After the erection of the precast sections (Figure 9.35), the splices are usually post-tensioned by the first group of continuity tendons. When the composite slab has been cast and the second group of continuity tendons have been post-tensioned, the temporary towers are removed.

9.3.2 Long-span beam structures with composite haunches

Recently the authors have developed precast beam structures that utilise the double composite action of the concrete. While the composite deck slab contributes to the resistance of the deck to the positive moment at the span sections, the composite pier haunches contribute to the resistance of the deck to the negative bending moments at the supports. This solution is suitable for multiple-span girder structures that require one longer span. This solution was formulated for the structure used in the bridge at Višňová, Slovakia (see Figure 9.17).

As the precast beams of typical spans are set on the pier caps formed by inverted-T beams, the longer span piers have a similar arrangement. Instead of the pier caps, slabs of a variable depth in both the longitudinal and the transversal directions are used (Figures 9.36 and 9.37). The lines of precast beams are assembled from span and pier segments.

The studied structure has three spans (lengths 58.50 + 75.00 + 58.5 m), and the girders have a depth of 2.00 m. The haunches are 2 × 17.50 m long and their maximum depth is 2.00 m.

The static function of the structure is developed from the process of the construction. First, the piers and support haunches are cast and post-tensioned. Simultaneously, the pier and span sections of the beams are cast and post-tensioned by the first group of tendons (Figure 9.38(a)). While the span beams are supported at their ends, the pier beams are supported at two points situated near their centres.

The pier beams are then erected and the longitudinal joints between their bottom flanges and haunches are reinforced and cast (Figure 9.38(b)). By removing the scaffolding, the pier beams and haunches have become composite structural members. Cantilever moments are resisted by the tension capacity of the prestressed girders and by the compression capacity of the haunches.

Next, the span girders of the side and central spans are progressively erected. During erection of the girders, the stresses in and the stability of the piers is carefully checked (Figure 9.38(c)). As soon as all the girders have been erected, the joints between them are cast and the first part of the continuity tendons is post-tensioned (Figure 9.38(d)). Once the composite deck slab has been cast, the second part of the continuity tendons is post-tensioned.

Figure 9.38 Construction sequence

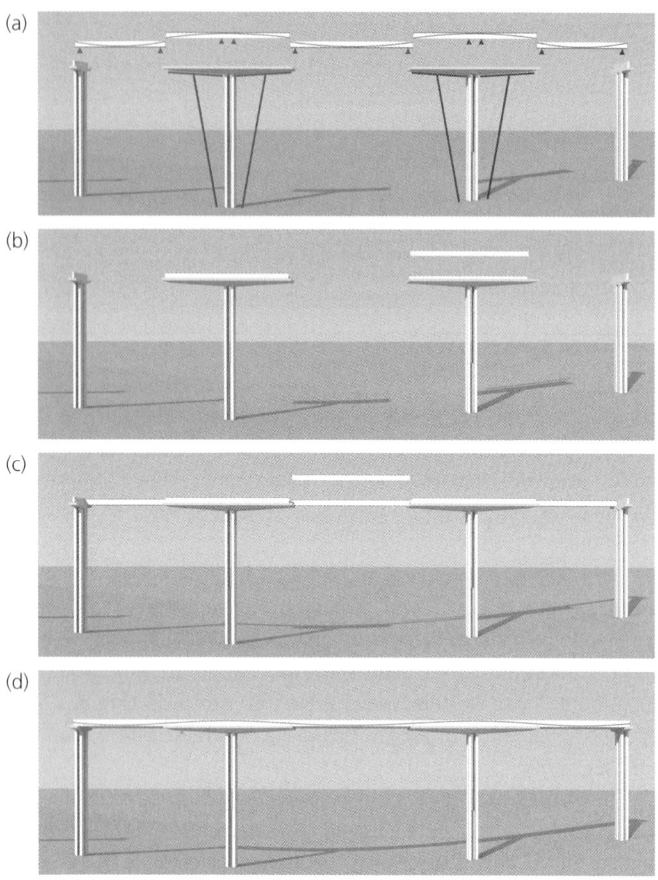

9.4. Precast overpasses

Prefabricated overpasses formed of continuous beams or strutted frames have been used for motorway crossings in several countries. Decks assembled of precast lines of T- or U-beams and a composite slab have been used in many countries. The decks of the bridges constructed in the Czech Republic (Figure 9.39) have been assembled of box girders that were cast with short cantilevers (Figure 9.40). The girders are mutually connected by longitudinal concrete joints reinforced by transverse loops that overlap the transverse reinforcement of the short cantilevers (Figure 9.40(c)) and by transversally prestressed end diaphragms. Detailed static analyses and tests have proved that this structural arrangement can guarantee the integrity of the structures and the good distribution of point loads.

Precast beams with a height of 1.20 and 1.60 m and widths of 2.00–2.40 m have been used for spans up to 45.00 m (Figure 9.41(a)). The decks are assembled of precast lines composed of segments having a maximum length of 14.0 m and weight of 40.0 t. The girders are supported by front and rear precast struts. The joints between the

Figure 9.39 Typical overpass

Figure 9.40 Cross-section: (a) bridge; (b) precast girder; (c) longitudinal joint

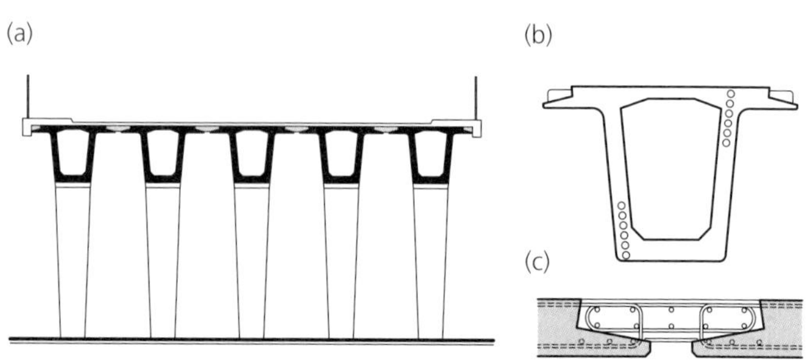

girder segments and between the girders and the struts are composed of concrete. During construction, the precast struts are prestressed to the girders. As the rear struts are loaded by the soil above them, they are fully prestressed to avoid cracking.

Bridges of this type have been used for skew crossings with skews of up to 50° (Figure 9.41(b)). The girders together with the struts form precast lines that are mutually staggered according to the angle of crossing. The transversely prestressed end diaphragms, which also support the approach slabs, guarantee the lateral strength of the bridge.

Figure 9.41 Typical overpass: (a) elevation; (b) plan

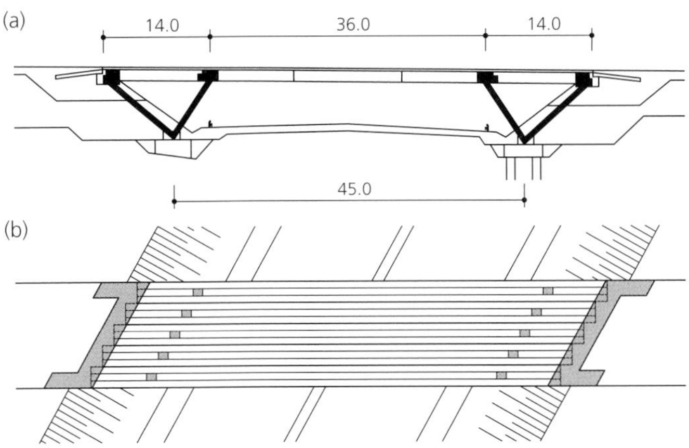

Figure 9.42 Precast members on temporary towers

During construction, the precast members are placed on temporary towers, which also support the precast struts (Figure 9.42). Then the cables are pulled through the ducts. After careful adjustment of the geometry, the joints between the girder segments and the girders and the struts are cast. When the concrete of the joints has achieved sufficient strength, the longitudinal girder cables and short cables connecting the struts to the girders are post-tensioned. The blocks for the top footings are cast. The erection is completed by the casting of the longitudinal joints between the girder lines and by the casting of the end diaphragms. Once the diaphragm tendons have been post-tensioned, the temporary towers are removed.

The same precast members have also been used for the construction of continuous beam structures.

9.5. Precast arch structures

Precast beams have been also widely used in a range of structural systems formed of continuous girders, frames and arches. On many occasions, the precast beams have been combined with cast-in-place structures for the spans or part of the spans built across obstacles where access is difficult. The wide variety of possible precast structures is demonstrated here using two examples of arch structures built in the USA.

9.5.1 Wisconsin Avenue Viaduct, Milwaukee, Wisconsin, USA

This bridge was designed to replace the existing old concrete arch structure. The structural solution addresses two criteria: (*a*) the public acknowledged the need to replace the existing structure, but expressed the desire to retain a nostalgic design, appropriate to its surroundings; and (*b*) the state of Wisconsin advocated a simple beam structure with a replaceable deck in order to adhere to its rigid maintenance policies and budget. The design of the new bridge fulfils both requirements by creating a simple and economical structure (Stráský *et al.*, 2001; Wanders *et al.*, 1994).

The viaduct has a total length of 443.80 m and is composed of 11 spans (lengths 22.86 + (8 × 48.16) + 12.80 + 22.86 m) (Figure 9.43). The typical 48.16 m long span is formed by two parallel, precast

Figure 9.43 Wisconsin Avenue Viaduct, Milwaukee, Wisconsin, USA: elevation

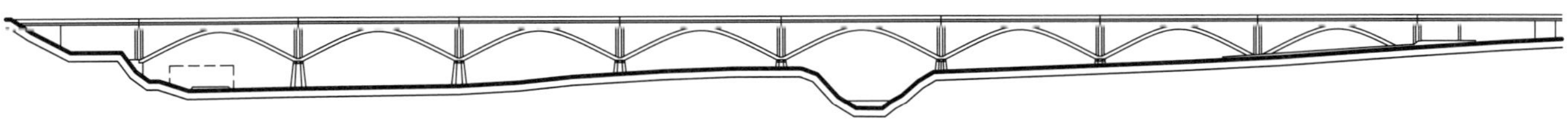

arches that support the deck, which is formed of precast beams and a composite slab (Figure 9.44). Each arch is assembled from two precast segments connected at the crown with a post-tensioned concrete joint. The precast segments have a trough cross-section, and were concentrically post-tensioned for handling before erection. Once erected, the troughs were filled with concrete, and combined with the precast arch segment to form the completed arch ribs. The precast beams are supported by cast-in-place diaphragms situated above the spandrel walls and at the crown at the mid-span.

The bridge was designed to allow progressive erection from one abutment to the other, without the need to temporarily strengthen the piers for excessive unbalanced lateral thrust (Figures 9.45 and 9.46). The sequence of construction generally followed distinct, overlapping stages that were designed to keep the unbalanced thrust within acceptable limits. Except for the temporary use of falsework support while erecting the arch elements at the crown, no further use of falsework was needed because the arch ribs served as the platform for the construction of the deck. Replacement of the slab deck is foreseen after about 30 years of service.

Figure 9.44 Bridge in Figure 9.43: cross-section at a quarter of a typical span

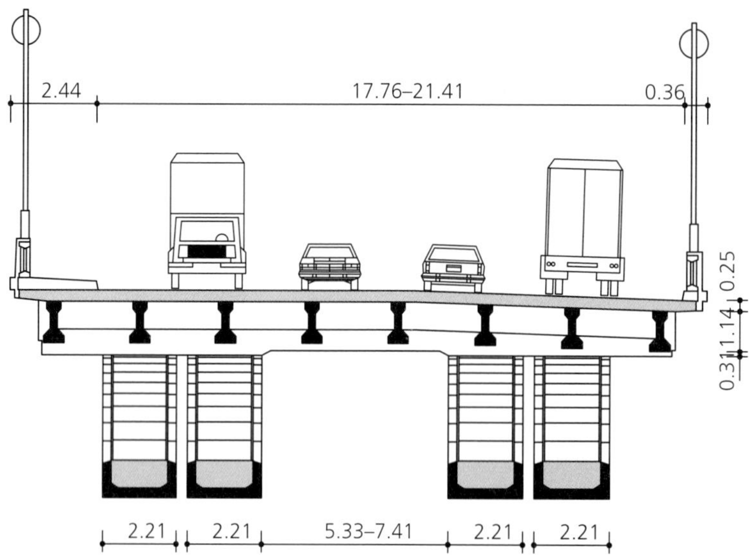

Figure 9.45 Bridge in Figure 9.43: progressive erection

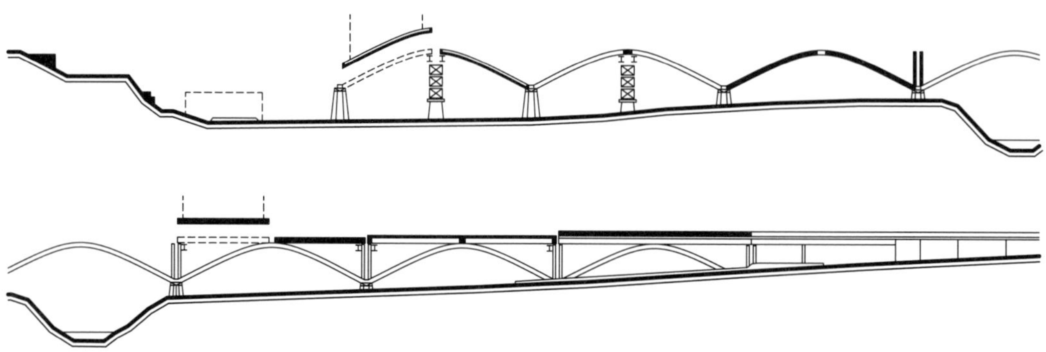

Figure 9.46 Bridge in Figure 9.43: progressive erection

Figure 9.47 Bridge in Figure 9.43: casting of the arch fill

Figure 9.48 Bridge in Figure 9.43: erection of the precast beams

The detailed construction sequence was as follows:

(*a*) construction of the substructure, casting and post-tensioning of the arch segments
(*b*) erection of the arch segments
(*c*) casting of the joints between the arch segments at the mid-span, casting and prestressing of the beams
(*d*) post-tensioning of the short continuity crown tendons, removal of the temporary tower at the mid-span
(*e*) casting of the arch fill (Figure 9.47) and the pier
(*f*) erection of the beams (Figure 9.48)
(*g*) casting of the diaphragms and the composite slab
(*h*) removal of the temporary beam support at the mid-span and above the piers
(*i*) casting of the sidewalks and railings.

During construction the structures utilised different static systems – boundary conditions changed, new structural members were assembled and/or cast, post-tensioning was applied, and temporary support elements were removed. Therefore, a detailed time-dependent analysis was mandatory. The time-dependent analysis, in which the arch and the deck were modelled as a series of parallel elements, was performed for 70-time modes.

The conceptual design was developed by the first author, and the final design is work was done by Charles Redfield (Mill Valley, California) and CH2M Hill (Milwaukee, Wisconsin).

9.5.2 Rich Street Bridge over the Scioto River, Columbus, Ohio, USA

This bridge is part of a larger reconstruction of the Scioto riverfront (Shanks *et al.*, 2013). The arch bridge has a total length of 171.30 m and is composed of five spans (lengths 22.94 + 41.40 + 42.62 + 41.40 + 22.94 m) (Figure 9.49). The bridge structure consists of four individual post-tensioned lines of precast arch ribs and deck

Figure 9.49 Rich Street Bridge over the Scioto River, Columbus, Ohio, USA

Figure 9.50 Bridge in Figure 9.49: elevation

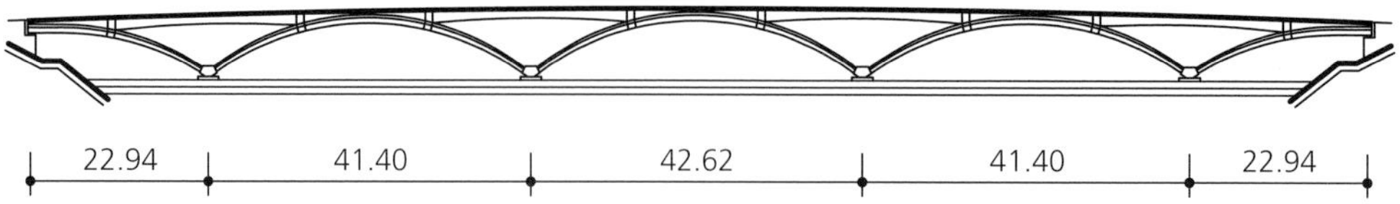

beams made of sand–lightweight concrete. The arch bridge is formed of a self-anchored arch structure in which the horizontal component of the arch force is resisted by a prestressed deck (Figure 9.50).

At the abutments, the superstructure is seated on laminated elastomeric bearings with a sliding surface that permits free longitudinal movement. Each pier consists of a reinforced-concrete cap supported on four 1.68 m diameter drilled shafts. The superstructure is fully integral with the pier cap at a single fixed pier, and is seated on laminated elastomeric bearings at the expansion piers.

The arch ribs are made of three precast segments, two identical arch legs and a keystone (crown) segment (Figures 9.51 and 9.52). At the piers, the arch rib segments are linked by cast-in-place arch blocks. The riding surface of the bridge is formed by a 254 mm deck slab with a 38 mm thick overlay made of concrete modified with microsilica. The deck slab is composite with both the arch crown segments and the deck beams.

The arch and deck segments were connected with short closure pours and by longitudinal post-tensioning. Four 19-strand tendons were placed at each arch rib, passing through a 1 m radius bend at each pier location. Each set of rib tendons passes through one arch block, with anchorages located at the abutments and in the cast-in-place closure beyond the arch crest in each span. Full-length longitudinal tendons are used in the beam segments and in the cast-in-place deck slab. These tendons are tensioned from both ends of the bridge.

To reduce the weight of the segments to shipping weight of 100 t, the interior arch crest segments have trough cross-sections that were later filled with concrete at the construction site. The geometry of the arch ribs is defined by two circular arcs, one forming the top of the rib and the other the bottom of the rib. The resulting shape is a rib of variable thickness, the thickest section being at the crest and the thinnest at the base. The edges of the ribs are formed by two intersecting surfaces oriented at angles of 30° and 45° from the top surface of the rib.

The precast segments were erected onsite on falsework towers, and tied together by cast-in-place closure pours. Girder closures were

Figure 9.51 Bridge in Figure 9.49: arch segments on temporary supports

Figure 9.52 Bridge in Figure 9.49: precast members

provided at the interface of the arch and girder segments. At the piers, the arch rib segments were linked by a cast-in-place arch block. Sleeves were embedded in the closure and arch block to allow continuous post-tensioning between the segments. Before construction of the deck, lateral support of the ribs and girders was provided by the falsework towers.

A large number of utilities are carried on the bridge. They are carried between the arch ribs just below the deck, and are concealed by grey fibreglass grating panels.

Frederick Gottemoeller (Bridgescape, Columbia, Maryland) developed the overall concept for the bridge and the architectural details, and the bridge was designed by engineers at Burgess & Niple (B&N, Columbus, Ohio) with Leonhardt, Andrä und Partner (Stuttgart, Germany).

REFERENCES

CEN (European Committee for Standardization) (2005) EN 1992-2:2005. Eurocode 2 – Design of concrete structures. Part 2: Concrete bridges: design and detailing rules. CEN, Brussels, Belgium.

CEN (2006) EN 1993-2:2006. Eurocode 3 – Design of steel structures. Part 2: Steel bridges: design and detailing rules. CEN, Brussels, Belgium.

fib (International Federation for Structural Concrete) (2001) *Precast Bridges*. State-of-the-Art Report, Task Group 6.3.

Rudie C, Poulson J, Ryzhikov V and Molas T (2008) Innovative solutions for rapid construction. *ASPIRE* Summer: 22–24.

Shanks JC, Butz TM and Gottemoeller F (2013) Unifying past and present. *Civil Engineering* **83(4)**: 60–64.

Stelmack TW, Davidson TE and Forars K (2016) SR 417 and Boggy Greek Road interchange. Florida's first curved precast concrete U-girder. *ASPIRE* Winter: 14–17.

Stráský J, Navrátil J and Suský S (2001) Applications of time-dependent analysis of in the design of hybrid bridge structures. *PCI Journal* **46(4)**: 56–74.

Troyano LF (2003) *Bridge Engineering. A Global Perspective*. Thomas Telford, London, UK.

Wanders SP, Maday MA, Redfield C and Stráský J (1994) Wisconsin Avenue Viaduct – design-construction highlights. *PCI Journal* **39(5)**: 20–34.

West C (2019) Girders achieve record length. *ASPIRE* Fall: 56–57.

ICE Publishing

Strášký and Nečas
ISBN 978-0-7277-6385-3
https://doi.org/10.1680/dcpb.63853.183
ICE Publishing: All rights reserved

Chapter 10
Span-by-span construction

Span-by-span construction is usually used for long bridges with spans of up to 60 m (Dyckerhoff and Widman, 1973; Troyano, 2003; Wittfoht, 1972). Bridges with deck widths up to 15 m and spans up to 45 m usually have a solid or double-T cross-section, while for longer spans they usually have a box section (Figure 10.1). The decks of wider bridges are usually formed of a one-cell box with large overhangs supported by ribs or precast struts. Multi-cell decks are commonly used on the west coast of the USA. Channel sections are mainly used in the construction of railway bridges. Span-by-span construction is commonly used for structures formed of continuous girders that are supported by vertical or inclined piers and/or arches.

10.1. Span-by-span construction of cast-in-place structures

Only relatively short bridges (up to about 100 m) are concreted at one time on stationary scaffolding that runs across the entire length of the structure. These structures are usually prestressed by continuous tendons that are post-tensioned from both ends. In the case of longer bridges, the structures are usually cast incrementally, span-by-span, with short cantilevers protruding into the next span. The joints between the cast parts are usually situated at points where the bending moments due to the self-weight are zero, that is, approximately one-fifth of the span length.

10.1.1 Static analysis

The static analysis has to consider the stage casting of the deck and the corresponding changes in the static system. To illustration the problem, Figures 10.2 and 10.3 show the bending moments originating in a deck that was cast incrementally (span-by-span) on stationary scaffolding, and was incrementally prestressed by continuous tendons all coupled at the construction joints. The procedure is as follows.

- *Stage 1*. The static effects due to the self-weight and prestressing are determined on a structure formed by a simple beam with an overhang.
- *Stage 2*. The static effects due to the self-weight and prestressing of the second cast part are determined on a structure formed by a two-span continuous girder with an overhang. The resultant static effects are equal to the sum of the effects determined in the first and second stages.
- *Stage 3*. The static effects due to the self-weight and prestressing of the third cast part are determined on a structure formed by a three-span continuous girder with an overhang. The resultant static effects are equal to the sum of the effects determined in the second and third stages.

Figure 10.1 Typical cross-sections: (a) solid; (b) double-T; (c) box; (d) one cell box with large overhangs; (e) multi-cell; (f) channel

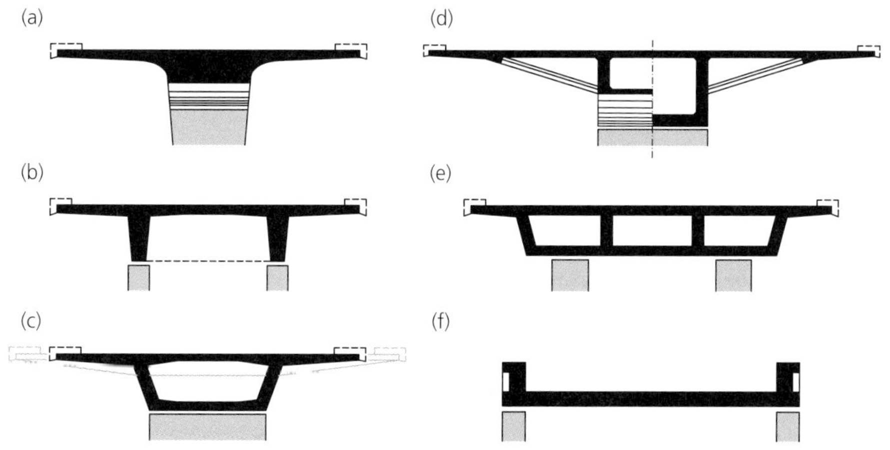

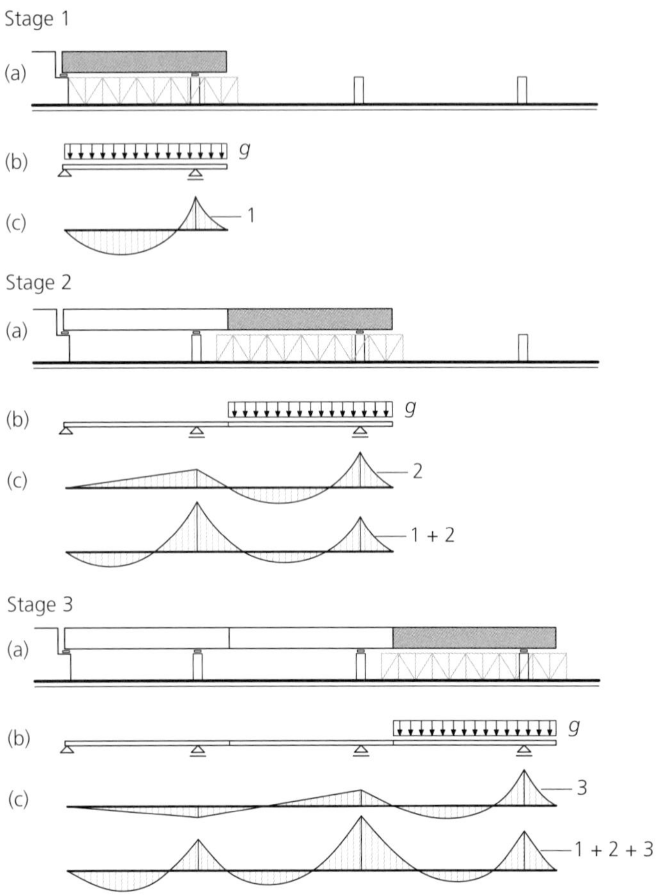

Figure 10.2 Bending moments due to the self-weight: (a) scheme of the structure; (b) loading and static system; (c) bending moments

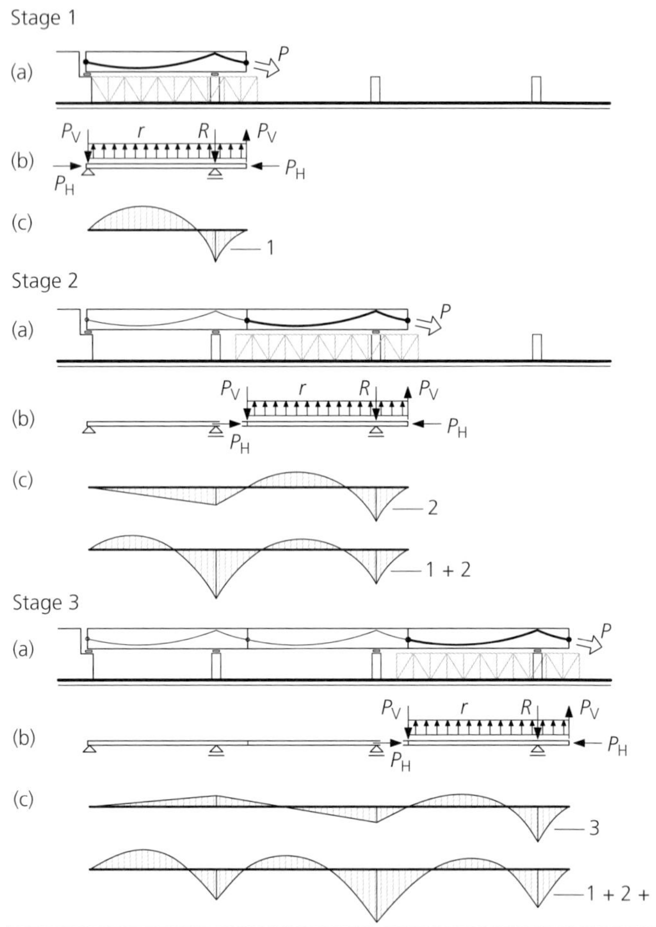

Figure 10.3 Bending moments due to prestressing: (a) scheme of the structure; (b) loading and static system; (c) stage and resulting bending moments

This procedure is repeated for all the subsequent spans.

Figure 10.4 shows the tension forces in prestressing tendons connected by compact couplers. To ensure that the couplers do not move, it is necessary to guarantee that the forces in the ends of the tensioned tendons are smaller than the anchor forces in the previous tendon.

Due to the creep of concrete, the structure has a tendency to function the same as a similar structure that is cast in one construction stage. To get an idea about this redistribution, Figure 10.5 shows a bridge deck that was cast in five stages on stationary scaffolding. The bridge deck of the double-T cross-section has five spans (lengths 30 + (3 × 42) + 30 m). Figure 10.5(b) shows the courses of bending moments at each stage, and Figure 10.5(c) shows the courses of the bending moments when the deck had been completed (stage 5), the bending moments in the structure cast at the same time, and the bending moments after 100 years.

If the bridge deck is progressively cast in a movable scaffolding, it is necessary to consider that the rear scaffolding support loads the

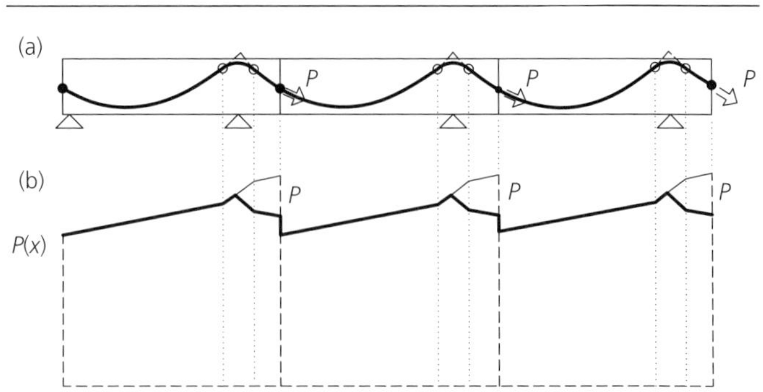

Figure 10.4 Tension forces in a coupled tendon: (a) scheme of the structure; (b) tension forces

overhangs of the previously cast span (Figure 10.6). As this load gradually increases during the casting, the bending moment in the previously cast bridge part is changing. Figure 10.7 shows the

Figure 10.5 Bending moments: (a) scheme of the structure; (b) moments during stage casting; (c) resulting moments

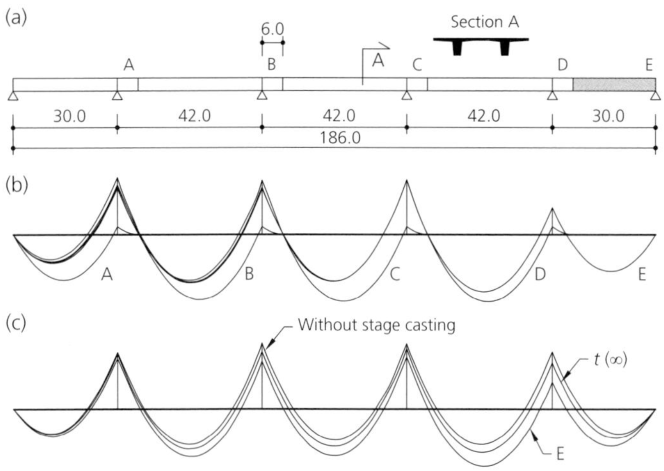

Figure 10.6 Movable scaffolding

Figure 10.7 Scheme of the structure and bending moments before and after casting of: (a) the second span; (b) the third span; (c) the fourth span

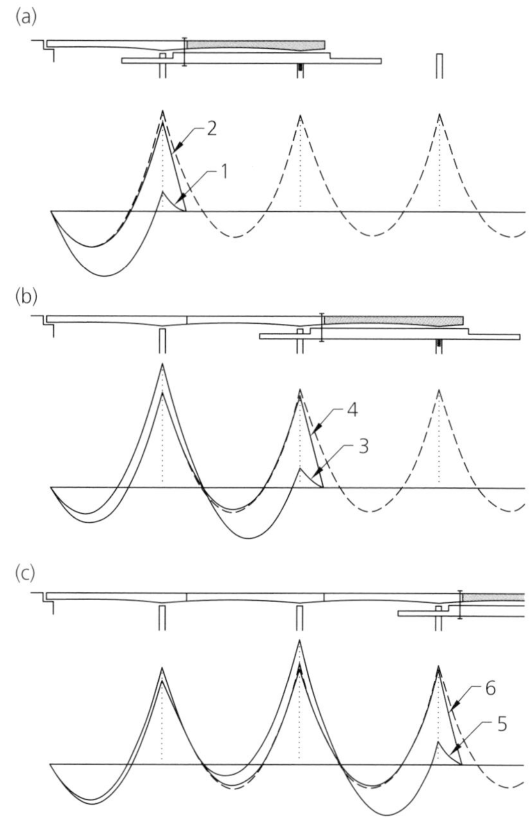

courses of the bending moments originating in the deck of the motorway viaduct at Knínice in the Czech Republic (see Figures 4.44, 10.14 and 10.15) for each stage of the erection before and after the erected span is cast. The course of the bending moments of the completed structure is shown by the dashed line.

Although the joints between the progressively cast parts are situated near the points where the moments due the self-weight are zero, the resulting course of the moments after construction can be different from the course of the bending moment of a structure that was cast all at the same time. The reason for this is the staged construction and the loading of the overhangs by the movable scaffolding. Therefore, a detailed time-dependent analysis of the structure is required. The analysis must also take into the account the stage tensioning of the tendons. The deformation of the structure is also significantly influenced by heavy loading of the young concrete of the overhangs.

10.1.2 Construction joints

In the first applications of span-by-span construction, all tendons were run in parallel and were anchored close to each other (Figure 10.8(a)). Although the cables have been anchored at points where the moments due the self-weight are zero, both vertical and horizontal cracks have been discovered at joints after several years.

The explanation for this is clear from Figures 10.9 and 10.10. Figure 10.9(a) shows the principal stress trajectories in a wall supported at one edge and loaded by a point load acting at the wall centre. The point load P represents an equivalent load created by the prestressing tendon at its anchor. The figure shows a non-uniform stress distribution, and the fact that large tensile stresses originate behind the anchor. Figure 10.9(b) shows the normal stresses $\sigma_{k,p}$ in the section in front of the anchor, and the normal stresses $\sigma_{k,z}$ beyond the anchor. The tension force (the integral of the tension stresses) has a value as large as 0.3 of force P.

Figure 10.8 Layout of the prestressing tendons

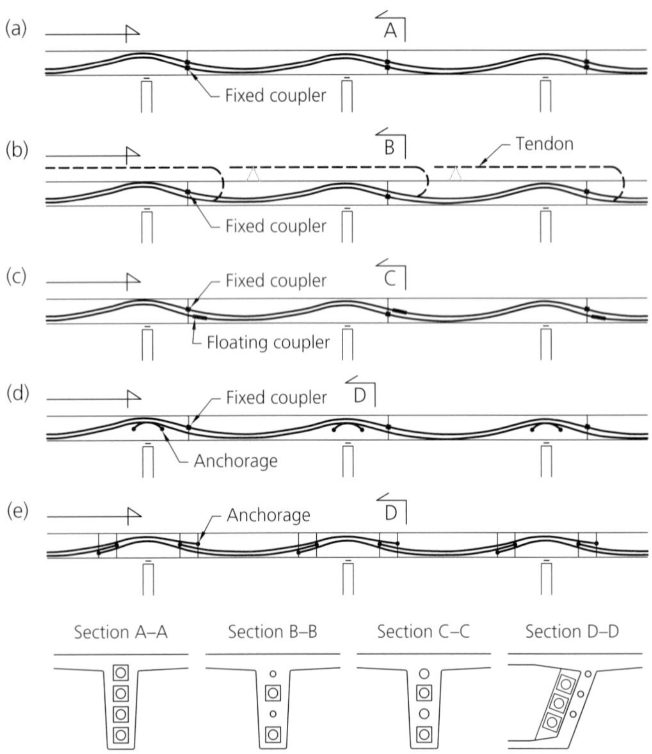

Figure 10.9 Stresses in the anchor area

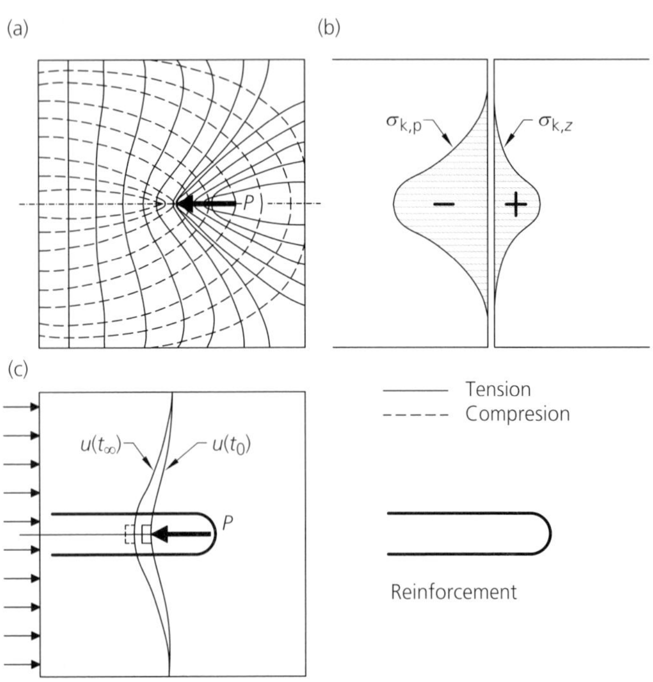

Figure 10.10 Stresses in the incrementally prestressed beam

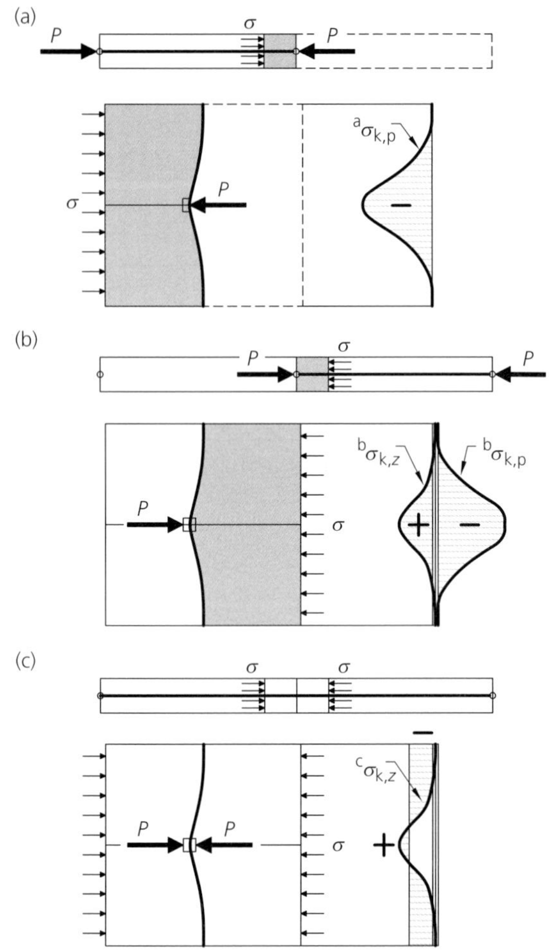

Figure 10.9(c) shows the deformation of the wall in the section in which the anchor is located. Due to the stress concentration under the anchor, the plastic deformation due to the concrete creep increases the total deformation over time. In order to avoid tearing off the wall part behind the anchor, the wall must be properly reinforced by reinforcing bars anchored in the part of the wall in front of the anchor.

Figure 10.10 shows the deformations and stresses originating in the beam of height b and thickness d during its two-stage construction. For each stage the top drawing shows the whole structure, and the drawing below shows the cut-out of the structure close the anchor of the tendon, which functions as a wall.

First, the left part of the beam is cast and centrically prestressed by a force P (Figure 10.10(a)). An uneven distribution of the normal stresses $^a\sigma_{k,p}$ originates in the section in front of the anchor. At a distance corresponding to the height b, the beam is stressed by a uniform compression stress $\sigma = P/bd$.

Then the prestressing tendon is coupled, and the right part of the beam is cast. After the concrete has achieved sufficient strength, the beam is prestressed by force *P* (Figure 10.10(b)). The action of the cable can be expressed by the force *P* acting at the coupler and in the anchor. An uneven distribution of the compression stresses ${}^{b}\sigma_{k,p}$ originates in the section to the right of the coupler, and an uneven distribution of the tension stresses ${}^{b}\sigma_{k,z}$ originates in the section to the left of the coupler.

The resulting stresses in the structure (Figure 10.10(c)) is given by the sum of the stresses determined in the two stages. If the structure is analysed as a beam, the part of the structure between the two cut-outs is subjected to uniform pressure, but, in reality, the resulting stresses are given by the sum of the stresses determined in the stage analyses of the two cut-outs modelled as walls. The resulting stresses are ${}^{c}\sigma = {}^{a}\sigma + {}^{b}\sigma$. Obviously, a concentration of compression stresses originates to the right of the coupler, and tension stresses ${}^{c}\sigma_{k,z}$ originate to the left of the coupler.

This explains why cracks have formed in the joints. It is clear that beam analysis cannot capture the complex stress distribution in the area of the cable couplers. Therefore, these areas must be carefully analysed and/or the design principles given in Eurocode 2 Part 2 (CEN, 2005) must be used. According to these principles, only half of the cables can be anchored at the joint, and the anchors must be evenly distributed over the cross-section. The solution presented in Figure 10.8(a) can be used only if a detailed analysis of the joint area modelled as a three-dimensional structure assembled of shell or solid elements proves that, for a characteristic loading, there is a minimum compression stress of 3 MPa in the joint.

10.1.3 Layout of prestressing tendons

Bridges in which the decks are formed by spine girders with large overhangs or by double-T girders are usually prestressed by internal tendons. There are two options for the arrangement of the tendons.

In the first option, half of the tendons are connected by compact couplers at the construction joints, and the other half, the lengths of which correspond to the length of two spans, are bent and placed above the cast span (see Figure 10.8(b)). When the next span is cast, the previously bent tendons are coupled, and the previously coupled tendons are bent.

This arrangement has been used in many structures. The 840 m long motorway viaduct near to Hladke Životice in the Czech Republic, which was built in 2009, can serve as an example (see Chapter 4, Figures 4.40 and 4.41). The bridge deck is formed by a continuous beam of a double-T cross-section. The girder depth varies from 2.60 m at the supports to 1.70 m at the midspans.

The deck was progressively cast span-by-span in a formwork supported by underslung launching scaffolding (Figure 10.11). The construction joint was situated at a distance of 8.16 m from the piers. During lunching, the scaffolding was supported on the temporary towers, and during the casting it was supported by temporary towers situated on the footings of the front pier of the cast section and suspended on the cantilever of the previously cast section.

The deck is made from C35/45 concrete and is prestressed longitudinally by draped cables (19 strands of 0.6 in. diameter) and two straight cables (15 strands of 0.6 in. diameter) (Figure 10.12). Only half of the tendons were tensioned and coupled at the construction joint (Figure 10.13). The second half were temporarily bent and placed on the already cast section.

In the second option, half of the tendons are connected by compact (fixed) couplers at the construction joints, and the other half are connected by floating couplers at a distance longer than the deck depth (see Figure 10.8(c)). When the next span is cast, the tendons previously connected by compact couplers are coupled by floating couplers, and the tendons previously connected by floating couplers are coupled by compact couplers.

This arrangement has been used in many structures cast using an underslung and overhead movable scaffolding system. An example is the viaduct on the D8 motorway at Knínice, Czech Republic, close to the border between the Czech Republic and Germany (see Chapter 4, Figures 4.44 and 4.45). The deck of the viaduct, which has a typical span of 42 m, is formed by a spine girder with large overhangs, giving a total width of 14.80 m. The spine girder has a solid cross-section of variable depth, from 2.6 m at the supports to 1.4 m at the midspan. The deck, which was designed as a partially prestressed structure, is post-tensioned in both the longitudinal and transverse directions of the bridge.

Figure 10.11 Motorway bridge near Hladke Životice, Czech Republic: construction

Figure 10.12 Bridge in Figure 10.11: movable scaffolding and tendon layout

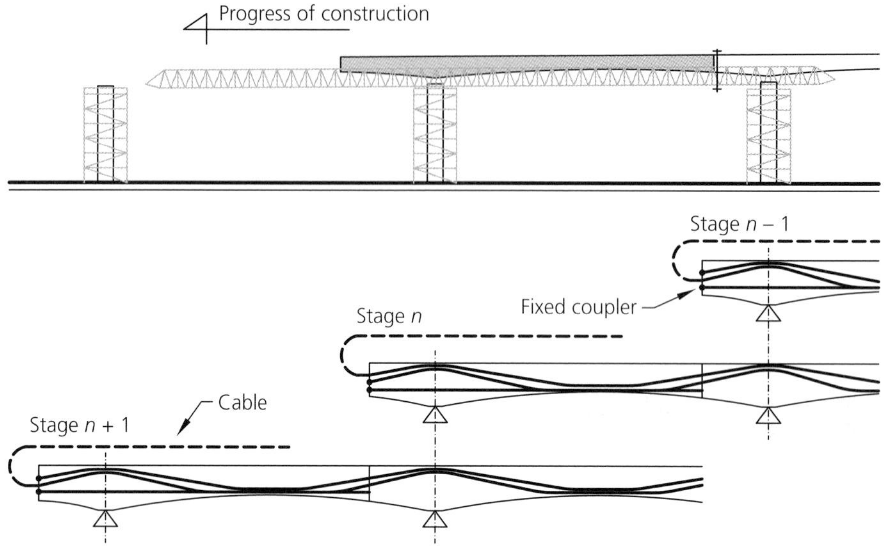

Figure 10.13 Bridge in Figure 10.11: bent tendons

The joints between the incrementally cast spans (see Figures 10.6 and 10.7) are situated at a distance of 8.5 m from the piers. The deck is prestressed by prestressing tendons progressively coupled in compact and floating couplers (Figure 10.14). In the joints, only -half of the longitudinal tendons are post-tensioned and coupled. The other half of the tendons are coupled in the floating couplers (Figure 10.15) situated at a distance of 2.60 m from the construction joint.

The same arrangement can be used for the construction of a deck formed by box girders. However, these decks can be also prestressed by tendons laid out in another arrangement, where half of the tendons are anchored and connected by compact couplers at the construction joints and the other half overlap above the piers (Figure 10.8(d)). If internal tendons are used, the tendons are anchored at the strengthened part of the webs. If external cables are used, the tendons are usually anchored at anchor blocks situated beyond the upper part of the pier diaphragms. This solution is commonly used in the construction of progressively erected decks. The overlapping tendons are usually tensioned after the movable scaffolding has been moved to the next span.

Figure 10.14 Motorway viaduct at Knínice, Czech Republic: layout of the tendons

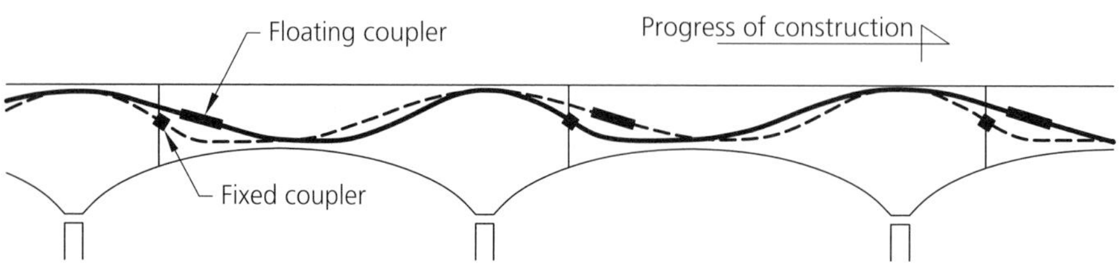

Figure 10.15 Bridge in Figure 10.14: floating couplers

The first solution described above is illustrated by the motorway viaduct near Bitarová in Slovakia (see Figures 4.46 and 4.47). The deck is formed of a typical single-cell box girder that was incrementally cast, span-by-span, in a formwork supported by movable scaffolding (Figure 10.16). The joints between the casting sections are at a distance of 11.70 m from the support.

The arrangement of the deck and the layout of the prestressing tendons were determined by the technology of the construction. The contractor decided to cast each whole span as one section, without interruption. Therefore, the internal movable formwork did not allow for pier diaphragms. The box girder webs are only locally widened above the piers.

Figure 10.16 Motorway viaduct near Bitarová, Slovakia: movable scaffolding

Figure 10.17 Bridge in Figure 10.16 – layout of the prestressing tendons: (a) elevation; (b) cross-section at the midspan; (c) cross-section above the piers

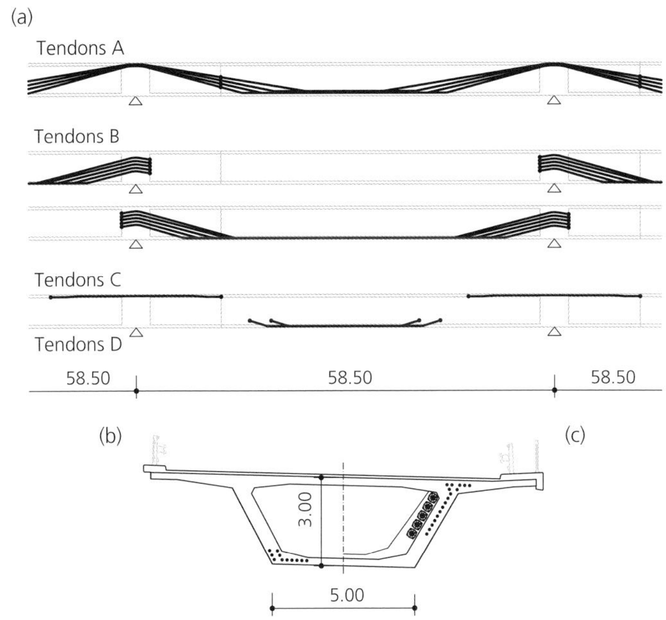

Half of the prestressing tendons (A), which are situated in the girders' webs, are anchored and coupled at the construction joints (Figures 10.17(a) and 10.18), and the other half (B) mutually overlap above the piers and are anchored on the opposite sides of the widened webs (Figure 10.19). The web tendons are supplemented by short tendons (C and D) situated at the top slabs above the piers and at the bottom slabs at the spans.

Figure 10.18 Bridge in Figure 10.16: external and internal formwork

Figure 10.19 Bridge in Figure 10.16: overlapping of the tendons above the piers

A combination of internal tendons and external cables was used in the construction of the approach spans of the cable-stayed motorway bridge across the River Odra and Antošovice Lake, near Ostrava in the Czech Republic (see Chapter 4, Figures 4.83–4.85). The typical span length of the approach viaducts bridging the Antošovice Lake is 39.00 m (Figure 10.20). The viaducts are formed by two-cell box girders (depth 2.20 m, width 14.80 m) without traditional overhangs (Figure 10.21).

The contractor decided to cast the deck span-by-span in formworks suspended on the overhead movable scaffolding (Figure 10.22(a)). During the construction, the movable scaffolding was shored up by pier segments that were supported by temporarily fixed bearings.

Figure 10.20 Bridge across the River Odra and Antošovice Lake, near Ostrava, Czech Republic: approach spans

Figure 10.21 Bridge in Figure 10.20: cross-section

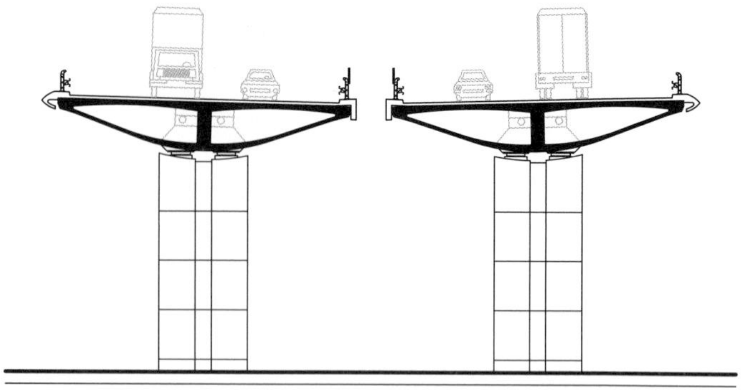

During the construction of the common spans, the internal tendons situated in the web (Figures 10.22(b) and 10.23) and the external cables (Figures 10.22(c) and 10.24) situated on both sides of the central web were post-tensioned. The draped internal tendons were supplemented by straight tendons situated at the edges of the deck. The draped tendons were coupled by compact couplers situated at each joint. Only half of the straight tendons were tensioned and coupled at the pier joints; the other half were connected by floating couplers (Figure 10.22(b)).

The external cables overlap at the pier segments. Each external cable is situated in two spans, and were post-tensioned from the face of the erected structure. This means that only half of the cables are anchored at the pier segments. The cables are bent in the pier diaphragms and span deviators.

Figure 10.22 Bridge in Figure 10.20: movable scaffolding and tendon layout

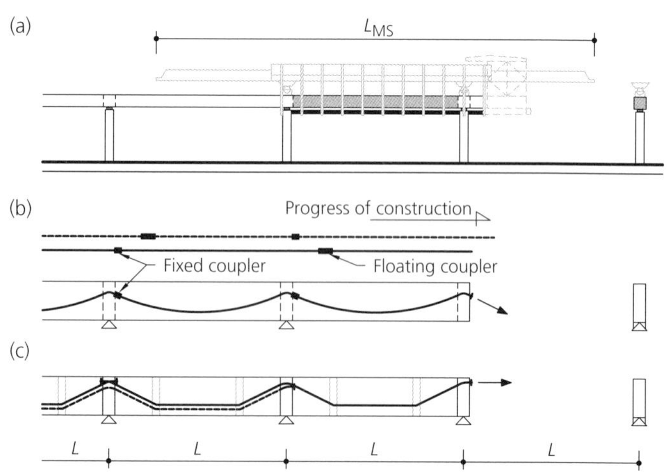

Figure 10.23 Bridge in Figure 10.20: internal tendons

Figure 10.24 Bridge in Figure 10.20: external cables

A combination of internal tendons and external cables has also been successfully used in the construction of several bridges with progressively erected decks. The advantage of this system is demonstrated by the design of the viaduct across the Hostovsky Creek Valley built on the Expressway R1 near Nitra, Slovakia (see Chapter 4, Figures 4.60 and 4.61). The 25.66 m wide bridge deck was progressively erected in both the longitudinal and transverse directions. First, the basic box girder was cast in formwork suspended on an overhead launching gantry, and then

Figure 10.25 Motorway viaduct across Hostovsky Creek Valley, near Nitra, Slovakia: progressive construction

the precast slab struts were erected and the overhangs were cast (Figures 5.90 and 10.25).

The deck is longitudinally prestressed by internal bonded tendons situated within the basic cross-section, and by external non-bonded tendons situated inside the central box (Figure 10.26). The six bonded tendons are situated in the webs that have a constant width of 500 mm, and are coupled in each construction joint. In the 69 m long spans these tendons are supplemented by 2 × 2 tendons situated in the strengthened part of the bottom slab. Above the supports an additional 2 × 6 straight tendons are situated at the top slab. External prestressing is provided by 2 × 4 external cables. The tendons are continuous across three spans, and are anchored at the pier diaphragms. They are deviated at the pier and span deviators.

If the use of couplers is prohibited, the tendons have to be anchored at each span at two web blisters. Usually, half of the tendons are anchored at the blisters situated at the construction joints (see Figure 10.8(e)). The other half of the tendons are anchored at blisters situated symmetrically to the joint blisters. When the span is cast, the tendons anchored at the joint blisters are post-tensioned. The other half of the tendons are post-tensioned when the scaffolding is moved to the next span. As the tension originates beyond the anchors and the webs are transversely bent by eccentrically anchored tendons, the stresses at the blister areas have to be checked carefully.

For example, this arrangement was used in design of the western approach viaducts to the Kayak Bridge, which is part of the new bypass system under construction around Bratislava, Slovakia (see Chapter 5, Figures 5.83 and 5.84). The viaduct has a total length of 785 m and is formed of a continuous girder of 12 spans (lengths 53.0–67.5 m) (Figure 10.27). The bridge deck is formed of a

Figure 10.26 Bridge in Figure 10.25 – arrangement of the prestressing tendons: (a) longitudinal section; (b) cross-section; (c) continuous bonded tendons, bonded tendons at supports, external unbonded tendons

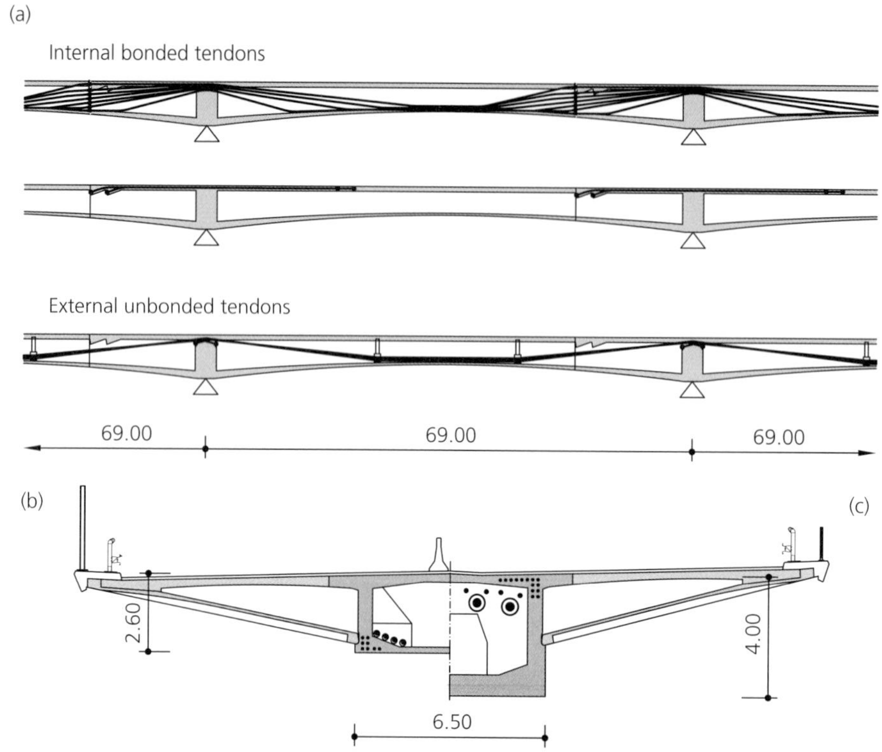

Figure 10.27 Western approach viaducts to the Kayak Bridge, Bratislava, Slovakia: progressive construction

one-cell box girder with large overhangs supported by precast struts (Figure 10.28). The bridge deck, which is 35.00 m wide and 4.30 m deep, was progressively erected in both the longitudinal and transverse directions. First, the basic box girder was cast in the formwork suspended on an overhead launching gantry, and then the precast bar struts were erected, and the overhangs were cast. The deck is prestressed by two groups of two tendons (see Figure 10.28) that overlap in the web blisters, where they are also anchored (Figure 10.29). The tendons anchored at the joint blisters are post-tensioned before the scaffolding is moved to the next span, and the tendons anchored at the span blisters are post-tensioned once the top slab has been cast.

10.1.4 Deck above the intermediate supports

When designing a continuous prestressed concrete bridge, it is necessary to carefully check the stresses in all the sections of the progressively erected deck. Critical tension stresses originate not only at the bottom fibres of the span section and at the top fibres of the support section, but also in the bottom fibres of the support sections. These stresses are crucial, mainly in deep decks composed of double-T sections (Leonhardt, 1979).

As the position of the centroid axis of these sections is close to the top fibres of the deck, the bottom fibres are stressed by very large compression stresses due to the service load. To eliminate tension stresses at the top fibres and to reduce compression at the bottom fibres, the pier sections are prestressed by tendons situated as close as possible to the top fibres.

Figure 10.28 Bridge in Figure 10.27 – arrangement of the prestressing tendons (PT): (a) elevation; (b) cross-section; (c) elevation; (d) cross-section

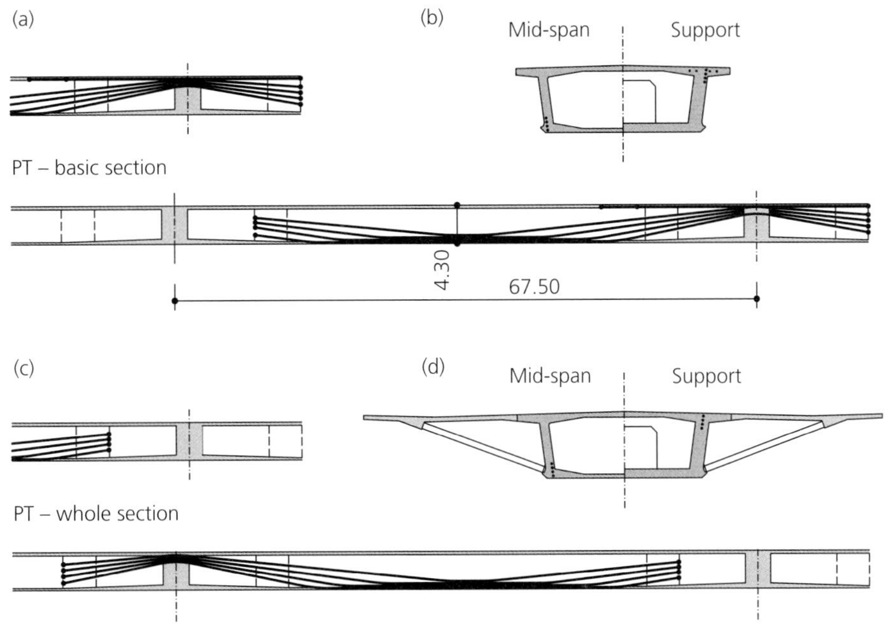

Figure 10.29 Bridge in Figure 10.27: anchoring of the tendons at the joint blisters

To guarantee that there is no tension at the bottom fibres at the midspan section, the prestressing force in the continuous tendons is relatively large and creates tension stresses at the bottom fibres of the support sections. These stresses increase due to the effects of temperature changes (Figure 10.30).

If the design is done according to Eurocode 2 Part 2 (CEN, 2005), the variation in the prestress expressed by favourable and unfavourable values of the coefficients r_{sup} and r_{inf} should be considered. The value of these coefficients is given in the UK National Annex to the

Figure 10.30 Double-T deck above the support

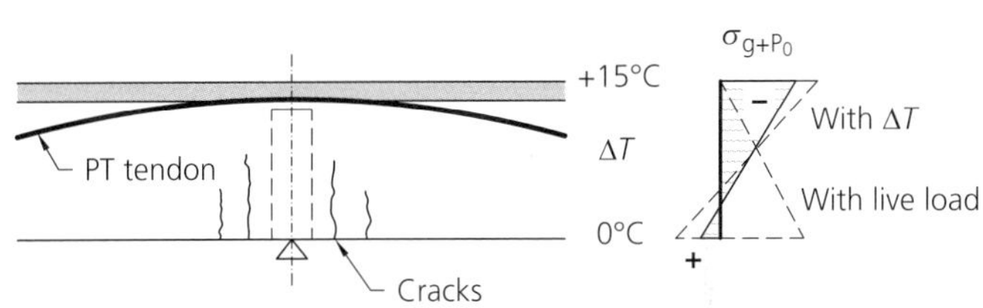

Eurocode. The annex sets both coefficients $r_{sup} = r_{inf} = 1.0$; however, other countries use the values $r_{sup} = 1.1$ and $r_{inf} = 0.9$. As it will be seen from the next example, these values significantly influence the structural arrangement of bridges.

The viaduct across Strazov Pond and the reservoir Hricov of the River Vah near Žilina, Slovakia (see Chapter 4, Figures 4.68 and 4.69) has approach spans formed by a double-T girder with a constant depth of 3.0 m. The double-T girders are indirectly supported by pairs of pot bearings situated on slender piers that have an elliptical cross-section (Figure 10.31).

While the main spans were segmentally cast in symmetrical cantilevers starting from the pier tables, the remaining spans were progressively cast, span-by-span, in formworks supported by stationary or movable scaffolding. The major portion of the approach viaducts

Figure 10.31 Viaduct across Strazov Pond and the Hricov reservoir, near Žilina, Slovakia: cross-section

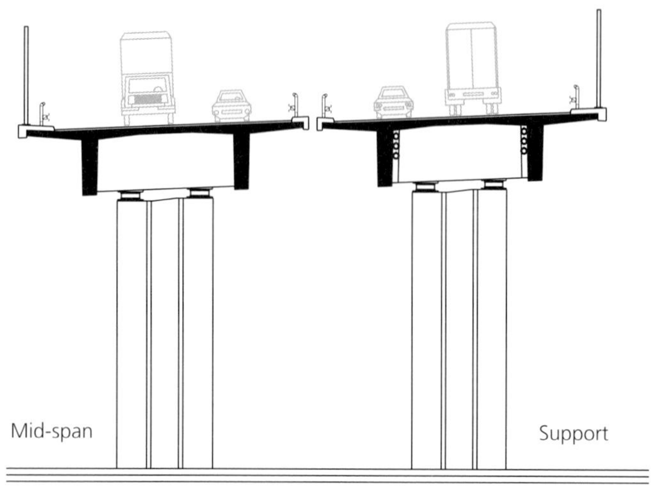

with typical span lengths of 42 m were erected using two types of overhead movable scaffolding, and this required advance construction of the pier tables (Figure 10.32).

The relatively deep section allows the design of very slender girder webs. However, slender webs mean it is difficult to situate compact and/or floating couplers. Therefore, the deck is prestressed by longitudinal prestressing tendons that are not coupled, but are led over two spans, overlap in the short parts of the girder close to the piers, and are anchored in the construction joints and in the anchor blocks of the pier cross-beams (Figure 10.33(a)).

The layout of the prestressing tendons corresponds to the common arrangement. At the midspan, the tendons are close to the bottom

Figure 10.32 Bridge in Figure 10.31: progressive construction

Figure 10.33 Bridge in Figure 10.31 – stresses in the deck: (a) arrangement of the prestressing tendons; (b) bending moments; (c) stresses in the deck

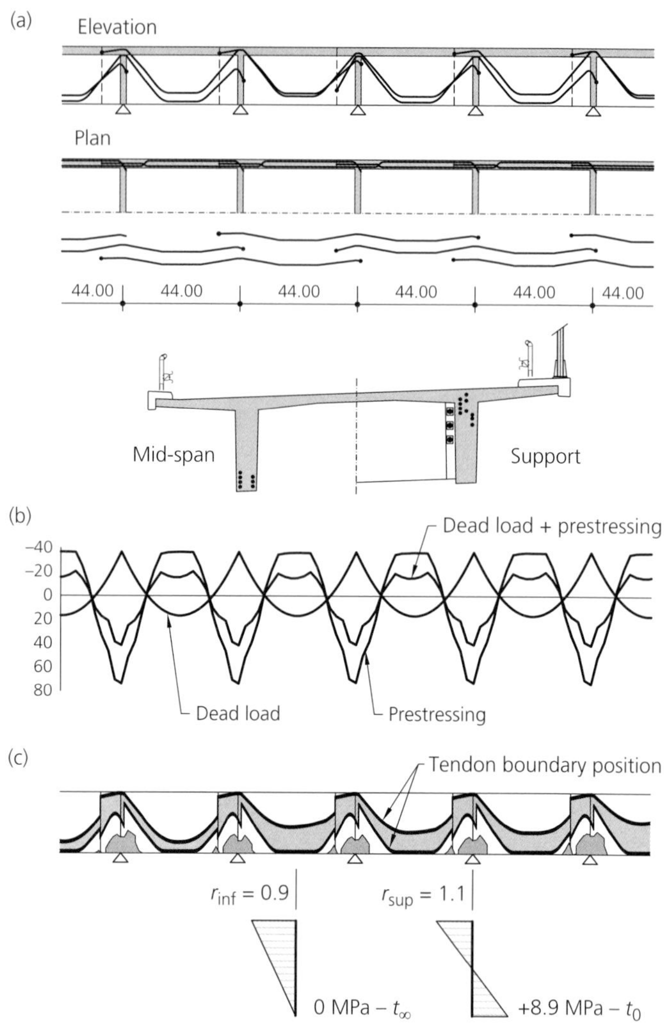

fibres, and above the piers they are close to the top fibres. For frequent loading, Eurocode 2 Part 2 (CEN, 2005) does not allow tension stresses at a distance of 100 mm from the tendons. Therefore, at the midspan section, no tension stresses are allowed at the bottom fibres of the deck. The same is true for the top fibres of the pier sections. Due to these constraints, the deck is strongly prestressed by tendons that create tension stresses at the bottom fibres at the pier sections. These tension stresses, the maximum values of which originate during prestressing, have to be resisted by reinforcing bars protruded from the pier tables (Figure 10.34).

As the Slovak National Annex requires the use of favourable and unfavourable values of the coefficients, r_{sup} = 1.1 and r_{inf} = 0.9, respectively, the reinforcement is very heavy – 4 × 4 = 16 ø B32 are

Figure 10.34 Bridge in Figure 10.31: reinforcement of the bottom fibres

placed at each girder. The detailed position of the tendons was also influenced by the requirement for there to be no tension stresses at a distance of 100 mm. It was checked that all sections of the viaduct met the Eurocode requirements. Figure 10.33(c) shows the boundary positions of the tendons and the locations where the tension stresses originate.

10.2. Incrementally launched structures

As described in Chapter 5, incrementally launched structures are cast segment-by-segment in forms situated behind the abutments, and then incrementally launched into the design position (Figure 10.35).

10.2.1 Static analysis

During launching, the structure is subjected to bending moments of both signs. The maximum bending stresses occur at the front

Figure 10.35 Bridge over Rybny Creek, Czech Republic

Figure 10.36 Bending moments: (a) originating in the deck at different construction steps; (b) envelope of bending moments

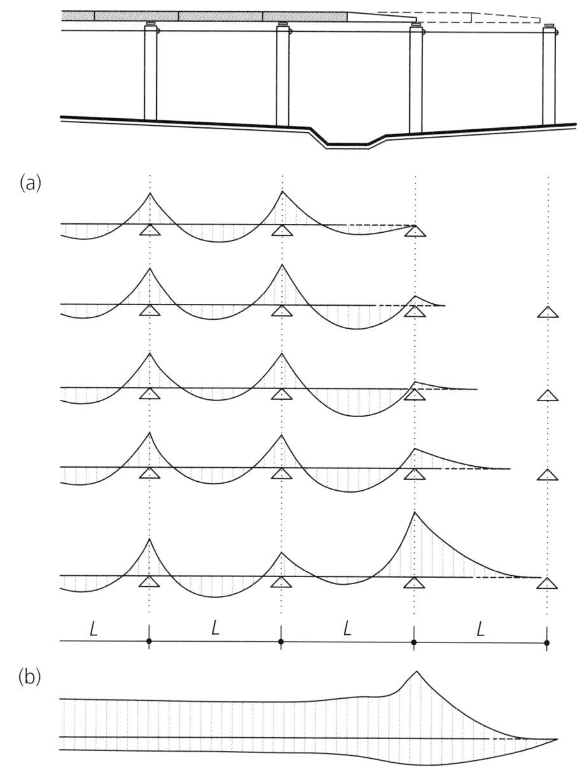

cantilever. Between the internal supports, the bridge deck of self-weight g is subjected to bending moments, the values of which correspond to the bending moments originating in the fixed beam: $-{}^{1}/_{12}\, gl^2$ or $+{}^{1}/_{24}\, gl^2$. As, during launching, these moments originate at each section, the structure is centrically prestressed. Figure 10.36 shows the bending moments originating in a structure launched with a steel nose. Figure 10.36(a) shows the bending moments originating in the deck during several construction steps, and Figure 10.36(b) shows the envelope of the bending moments.

As the bridge deck is cast in the design geometry, and deck is seated on bearings in the design geometry, and the final state of the stresses corresponds to the state of the stresses in a structure cast at the same time on a stationary scaffolding. Therefore, the redistribution of the stresses due to concrete creep is caused only by the different ages of the incrementally cast segments.

10.2.2 Layout of prestressing tendons

During construction, the stresses in the launched structure have to be checked at each erection stage. As each section of the launched structure is stressed by bending moments of both signs, it is necessary to prestress the deck centrically.

Figure 10.37 Layout of prestressing (PT): (a) internal bonded tendons; (b) external unbonded cables; (c) external unbonded cables with temporary cables

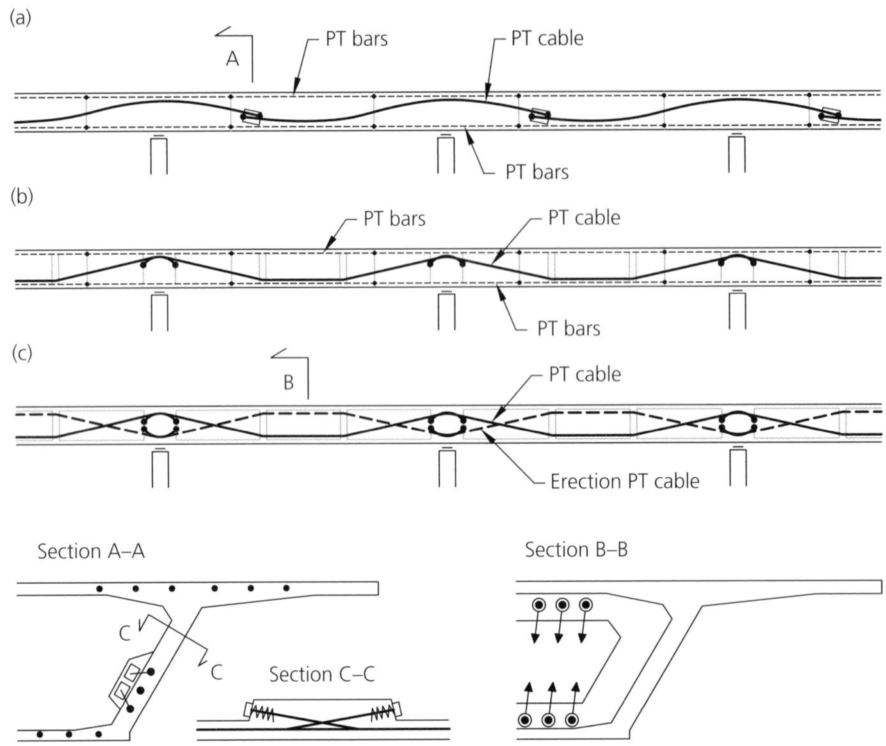

Usually, the centric prestressing is created by prestressing bars situated in the top and bottom slabs that are coupled at the construction joints. When the deck is launched into the final position, the prestressing is complemented by continuous tendons situated at the girder webs. The tendons usually overlap at the blisters where the tendons are anchored (Figure 10.37(a)).

As part of the development of the technique of external prestressing, another arrangement of prestressing cables has been developed in France. The launched deck is prestressed by two cable systems: final external cables, the layout of which corresponds to the course of the bending moments, and temporary external cables that have the opposite layout (Figure 10.37(c)). During construction, both systems are erected and tensioned. Then, when the launching is complete, the temporary cables are removed and used in another structure.

It is evident that the prestressing bars can be substituted by tendons, and continuity internal tendons that are anchored in the web blisters can be substituted by external cables that overlap at the pier diaphragms, where they are also anchored (Figure 10.37(b)). This method was used in the construction of the approach bridge to the viaduct across the Nitra River, which was built on the R1 expressway, near Nitra in Slovakia (see Chapter 4, Figure 4.70 and 4.71). The twin deck of the 813 m long approach bridge is composed of 20 spans (lengths 28.0–45.0 m).

The deck of this bridge was incrementally cast in segments of lengths ranging from 12.36 to 22.50 m. During launching (Figure 10.38), the deck was centrically prestressed by internal tendons situated at the top and bottom slabs. These tendons overlap at the blisters situated at the

Figure 10.38 Bridge on the R1 expressway across the Nitra River, near Nitra, Slovakia

span diaphragms, where they were also anchored. External tendons, which are bent at the span diaphragms, overlap at the pier diaphragms, where they are also anchored (Figure 10.39).

A combination of internal centric prestressing and additional external prestressing was also used in the design of the D8 motorway bridge across Rybny Creek, close to the border between the Czech Republic and Germany (see Chapter 4, Figures 4.55 and 4.56). The 356 m long, 31.10 m wide bridge is formed by a continuous box girder of seven spans (lengths 34–58 m) (see Figures 10.36 and 10.40).

To control the stresses in the piers, each pier was tied to the abutment by prestressing cables. The deck was incrementally cast and launched in 30 m long segments (Figure 10.41). The deck was produced in two steps: first, the bottom slab and the webs were cast

Figure 10.39 Bridge in Figure 10.38 – arrangement of the prestressing: (a) cross section at mid-spans; (b) cross section above supports; (c) elevation with internal bonded tendons; (d) elevation with external cables

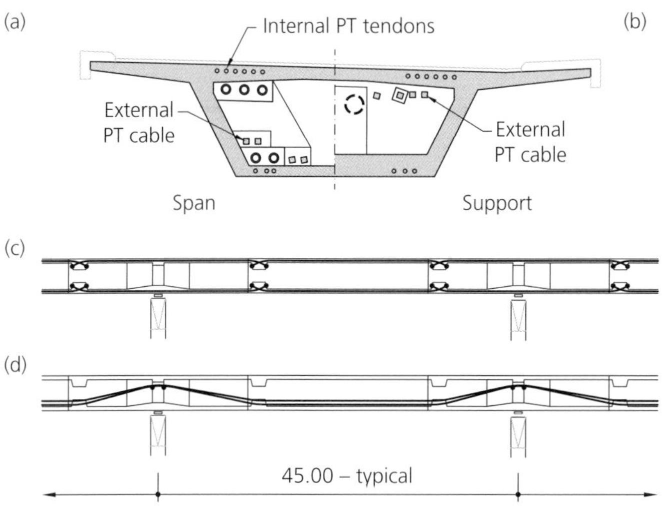

Figure 10.40 Bridge across Rybny Creek, Czech Republic: deck

in an adjustable casting bed; then, the external struts were erected, and the top slab was cast (Figure 10.42).

During launching, the box girder was progressively prestressed by coupled straight tendons uniformly distributed at the top and bottom slabs, and at the webs (Figures 10.43 and 10.44). These tendons create uniform compression stresses in the girder. The prestressing process and the arrangement of the couplers were designed in such a way that only half of the tendons were coupled in the joints. Due to the large positive bending moments that originate in the deck beyond the nose during launching, additional tendons were placed in the bottom slab of the first segment. These tendons were anchored at bottom plate's stiffeners of the launching nose to transfer stresses from the nose into the bottom girder's slab (Figure 10.45).

After launching, the girder was prestressed by continuous external cables anchored at the additionally cast end diaphragms. 2 × 8 cables are led inside of the box along the whole length of the bridge and are deviated at the pier and the span deviators. The deviators, together with stiffening ribs that transfer the radial forces of the tendons into the girder webs, were cast after the launch.

As stated previously, the launched deck is usually centrically prestressed by straight bars or tendons. The prestressing, which creates uniform compression in the whole deck, cannot balance the concentration of stresses originating in the top and bottom slabs during launching (Figure 10.46). Therefore, it is necessary to resist the peaks of the tension stresses, which originate above the webs, by reinforcing bars, and to check the width of the cracks.

Figure 10.41 Bridge in Figure 10.40: incremental construction

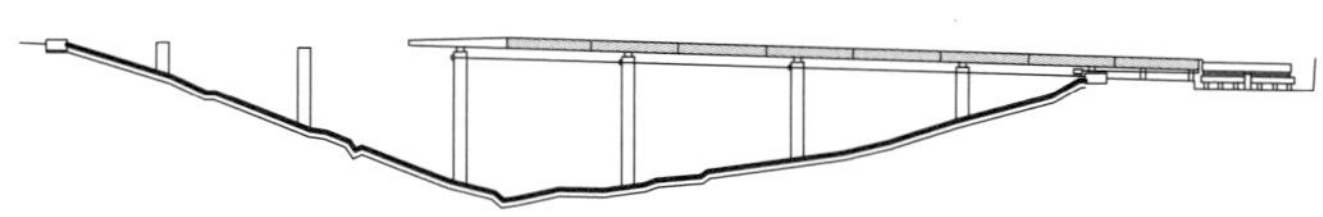

Figure 10.42 Bridge in Figure 10.40: Incremental casting of the deck

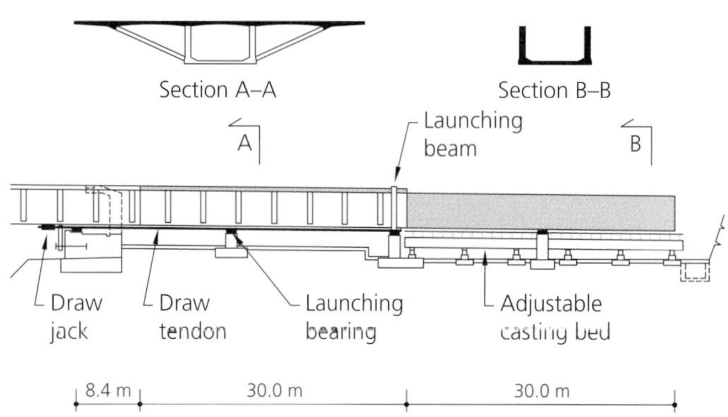

Figure 10.43 Bridge in Figure 10.40 – prestressing (PT): (a) elevation – internal bonded tendons; (b) elevation – external unbonded cables; (c) cross-section

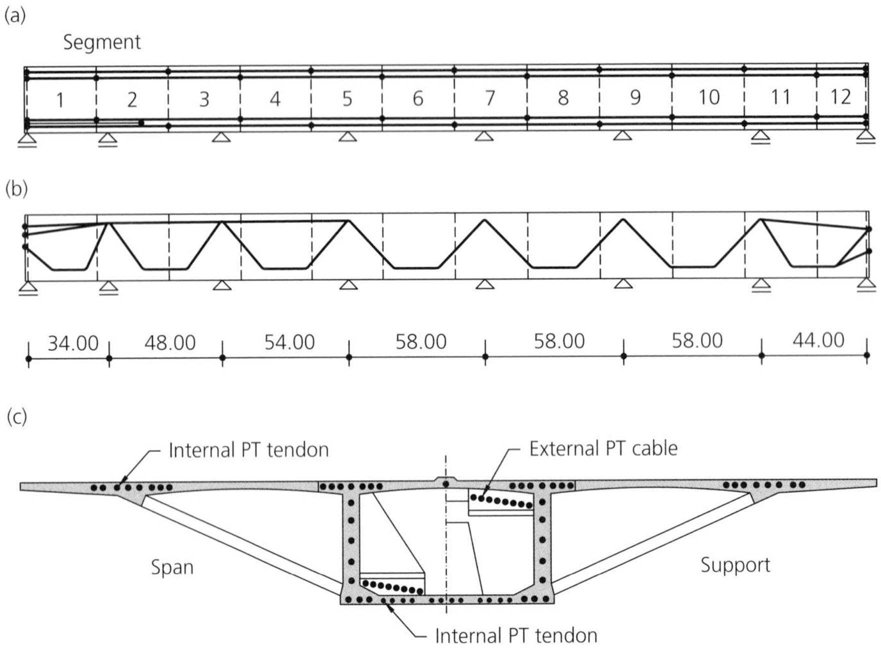

Figure 10.44 Bridge in Figure 10.40: internal bonded tendons

Figure 10.45 Bridge in Figure 10.40: launching nose

Figure 10.46 Bridge in Figure 10.40: distribution of the normal stresses at the top slab

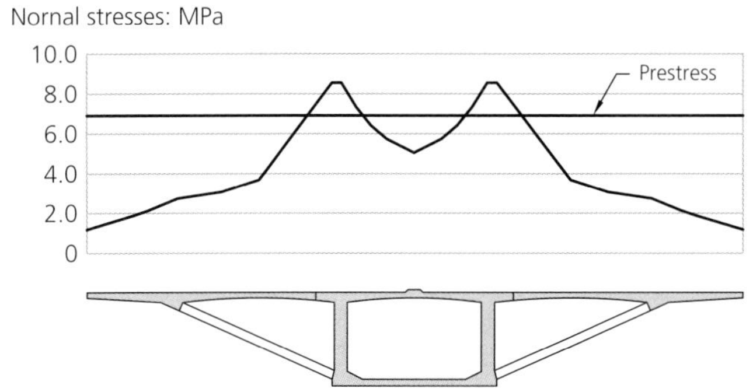

10.2.3 Connection of the deck to the steel nose

The safety of the launched structure also depends on the connection of the steel nose to the deck. This connection has to resist large shear forces of both signs. In the case of the Rybny Creek Bridge (see Section 10.2.2), the nose–deck connection was stressed by an upward force as large as 8.100 MN and downward force reaching 1.104 MN. Therefore, the steel nose was provided with two stiff corbels to transfer the shear forces from the nose into the webs. The strut-and-tie models shown in Figure 10.47 were developed to check the stresses originating in the webs.

The steel nose is relatively flexible compared with the stiffness of the launched concrete deck. Therefore, it is necessary to carefully check the deformation and the distribution of the stresses from the steel nose into the progressively cast deck. Figure 10.48 shows the deformation of the launched structure in the first steps of the erection.

Figure 10.47 Bridge in Figure 10.40 – strut-and-tie model for the connection of the steel nose to the deck: upward force, (b) downward force

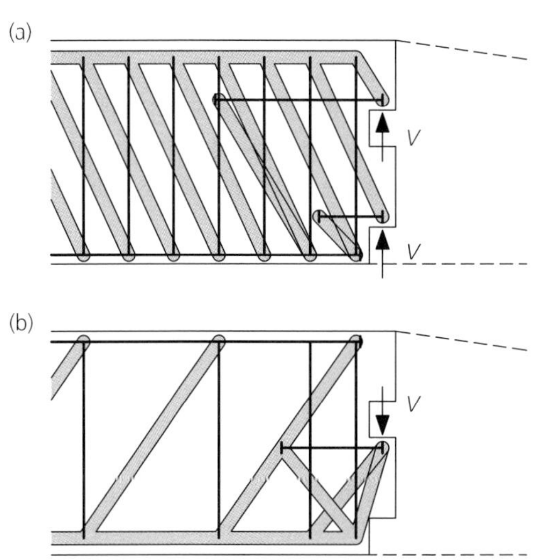

Figure 10.48 Bridge in Figure 10.40 – deformation of the launched structure: first span before (a) and after (b) reaches the first support; second span before (c) and after (d) reaches the second support

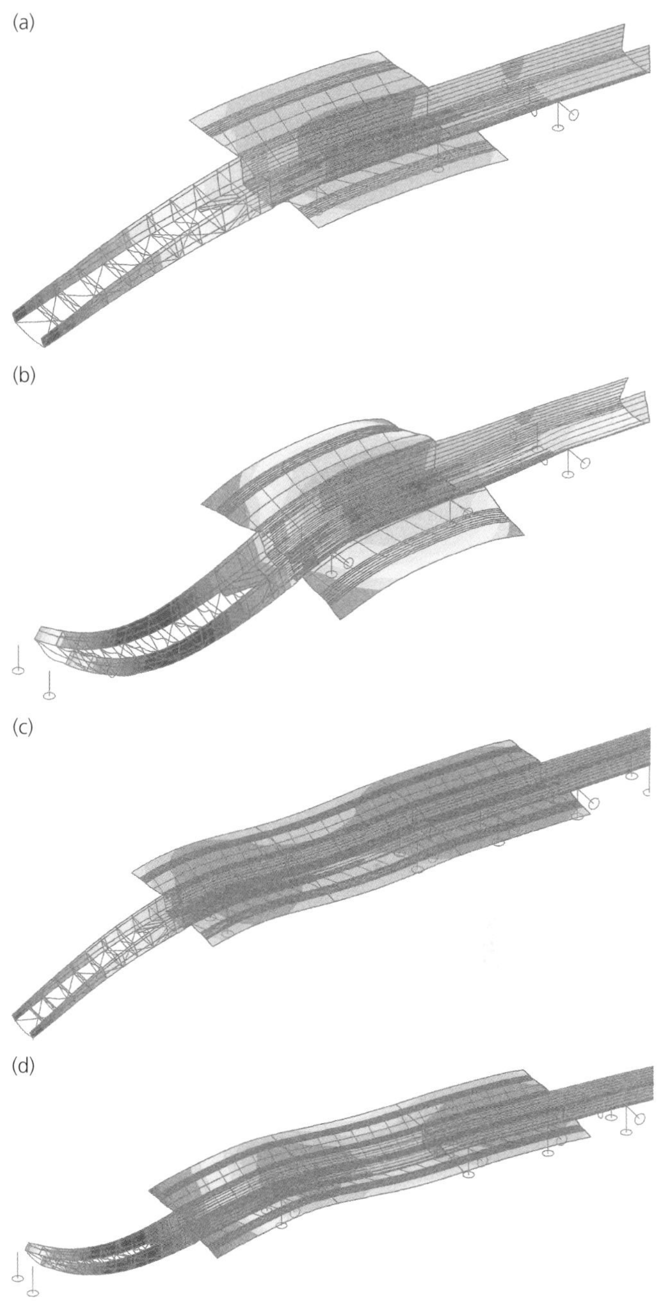

The deck is usually launched without the pier diaphragms and span deviators, which are cast after the launching has been completed. Their design is similar to the design of common structures (see Chapter 8).

10.3. Span-by-span erection of precast segmental structures

The first medium-span segmental bridges, which were built in the 1960s, were erected similarly to cast-in-place structures erected span-by-span with overhangs protruding into the next span (Anon., 1970). Usually, the bridge decks were prestressed by internal prestressing tendons that were coupled at the construction joints.

10.3.1 Structural arrangement

The development of external prestressing in the 1980s significantly contributed to the renaissance of span-by-span construction (Podolny and Muller, 1982). Currently, segmental bridges with spans up to 50 m are usually assembled of segments supported on or suspended from a movable gantry situated at the erected span. The segments form either a chain of simply supported girders or continuous girder structures. The decks usually have a trapezoidal single-cell box section. Wider decks are stiffened by internal struts. Wide bridges usually have a single-cell box section with large overhangs supported by struts. Railway bridges often have a channel cross-section, which has also been used in the design of several overpasses.

The box girders are usually prestressed by external cables that are anchored at the pier diaphragms, where, usually, they overlap. In the spans, the cables are bent either at several blisters situated at the bottom corners of the box section (Figure 10.49(a)) or at the span diaphragms (Figure 10.49(b)).

10.3.2 Static analysis

Similar to cast-in-place structures, the analysis has to follow the stages of construction. Figures 10.50 and 10.51 show the staged construction for the study design for the Jahra Road bridge in Kuwait. The bridge is formed by a chain of continuous girders in five spans (lengths 37.15 + (3 × 45.00) + 37.15 m). The girders are prestressed by external cables that are anchored at the pier segments and deviated at the span diaphragms (Figure 10.52). The analysis is as follows.

- *Stage 1*. The static effects due to the self-weight and prestressing are determined on a structure formed by a simple beam.
- *Stage 2*. The static effects due to the self-weight and prestressing of the second erected span are determined on a structure formed by a two-span continuous girder. The resultant static effects are equal to the sum of the effects determined in the first and second stages.
- *Stage 3*. The static effects due to the self-weight and prestressing of the third erected span are determined on a structure formed by a three-span continuous girder. The resultant static effects are equal to the sum of the effects determined in the second and third stages.

This procedure has to be repeated for all the subsequent spans.

Figure 10.49 Arrangement of the external cables

(a)

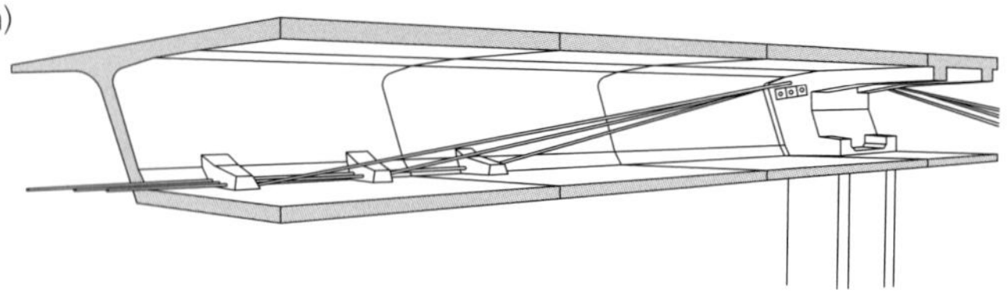

(b)

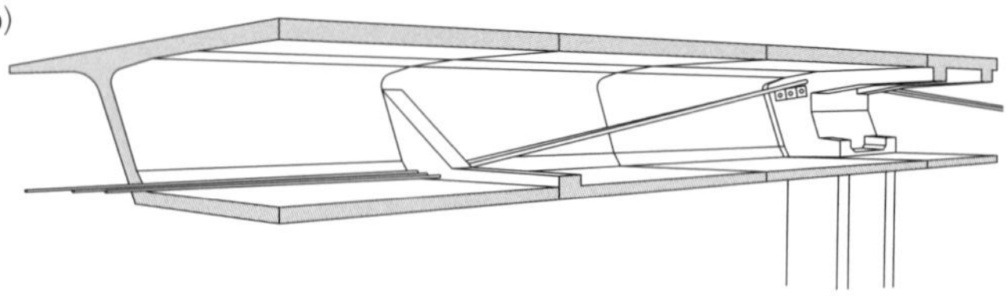

Due to concrete creep, a structure that is erected span by span has a tendency to reach a state of stress that corresponds to the function of a structure that was erected in one stage. To give an idea of the stress redistribution, Figure 10.53 shows the results of the analyses of the effects of the self-weight for the bridge presented in Figure 10.52. The bridge was erected span-by-span in five erection stages, with the erected segments being supported by movable scaffolding. Figure 10.53(b) shows the courses of the bending moments at each stage. Figure 10.53(c) shows the courses of the bending moments after the deck had been completed (stage 5), the bending moments in a structure cast at the same time, and the bending moments after 100 years.

10.4. Progressive erection of precast segmental structures

Medium-span bridges can also be erected as a progressive cantilever in the direction from one abutment to the other. The static effects in the erected cantilever can be reduced by temporary supports or by a temporary pylon and stays (Figure 10.54).

In span-by-span construction, the concrete joints situated between the individually erected spans mean that the final static system is composed of the sum of the individual erection stages.

10.4.1 Static analysis

In the progressive construction method, the precast deck is cast in the shape corresponding the designed bridge geometry. If the segments are correctly produced to the designed geometry, and if the vertical reaction corresponding to the erection stage is created when the pier segments are set on the bearings, the static function of the structure is the same as the function of a structure assembled in one stage on stationary scaffolding. Therefore, the structure is not subjected to a redistribution of the internal stresses due to changes in the static system.

Figure 10.50 Bending moments due to self-weight: (a) scheme of the structure; (b) loading and static system; (c) bending moments

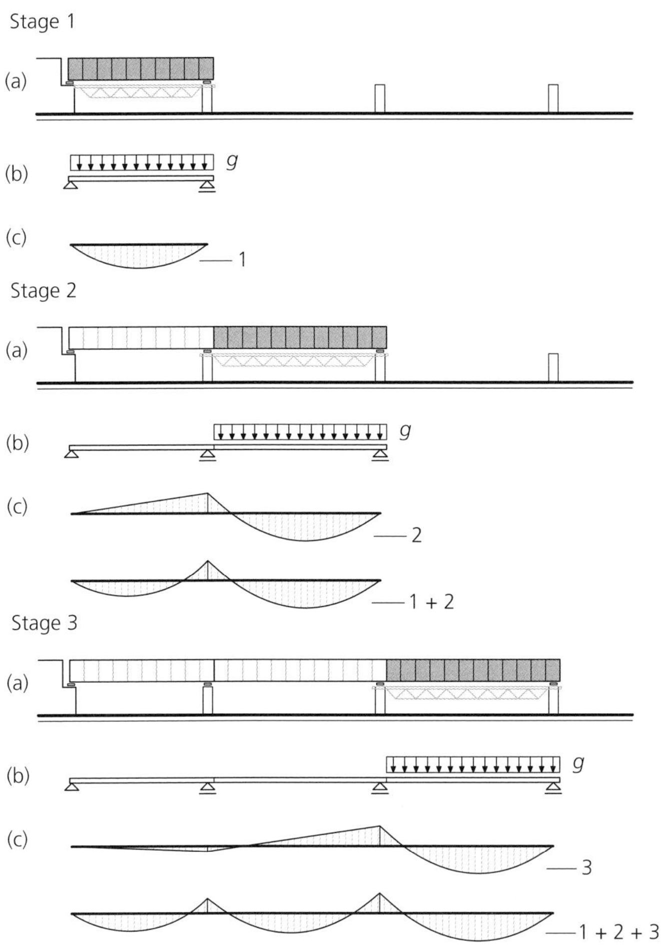

To illustrate this phenomenon, the five-span structure (span lengths 37.15 + (3 × 45.00) + 37.15 m) formed of the box girder shown in Figure 10.52(a) was analysed in this way. Figure 10.55 shows the courses of the bending moment in each erection stage. It is evident that the course of the bending moments at the final stage corresponds to the course of the bending moments of a structure erected in one stage on stationary scaffolding. Of course, there is a negligible redistribution of the moments due to the different ages of the individual progressively cast segments.

10.4.2 Layout of the prestressing cables – segment erection

The progressive erection method discussed above has been used in the design of several segmental bridges built in the Czech Republic (Stráský and Korenek, 1986, 1989). The bridges, which have a total length of 5 km, are composed of precast segments having a box or a

Figure 10.51 Bending moments due to prestressing: (a) scheme of the structure; (b) loading and static system; (c) bending moments

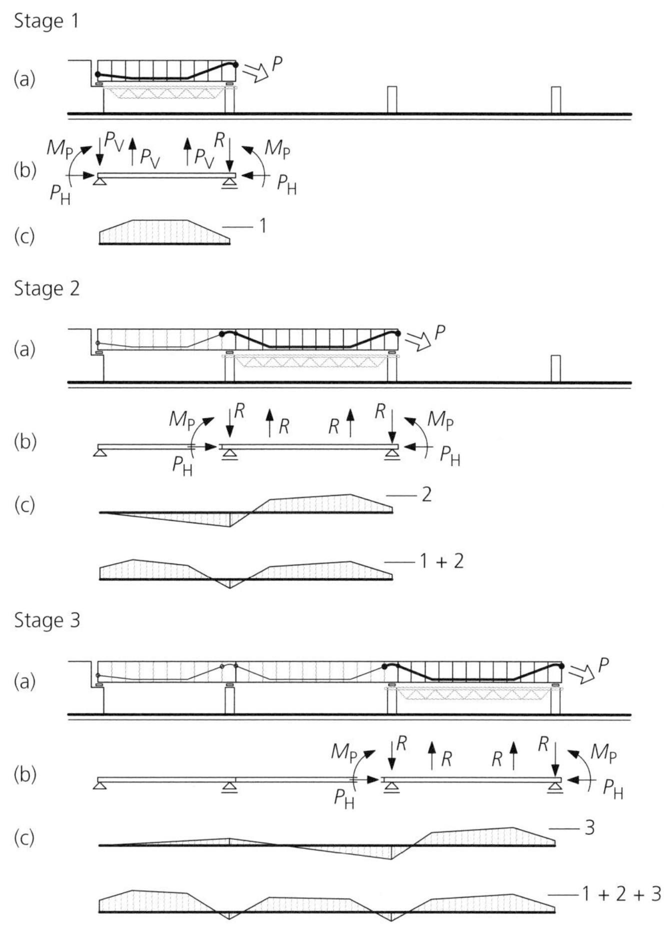

double-T cross-section (Figures 10.56 and 10.57). The decks are composed of 3.00 m long precast match-cast segments. The box girders have a depth of 1.40 m and are used for the bridges that have spans of up to 30 m. The double-T girders are 1.60 and 2.00 m deep,

Figure 10.52 Typical span: (a) cross-section; (b) elevation

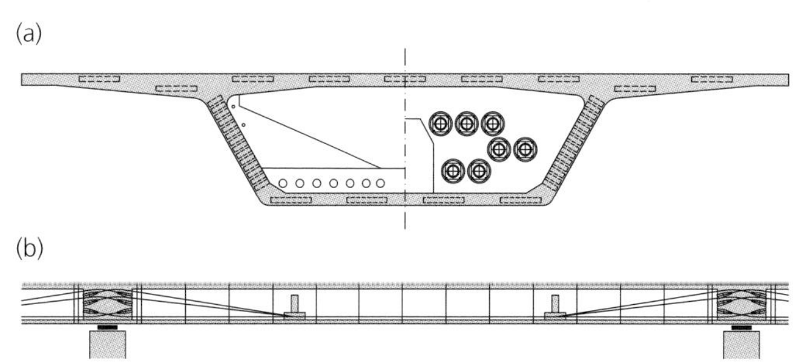

Figure 10.53 (a) Scheme of the structure; (b) bending moments during staged erection; (c) resulting bending moments

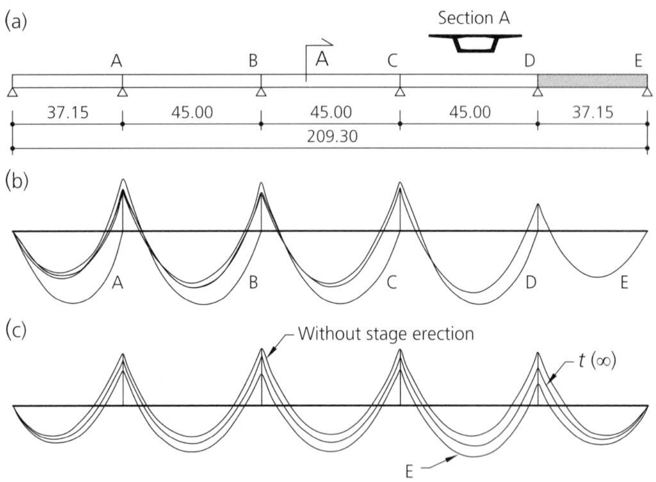

Figure 10.54 Progressive erection: (a) with temporary supports; (b) with temporary pylon and stays

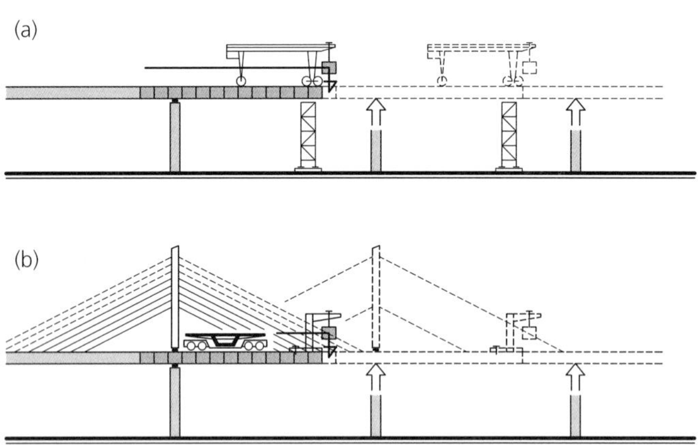

Figure 10.55 (a) Scheme of the structure; (b) bending moments during progressive erection

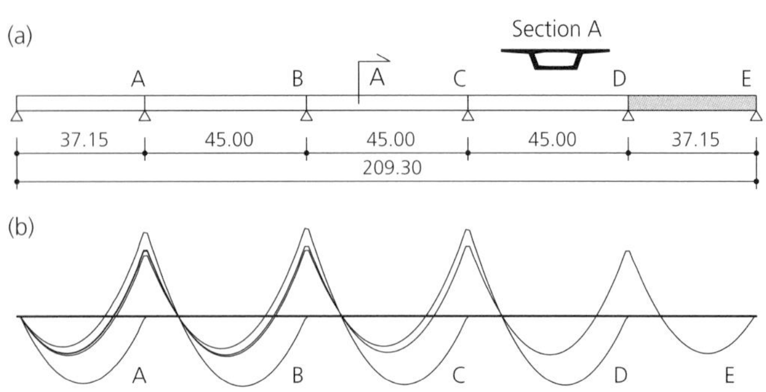

Figure 10.56 Cross-sections and plans: (a) box girder bridges; (b) double-T girder bridges

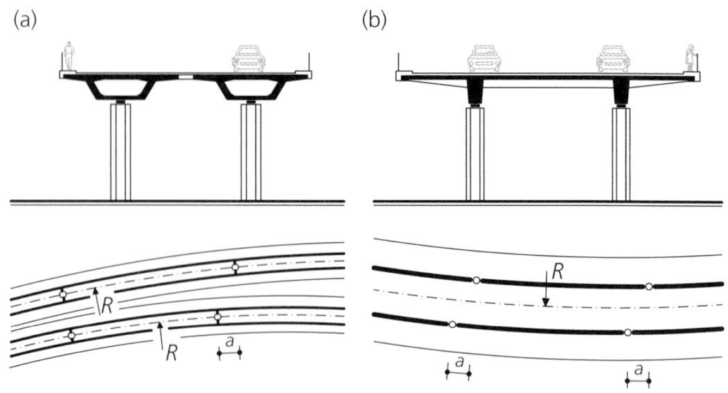

Figure 10.57 Brno viaduct

and are used for the bridges with spans of up to 42 m. The bridges that could have a skewed arrangement of the intermediate supports are built in space curves and with a super-elevation. The erected structure was progressively prestressed by two groups of tendons: straight tendons situated at the top slab close to the piers, and draped tendons situated in the webs (Figure 10.58).

This structural system was used for the first time in the construction of an urban viaduct in Brno (see Figure 10.57). The viaduct is formed by a twin bridge of eight continuous spans with lengths of $22 + (6 \times 30) + 22$ m. The deck of each bridge is assembled from two 6.80 m wide, 1.4 m deep box girders that are mutually connected by a transversally prestressed deck slab.

The girders are hinge connected to precast columns. During construction, the precast columns were temporarily fixed into the

Figure 10.58 Prestressing tendons: (a) erection; (b) box girder cross-section; (c) double-T girder cross-section; (d) elevation

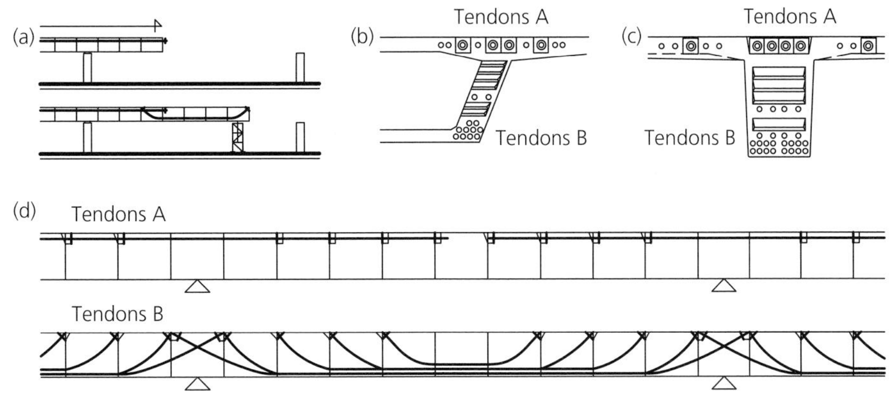

footing pockets. Once the pier segments had been erected and supported by temporary towers, the columns were drawn to the segments by prestressing tendons. Concrete hinges were then created by grouting the joints between the pier segments and the columns, and bottom parts of the column were cast in the pockets of the footings.

The segments were erected as a progressive cantilever, starting at the left abutment (Figure 10.59). First, the segments of the first span were erected on the scaffolding and were prestressed by draped tendons. Then, the structure was supported by temporary towers situated on the footings of the first span, and the geometry of the structure was adjusted. The precast columns were then connected to the deck (Figure 10.59(a)).

Then the second span was assembled. First, four segments were erected and prestressed by straight tendons (Figure 10.59(b)). Thereafter, the next three segments were erected and prestressed by erection prestressing bars and draped tendons (Figure 10.59(c)). When the second segment had been erected, the cantilever was supported by a temporary support (actually, it was loaded by a small force).

Then an additional two segments were erected and prestressed by erection bars (Figure 10.59(d)). After that, the end of the cantilever was supported by a temporary support, the deck was prestressed by draped tendons, and the force in the jack on the temporary support was adjusted. The first temporary support was released. Then the next two

Figure 10.59 Progressive erection

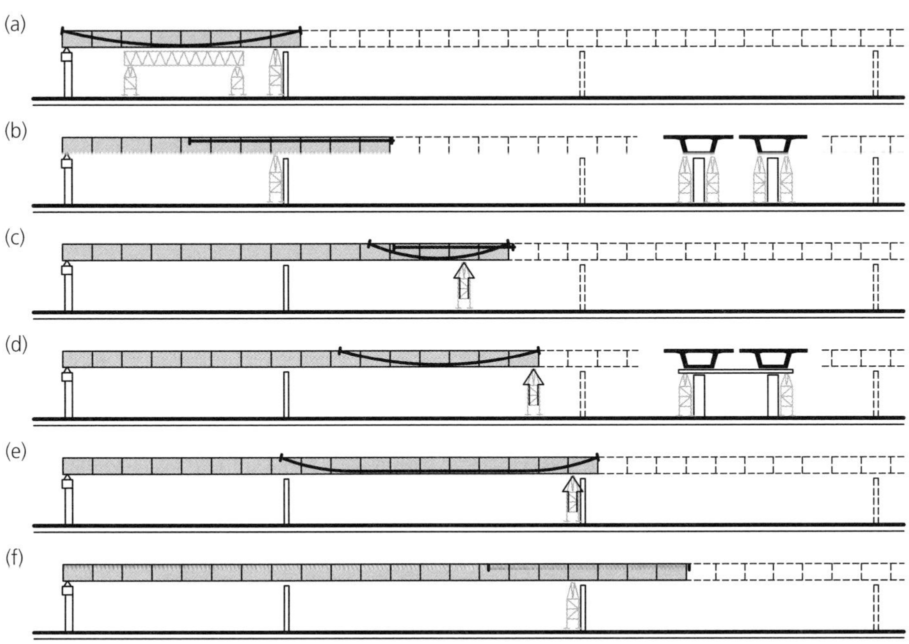

Figure 10.60 Bending moments during the erection of the: (a) second span; (b) third span (M'_p, secondary prestressing moment)

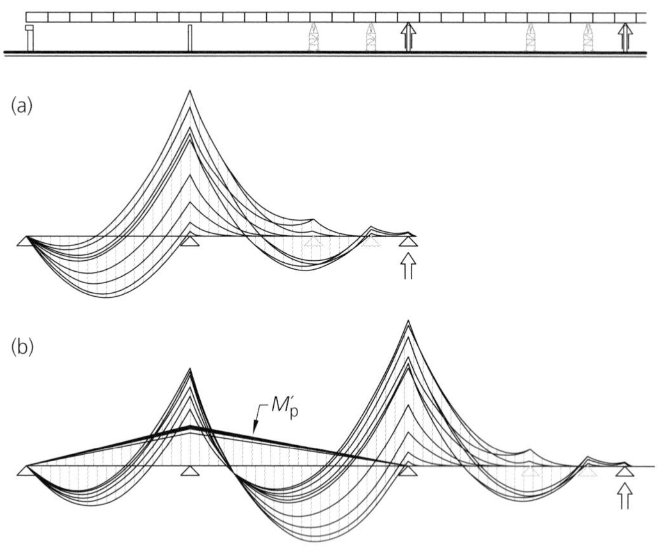

Figure 10.61 Prestressing: (a) cantilever tendons; (b) continuity tendons

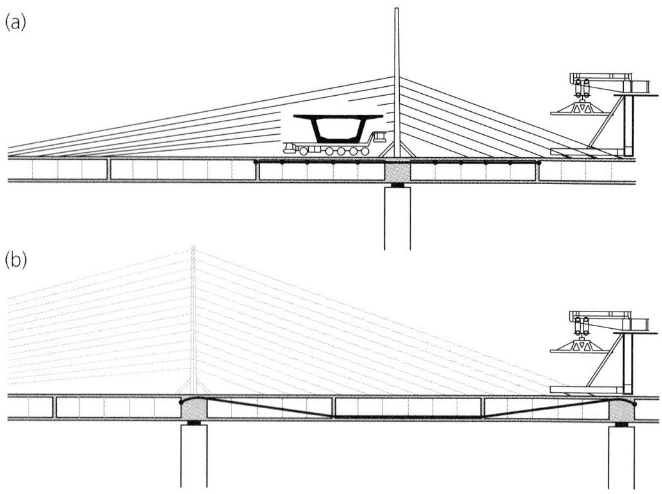

segments were erected and prestressed by erection bars (Figure 10.59(e)). After that, the cantilever was supported by the jacks located on the temporary towers situated on the pier footing, the force in the jacks was adjusted (the structure was loaded by a force corresponding to the reaction of the two-span girder loaded by the self-weight and the prestress), and the span support was removed. The precast columns were then connected to the deck. The erection of all the remaining spans was done in a similar way (Figure 10.59(f)).

The bending moments due to the self-weight and the secondary moments (M'_p) due to the prestress during the erection of the second and third spans are shown in Figure 10.60. From the figure it is evident how the bending moments change. It is clear that it is necessary to determine the layout of the prestressing tendons and the sequence of their tensioning that will balance the effects of the self-weight in each erection step.

Progressively erected bridges with cantilevers suspended on the temporary pylons behave in a similar way. During construction, the segments close to the piers are prestressed by straight cantilever tendons situated in the top slab, while the other segments are suspended on the temporary stay cables (Figure 10.61(a)). First, the erected segments are centrically prestressed to the already assembled structure by straight bars situated in the top and bottom slabs. The segments are then suspended on stay cables. The forces in the stay cables are determined such that they meet the condition of balancing the weight of the segments (i.e. the vertical components of the stay forces are equal to the weight of the segments). When the erection of all the segments has been completed, the pier segment is supported by hydraulic jacks, and span and continuity tendons are placed and post-tensioned (Figure 10.61(b)). The temporary stays are then released. Before the bearings are activated, the jacks situated on the piers load the structure by forces corresponding to the vertical reaction of the erection stage.

REFERENCES

Anon. (1970) Prestressed concrete in Britain, Western Avenue Extension. *Concrete* **4(6)**: 238–241.

CEN (European Committee for Standardization) (2005) EN 1992-2:2005. Eurocode 2 – Design of concrete structures. Part 2: Concrete bridges: design and detailing rules. CEN, Brussels, Belgium.

Dyckerhoff and Widman (eds) (1973) *Festschrift. Ulrich Finsterwalder 50 Jahre für Dywidag*. G. Braun, Karlsruhe, Germany.

Leonhardt F (1979) *Vorlesungen über Massivbau. Grundlagen des Massivbrückenbaues*. Springer, Berlin, Germany.

Podolny W and Muller J (1982) *Construction and Design of Prestressed Concrete Bridges*. John Wiley, New York, NY, USA.

Stráský J and Korenek M (1986) Short span segmental bridges in Czechoslovakia. *PCI Journal* **31(1)**: 106–132.

Stráský J and Korenek M (1989) Progressive placing of the segmental bridge across the Rokytka Valley near Prague. *Industria Italiana del Cemento* **56(10)**: 586–604.

Troyano LF (2003) *Bridge Engineering. A Global Perspective*. Thomas Telford, London, UK.

Wittfoht H (1972) *Triumph der Spannweiten*. Beton, Düsseldorf, Germany.

ice Publishing

Stráský and Nečas
ISBN 978-0-7277-6385-3
https://doi.org/10.1680/dcpb.63853.205
ICE Publishing: All rights reserved

Chapter 11
Cantilever construction

11.1. Deck

Cantilever construction is usually used for bridges with span lengths longer than 60 m. The method is described in many books, but the most comprehensive information can be found in Mathivat (1983) and Podolny and Muller (1982). Bridges with deck widths up to 20 m usually have a one-cell box section, while those with wider decks usually have a one-cell box section with large overhangs supported by ribs or struts (Figure 11.1). Two-cell box sections are also designed. However, as it is more complicated to analyse and construct these, they are seldom used. Cantilever bridges can be assembled of cast-in-place or precast segments.

Cantilever bridges are erected sequentially, segment-by-segment, in balanced cantilevers starting at the supports. In view of the cantilever moment, the deck usually has a variable depth that transfers the weight to the supports (Figure 11.2). The depth of the bottom slab is also variable, increasing near the supports.

The depth of the deck at the supports is usually 1/18 of the span length, and the depth at the midspans is usually 1/40. From an aesthetic point of view, a flat parabolic haunch following the curve $y = ax^{1.6}$ has proven its worth.

Contractors usually prefer box girders with vertical webs. However, structures with inclined webs are more aesthetically pleasing, as they allow the design of narrow piers (Figure 11.3). In addition, the inclination of the webs contributes to the reduction of deformations and stresses of the cross-sections.

Cantilever construction is used for bridges formed of continuous girders or frames that are supported by vertical or inclined piers. It is also used for the construction of cable-stayed bridges and arches.

The first cantilever bridges had decks frame-connected to the piers and expansion joints at the midspans. The cantilevers were mutually connected by joints allowing their mutual horizontal movement while also resisting the shear forces. Due to the effects of creep and shrinkage, these structures undergo large long-term deformations, and therefore they are no longer designed. Nowadays, the decks are formed by continuous structures. In the case of very long bridges, which require expansion joints, the joints are situated close to the quarter-span (Figure 11.4(a)), or steel girders resisting significant

Figure 11.1 Typical cross-sections: (a) one-cell box section; (b) one-cell box section with large overhangs supported by ribs; (c) one-cell box section with large overhangs supported by struts; (d) two-cell box section

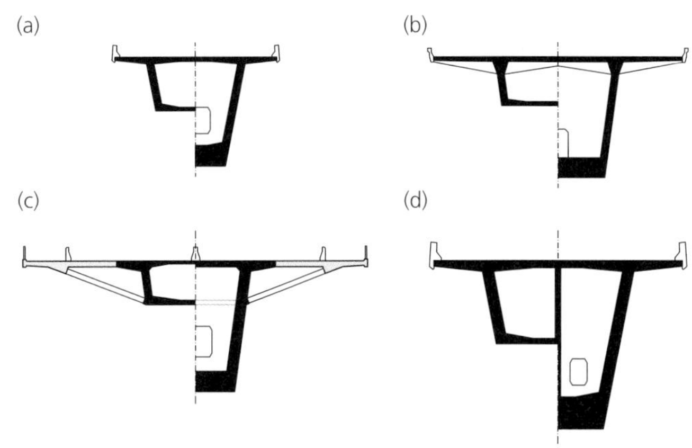

Figure 11.2 Typical cantilever: elevation

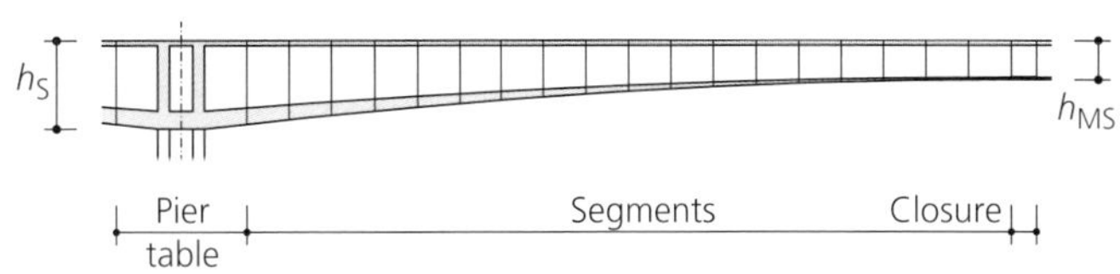

Figure 11.3 Typical cross-section: (a) vertical webs; (b) inclined webs

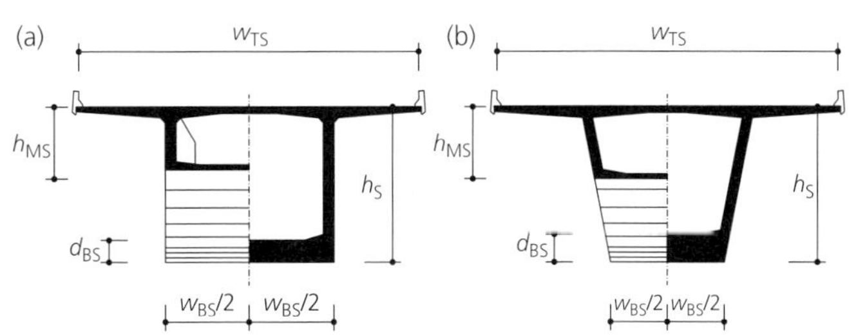

portions of the midspan bending moments are inserted into the box at the midspans (Figure 11.4(b)). Steel girders fixed into one cantilever and inserted inside the box of the neighbouring cantilever, which was the approach used in the construction of the Benicia–Martinez Bridge across the Sacramento River in California, USA (Murugesh and Cormier, 2007), provide a similar function. Figure 11.5 shows the erection of the midspan steel girder of the Skyway viaduct of the Bay Bridge between San Francisco and Oakland, California, USA (Brown, 2007) (see Chapter 4, Figures 4.86 and 4.87).

Figure 11.4 Expansion joints: (a) at quarter-span; (b) at midspan

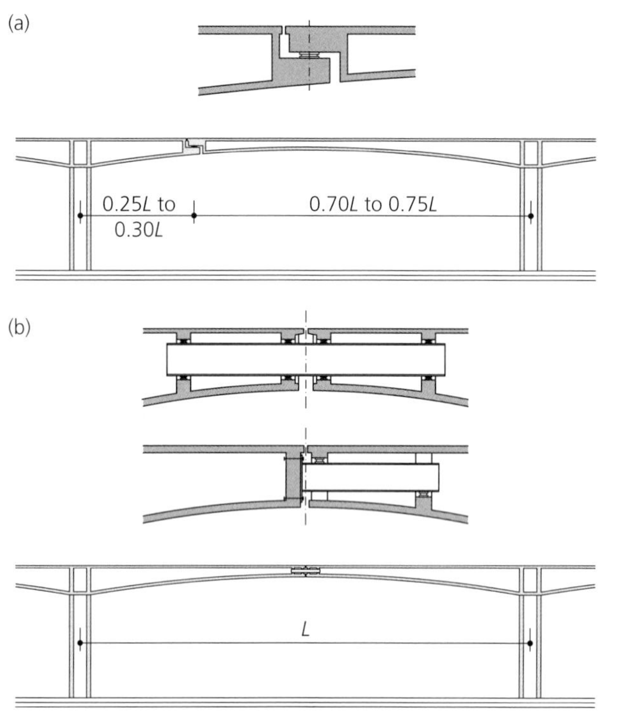

Figure 11.5 Skyway viaduct of the Bay Bridge between San Francisco and Oakland, California, USA: midspan steel girder

11.2. Supports – stability of the erected structures

The arrangement of the structures above the supports has been developed from the requirement to ensure stability during construction. The cast-in-place or precast cantilever must be checked for the possibility of uneven loading from the erected segment, the possibility of uneven distribution of the mass of the neighbouring cantilevers, wind loads and the random concentration of the live load on one side of the erected structure (Figure 11.6). In addition, the possibility of the falling of the erected segment should be considered.

As the uneven load and the corresponding bending moment can reach considerable values, sometimes a non-symmetrical pier table or/and erection support are proposed on one side only, and a procedure for the erection of symmetrical segments is determined. The segment is first cast on the shorter cantilever side of the pier table, and then on the longer cantilever side. Similarly, in structures with one-sided erection support, erection starts on the side of this support, followed by erection on the opposite side.

The need to ensure the stability of the structure during the erection of the deck affects the structural arrangement of the cantilever bridges. The deck can be supported by one row of bearings situated on slender piers (Figure 11.7(a)), or by one or two rows of bearings situated on stiff piers (Figure 11.7(b)), or it can be frame connected with stiff piers (Figure 11.7(c)). The stability of the erected deck can also be ensured by a pair of slender piers (Figure 11.7(d)).

11.2.1 Bridge decks supported by one row of bearings

If the bridge deck is supported by slender piers, it is necessary to stiffen the structure by temporary supports during the erection. The simplest way is to support the erected structure with twin temporary supports situated on both sides of the piers. The bearings

Figure 11.6 Radom loads applied to erected cantilevers. (According to Mathivat (1983))

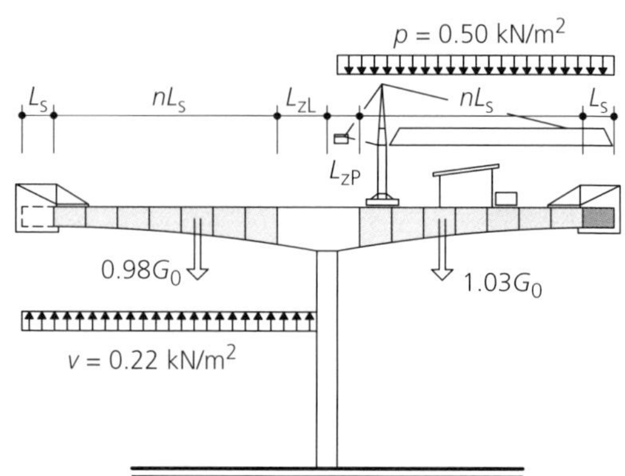

Figure 11.7 Support of a cantilever bridge during erection of the deck

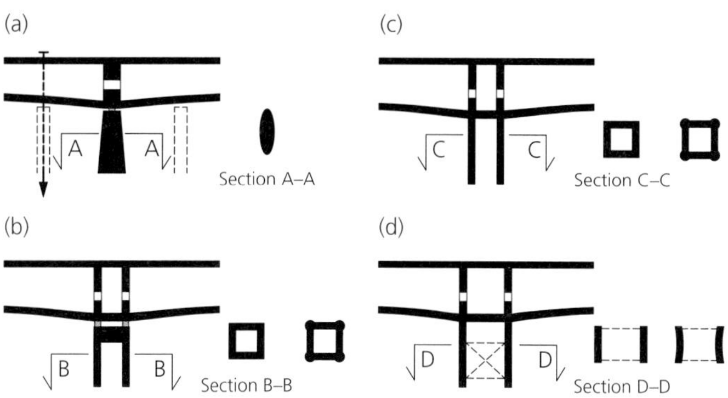

are activated when the closure between neighbouring cantilevers is cast. In this case, the temporary supports have to resist both the whole weight of the erected structure and the pair of forces caused by the unbalanced moment due to the progressive erection of segments.

This solution has been used in construction of many structures, for example in the design of the viaduct across the Nitra River, on the R1 expressway in Slovakia (see Chapter 4, Figures 4.70 and 4.71). The main bridge, with span lengths of 50.0–85.0 m, was erected in symmetrical cantilevers that were supported by two temporary concrete walls situated on both sides of the piers at a distance of 8.0 m (Figures 11.8 and 11.9). The walls were supported by the pier footings. The contractor cast symmetrical segments simultaneously. However, the stability of the erected cantilever was checked for the possibility of a non-symmetrical loading by one half of the weight of the cast segment.

Figure 11.8 Construction of the viaduct across the Nitra River, Slovakia

Figure 11.9 Bridge in Figure 11.8: bending moments

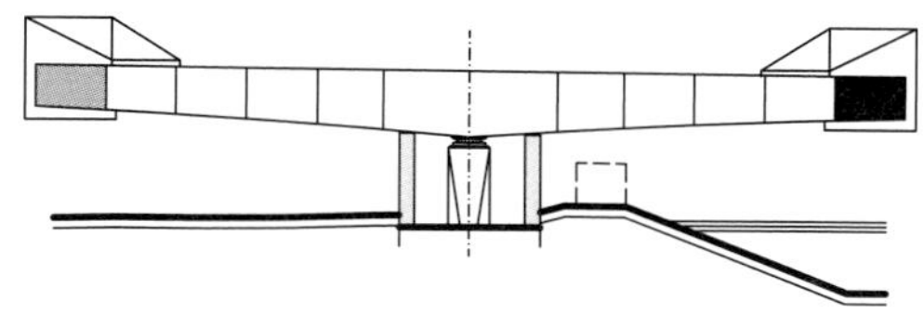

The construction of temporary supports for tall bridges is very expensive. Therefore, during the erection of segments, unbalanced moments are resisted by the piers. The pier segments are usually supported by pairs of hydraulic jacks situated on the pier caps. When all the segments have been erected, the geometry of the erected structure can be adjusted. After this adjustment, the closure joints are cast and prestressed, the joints between the bearings and pier segments are grouted, and the hydraulic jacks are released.

The above approach is also used in the construction of segmental bridges, which are, according under licence from Freyssinet, built by the Czech firm SMC, Prague, and the Slovak firm Doprastav, Bratislava. Bridge No. 203, a 762.4 m long viaduct built on the R1 expressway near Nitra in Slovakia, serves as an example. The twin bridge, with typical span lengths of 61.0 m, was assembled of precast, match-cast segments of width 12.25 m and depth 3.0 m. The segments were erected using two launching gantries (Figure 11.10). Due to the possibility of adjusting the geometry, the erected structure is not tied to the piers. To guarantee the required stability of the erected structure, the hydraulic jacks were situated at a sufficient distance (4 m). Therefore, the pier caps have large dimensions (Figure 11.11) and, because the jacks carry the whole dead load, the pier caps are heavily reinforced.

Figure 11.10 Construction of the Bridge No. 203 on the R1 expressway near Nitra, Slovakia

Figure 11.11 Support of the erected cantilevers

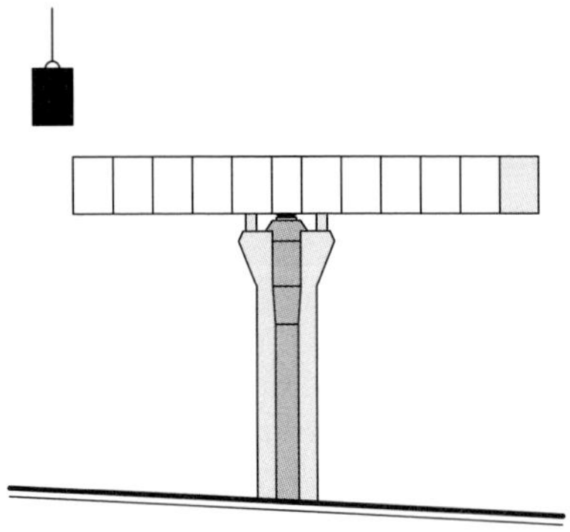

It is illogical to design the piers for a load that occurs only during construction, and it is not reasonable to construct twin temporary supports to resist the whole load of the erected structure. Therefore, other, more economical, solutions can be used.

It is useful to support the deck by definitive bearings, and to stabilise the erected cantilever by one temporary support situated on one side of the pier. Then, the deck has to be erected non-symmetrically. First, the segment situated on the side of the temporary support is cast, and then the segment on the opposite side. To guarantee the stability of the erected structure, the temporary supports have to be prestressed to the deck and the footings. The vertical reaction, which the temporary support resists, corresponds to the unbalanced moment divided by the distance of the support from the bearings.

This approach was being used for the Kayak Bridge, on the D4 motorway, which erection was recently completed near Bratislava in Slovakia (Figure 11.12) (Chalupec *et al.*, in press). The bridge has a

Figure 11.12 Construction of the Kayak Bridge, near Bratislava, Slovakia

Figure 11.13 Bridge in Figure 11.12: (a) cross section at mid-span; (b) cross section at piers; (c) partial elevation at piers; (d) elevation during construction

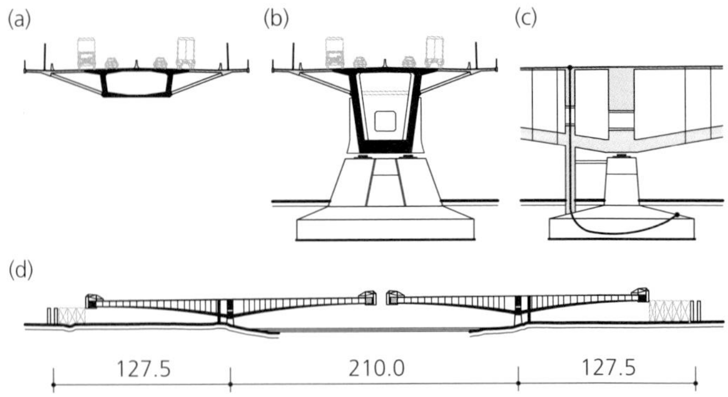

total length of 465 m and has three spans (lengths 127.5 + 210.0 + 127.5 m) (Figure 11.13(a)). The deck is formed of a 35.0 m wide one-cell box girder with additionally cast overhangs supported by precast struts (Figure 11.13(b)). The 16.30 m wide spine girder has a depth above the piers of 13.0 m and 4.3 m at the midspan. The length of the segments ranges from 3.90 to 5.00 m; the length of the pier tables is 20.00 m. The pier tables are supported by temporary concrete walls situated at the side spans at a distance of 8.60 m from the bearings (Figure 11.14). The walls are post-tensioned by prestressing tendons anchored at both the girder and the pier footings.

A similar principle was used in a design of the small precast segmental bridge in Plzen, Czech Republic (Stráský *et al.*, 2001). However, instead of the temporary supports, steel brackets and the bending stiffness of the piers were utilised. The 290.65 m long viaduct has seven spans with lengths ranging from 34.0 to 45.0 m. The bridge deck is formed of a three-cell box girder without

Figure 11.14 Bridge in Figure 11.12: temporary support

Figure 11.15 Bridge in Plzen, Czech Republic: cross-section

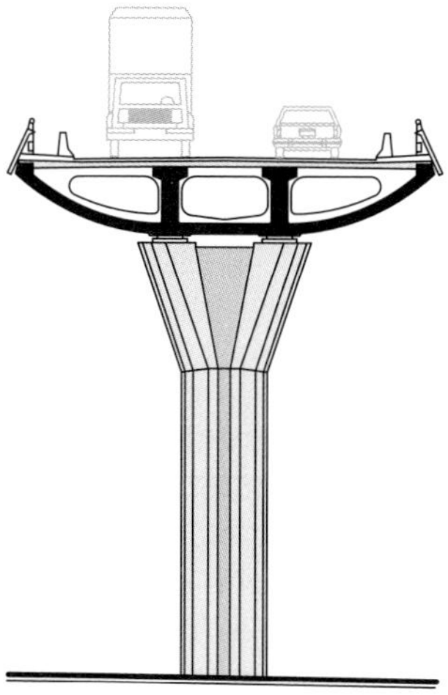

traditional overhangs (Figure 11.15). The girder is supported by two pot bearings situated on single supports formed by elliptical piers with caps. The 2.20 m deep box girder is assembled of precast, match-cast segments with an open cross-section and an additionally cast deck slab. Typical segments are formed of a curved bottom slab and two webs (Figure 11.16). Transverse ribs situated in the middle of their length stiffen the segments during their production and erection.

The erection of a typical span began with the erection of the steel brackets anchored at the piers. First, the pier segment and the first span segment were placed on the steel brackets and prestressed by the erection prestressing bars. Then, the pier segment was tied to the brackets. In this way, a non-symmetrical pier table was created. Next, the first opposite segment was prestressed by the erection prestressing bars, and cantilever tendons were placed and post-tensioned. The erection of typical segments was then continued (Figure 11.17). First, a segment was erected on the site of the steel brackets, followed by the

Figure 11.16 Bridge in Figure 11.15: precast segment

Figure 11.17 Bridge in Figure 11.15 – progressive segment erection: (a) on the bracket's side, (b) on the opposite site

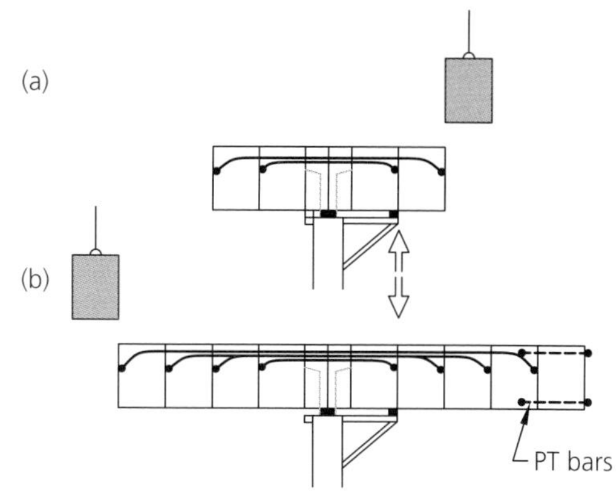

opposite one. When all segments had been erected (Figure 11.18), the closure joint was cast and prestressed. Finally, the composite deck slab was cast, and the deck was post-tensioned by external continuity cables. As the steel brackets resit only the vertical forces corresponding to the unbalanced moment divided by the distance from the bearings, they are light, and the requirements regarding their anchoring into the piers were easily accomplished.

Long segmental bridges are usually erected using launching gantries, which can also be used as a structural member to guarantee the stability of the erected cantilevers. The function of launching gantries can be illustrated by the construction of the recently opened viaduct across Roskilde Fjord in Denmark (Novák *et al.*, 2019). The viaduct has a total length of 1.36 km and is composed of 17 spans (lengths 57.35 + (15 × 83.00) + 57.35 m). The 19.70 m wide deck, which is assembled of precast, match-cast segments with a one-cell box section, was erected in

Figure 11.18 Bridge in Figure 11.15: erected cantilever

symmetrical cantilevers from the piers (Figure 11.19). The 3.50 m deep deck is strengthened at the piers to a depth of 4.00 m by short haunches. The deck is supported by two concrete hinges or bearings set on slender piers, which were designed for bending moments originating from the service load only (Figure 11.20).

Construction of a typical span started with the erection of short steel brackets, which served to support the launching gantry and the pier segments (Figures 11.21(a) and 11.22). After post-tensioning, the gantry was supported by an erection support that served for gantry launching. The first segments were then erected, post-tensioned and supported by hydraulic jacks (Figure 11.21(b)). In this way, a pier table was created that was able to resist the unbalanced moment due to the erection of the next three symmetrical segments. Next, the already erected structure was connected to the launching gantry by a stabilising arm (Figure 11.21(c)), creating a structural system that allowed the erection of the next four symmetrical segments. After the erection of these segments, the stabilising arm was moved into a new position and the next four symmetrical segments were erected. This procedure was repeated until all the segments had been erected. The closure joint was then cast and post-tensioned (Figure 11.21(d)).

Figure 11.19 Construction of the Roskilde Fjord viaduct, Denmark

Figure 11.20 Bridge in Figure 11.19 – pier: (a) cross-section; (b) elevation

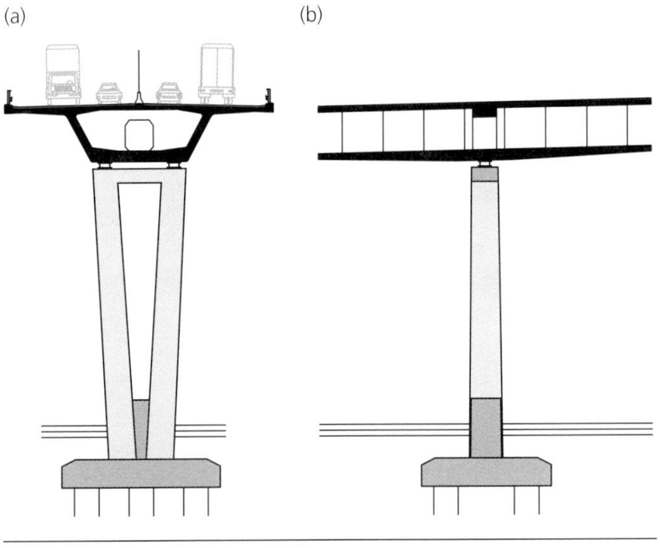

11.2.2 Bridge decks supported by two rows of bearings

The structural system in which the deck is supported by two rows of bearings both during construction and during service was used in the construction of the extradosed viaduct on the D1 motorway at Považká Bystrica, Slovakia (see Chapter 4, Figures 4.57–4.59) ((Stráský, 2010)). The 30.40 m wide viaduct deck is formed of a narrow box girder with large overhangs supported by V-shaped precast struts, and is suspended on seven low pylons. The deck is stiffened by A-shaped diaphragms that transfer the load from the pylons into two rows of bearings. To guarantee the stability of the deck during erection, the bearings were temporarily fixed, and the deck was anchored to the pier heads by post-tensioning of prestressing bars (Figure 11.23). The bearings on the two central piers are fixed; the remaining bearings are movable in the longitudinal direction of the bridge. The stability of the deck in the longitudinal (Figures 11.24 and 11.25) and transverse directions, both during construction and service, was carefully checked.

The decks of many cantilever bridges designed by French engineers are supported by two rows of neoprene or elastomeric bridge pads, which ensure the partial fixing of the deck into the piers and at the same time allow limited horizontal movement of the deck (Figure 11.26). During construction, the deck is supported by concrete wedges and it is prestressed by prestressing bars to the piers. When the closures between the individual cantilevers are cast, the prestressing bars are released, and the concrete wedges are substituted by neoprene pads.

The bridge pads consist of a sandwich of mild steel shims and rubber moulded as one unit. Bridge bearing pads resist the vertical load, accommodate limited rotation and provide lateral shear movement (Figure 11.27). The function of the pads can be expressed as the spring constant determined from the values of the Young modulus and shear modulus provided by the manufacturer. The stiffness of the deck fixing into the pier can be achieved by varying the thickness of the neoprene layer (Mathivat, 1983).

Elastomeric bearing pads have been used in the construction of many structures, the most interesting example being the Brotone Bridge, built across the River Seine in Normandy, France, which was built in 1977 (see Chapter 4, Figures 4.81 and 4.82) (Mathivat, 1983). The segments of the main and side spans were progressively cast in symmetrical cantilevers starting at the piers. During construction of the main span, the deck segments were progressively prestressed, and all the other segments were suspended on the pylon. Because the bending capacity of the pier and foundation of the pylon

Figure 11.21 Bridge in Figure 11.19: segment erection using a launching gantry

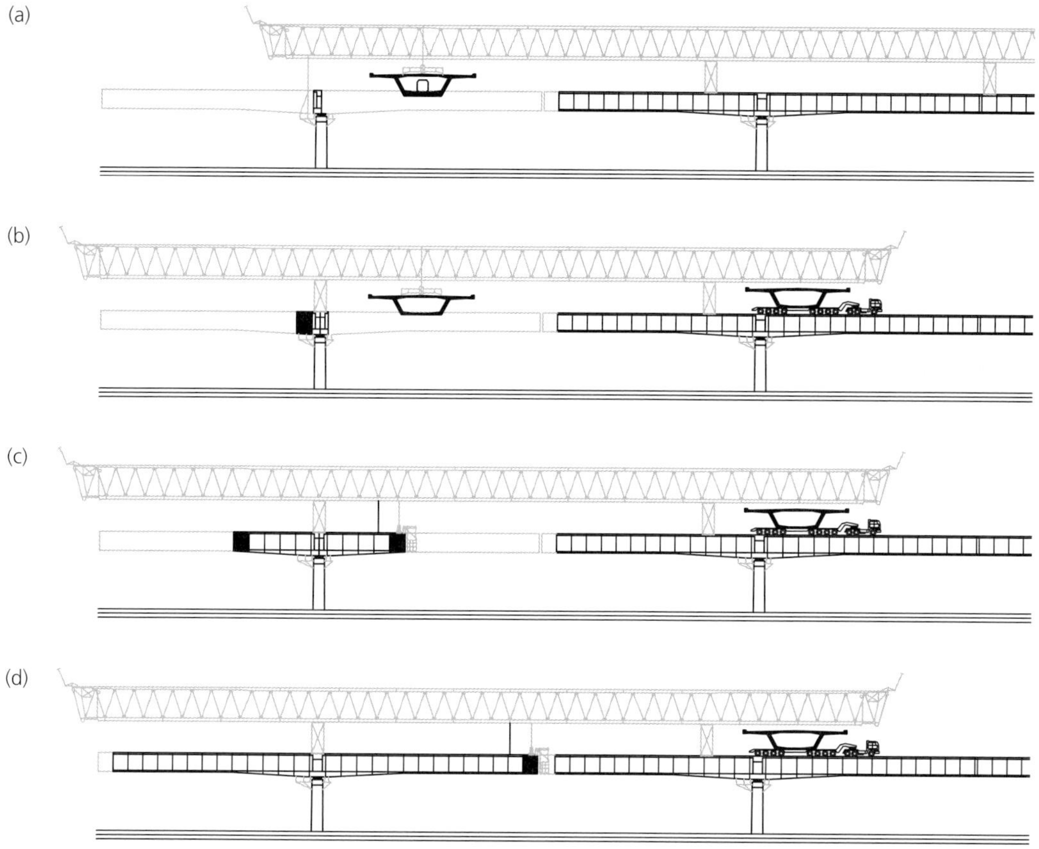

Figure 11.22 Bridge in Figure 11.19: support of the first segments

Figure 11.23 Viaduct at Považká Bystrica, Slovakia: (a) cross-section; (b) partial elevation

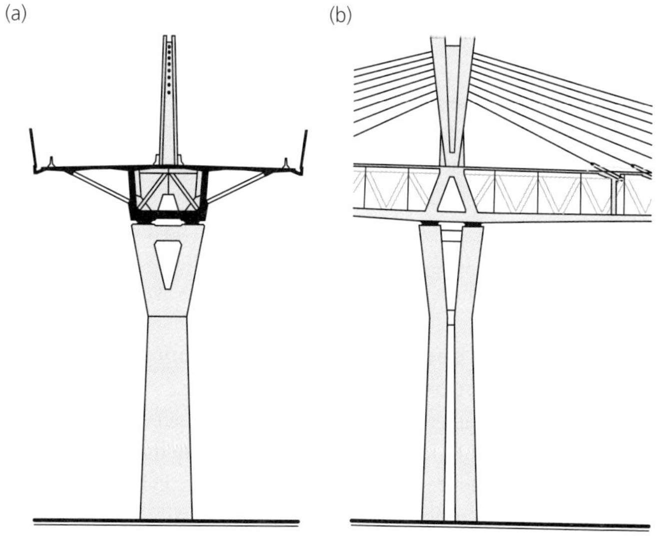

Figure 11.24 Bridge in Figure 11.23: cantilever construction

Figure 11.25 Bridge in Figure 11.23: deck supported by two rows of bearing

had to accommodate an unsymmetrical load due to the cantilever construction, the bending capacity was also utilised in the final structure to reduce the effect of the live load in the deck. Therefore, the pylon was constructed integrally with the deck at its base, and both the pylon and the deck were separated from the pier by a ring of neoprene bearings (Figures 11.28 and 11.29). In this way a flexible connection between the deck and the pylon piers was created.

11.2.3 Bridge decks supported by a monolithic connection to the piers

Hinge or frame connection of the deck to the piers represents the most economical solution, and also requires minimum maintenance. Therefore, it should be used whenever possible. Due to the frame connections, the bridges form multiple statically undetermined (redundant) structural systems. Therefore, they are predominantly

Figure 11.26 Support of the deck by neoprene pads: (a) elevation; (b) plan

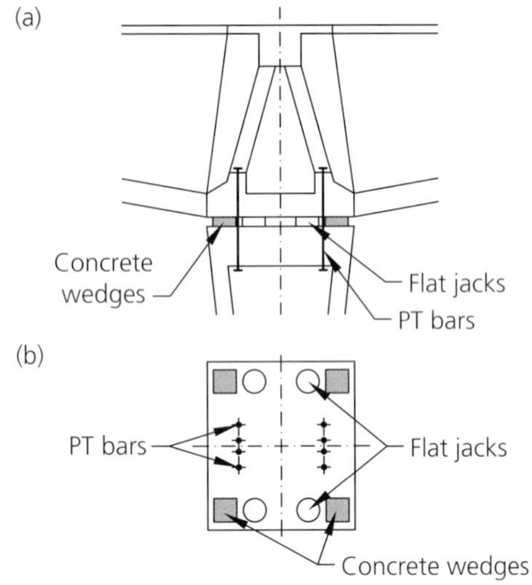

Figure 11.27 Static function of neoprene pads

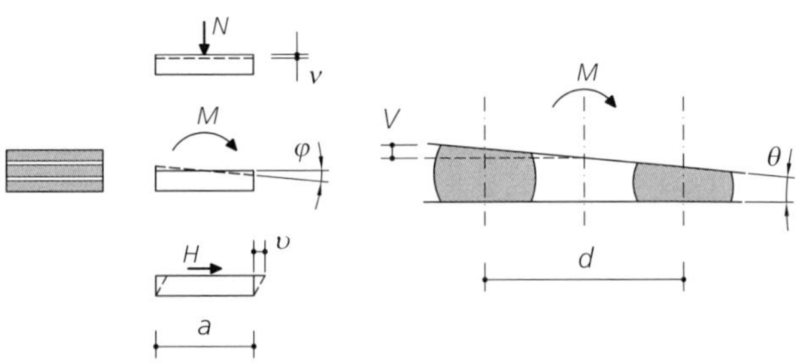

designed in seismic areas. An example of such a structure is the Benicia–Martinez Bridge across the Sacramento River in California, USA, which was built in 2007 (see Chapter 4, Figures 4.72 and 4.73) (Murugesh and Cormier, 2007).

The main bridge is formed by three frames of lengths 515.8 + 644.0 + 811.4 m, which are formed by multiple span structures having span lengths in the range 161.0–200.8 m. The frames are mutually connected by steel girders fixed into the end diaphragm of one cantilever and inserted into the box of the neighbouring cantilever (see Figure 11.4(b)). The 25.07 m wide one-cell box girder has a variable depth, and is frame connected to the piers (Figures 11.30 and 11.31). The piers are formed by four columns with an octagonal cross-section, which are mutually connected by transverse and longitudinal shear webs. The longitudinal bars, which are properly anchored in both the footings and the vertical pier diaphragms, are heavily confined by welded loops (Figure 11.32), which guarantee plastic behaviour during accidental loading. The footings are sup-

Figure 11.28 Brotone Bridge across the River Seine, Normandy, France

Figure 11.29 Bridge in Figure 11.28 – support of the deck by neoprene pads: (a) cross-section; (b) elevation

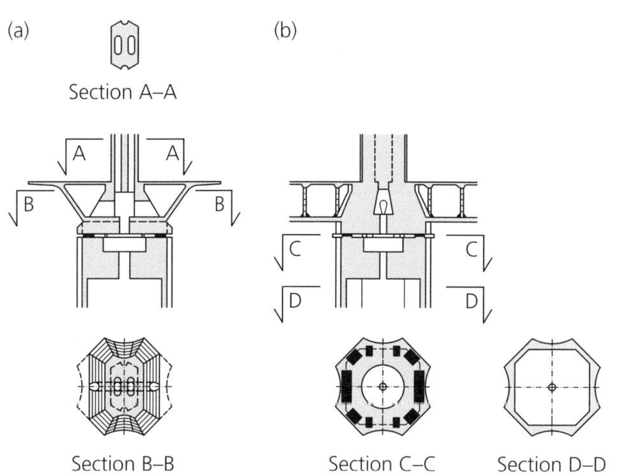

Figure 11.30 Benicia–Martinez Bridge across the Sacramento River, California, USA: pier table

Figure 11.31 Bridge in Figure 11.30: (a) bridge partial longitudinal section; (b) cross-section of the piers

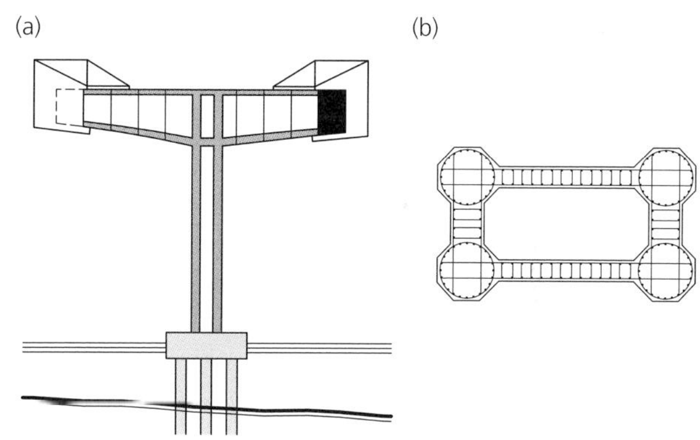

ported by driven steel pipes filled with concrete. The piers and non-symmetrical pier tables are made of normal weight concrete, and the box girder is made of lightweight concrete with a characteristic cylinder strength of 45 MPa.

A similar arrangement of the deck–pier connection was used in the construction of the Skyway viaduct of the Bay Bridge between San Francisco and Oakland, California, USA (Brown, 2007). The 1.3 mile long twin bridge is divided into four frames separated by expansion joints that resist both moment and shear (see Figures 11.4(b) and 11.5). The typical span consists of a cast-in-place pier table and progressively erected precast segments (Figure 11.33). The pier tables are frame

Figure 11.32 Bridge in Figure 11.30: anchoring of the longitudinal bars in the pier diaphragm

Figure 11.33 Skyway viaduct of the Bay Bridge between San Francisco and Oakland, California, USA: segment erection

connected to the piers, which are formed by four columns having a pentagonal cross-section mutually connected by transverse and longitudinal shear webs (Figures 11.34 and 11.35).

The moment-resisting connection of the deck with the piers guarantees the safe erection of cantilever bridges. However, due to the

Figure 11.34 Bridge in Figure 11.33 – pier: (a) deck's cross section near the pier; (b) cross section of the pier table; (c) cross section of the pier

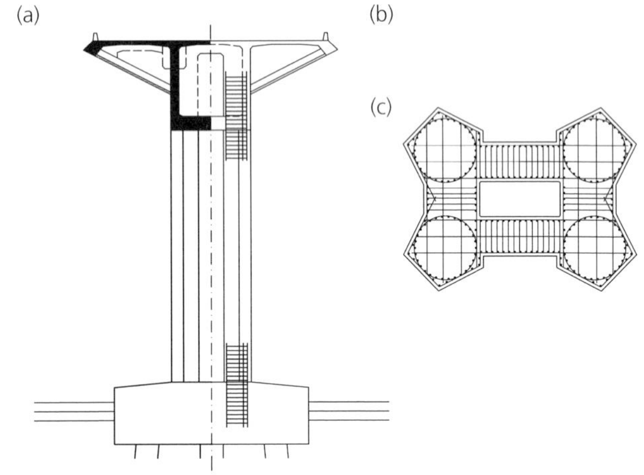

Figure 11.35 Bridge in Figure 11.33: reinforcement of the pier

effects of prestressing, temperature changes, and creep and shrinkage of concrete, this type of connection is mainly suitable for tall bridges. In low bridges, the stiff piers prevent free shortening of the deck, and therefore this type of connection can only be used in short structures. The secondary effects can be reduced by jacking of the cantilevers before the central closure is cast (Figure 11.36).

Figure 11.36 Jacking

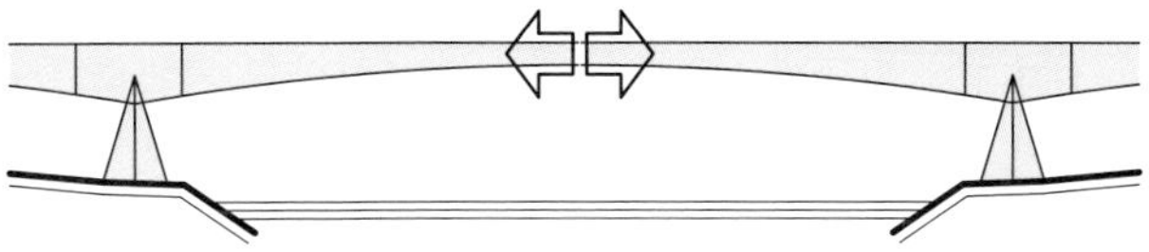

Figure 11.37 Static function of double-wall piers

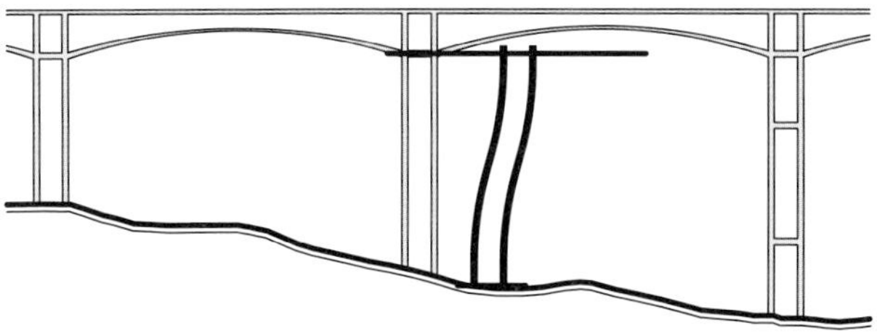

11.2.4 Decks supported by a pair of slender piers

For longer structures it is possible to substitute the box section of the piers by two slender walls that can easily accommodate the volume changes (Pauser, 1986) (Figure 11.37). During construction, the slender piers can be temporarily stiffened by steel truss members. The twin transverse walls can also be stiffened, by a central longitudinal wall (Figure 11.38). By changing the height of this wall, the required stiffness of the bridge structure can be achieved.

This approach was used in design of the cantilever structures built on the D3 motorway that connects the city of Žilina in Slovakia with Poland. The bridges at Valy and Rieka, in Slovakia, have total lengths of 592 and 502 m, respectively, and form continuous structures with span lengths of 24.50–92.00 m (see Chapter 4, Figures 4.96 and 4.97) (fib, 2018). The long spans were segmentally cast in symmetrical cantilevers (Figure 11.39). The 11.40 m wide decks of both bridges are formed of single-cell box girders supported by slender piers, which guarantee the stability of the bridges both during erection and during service. The piers have H-shaped cross-sections formed by two

Figure 11.38 Bridge at Valy, Slovakia

Figure 11.39 Bridge in Figure 11.38: segmental casting in symmetrical cantilevers

slender transverse walls that are mutually connected by a longitudinal wall. In low piers and the top portions of the tall piers, the longitudinal walls are omitted. The longitudinal distance of the transverse walls and the height at which the longitudinal walls were omitted were determined by means of extensive parametric studies.

Twin piers were also used in the construction of the main spans of the viaduct across Strazov Pond and the Hricov reservoir for the River Vah (see Chapter 4, Figures 4.68 and 4.69) (fib, 2018). The main spans were segmentally cast in symmetrical cantilevers starting from the pier tables (Figure 11.40). To guarantee the stability of the progressively cast structure during cantilever construction, and to give the bridges sufficient stiffness during service, the main spans are supported by twin piers, each formed of two transversally inclined slender walls that directly support the box girder webs (Figures 11.41 and 11.42). In the longitudinal direction of the bridge, the walls of the central pier are frame connected to the deck; the walls of other two piers are hinge connected to the deck. All the piers are founded on drilled piles.

The girders are stiffened by pier diaphragms that follow the pier walls in the upward direction. When designing the piers, the whole of the integral structural system, including the hinge connection of the side piers and the elastic supports of the piles, was considered. That is why relatively short twin piers integral with the deck could be used.

From a purely technical point of view, piers formed by twin walls represent a very elegant solution. However, from an aesthetic point of view they often look unnatural if the pier walls are very slender and the deck is very deep. The aesthetic impression can be significantly improved by designing a suitable curve of the pier cross-sections. With careful design, the stiffness of the piers can be designed to retain all the advantages of the solution. Greater wall stiffness can also help to eliminate the erection stiffening of the slender piers. An excellent example of this solution is the Qurashiyah Viaduct in Riyadh, Saudi Arabia (Srinivasan, 2004) (Figure 11.43).

Figure 11.40 Viaduct across Strazov Pond and the Hricov reservoir, near Žilina, Slovakia

Figure 11.41 Bridge in Figure 11.40: cross-section

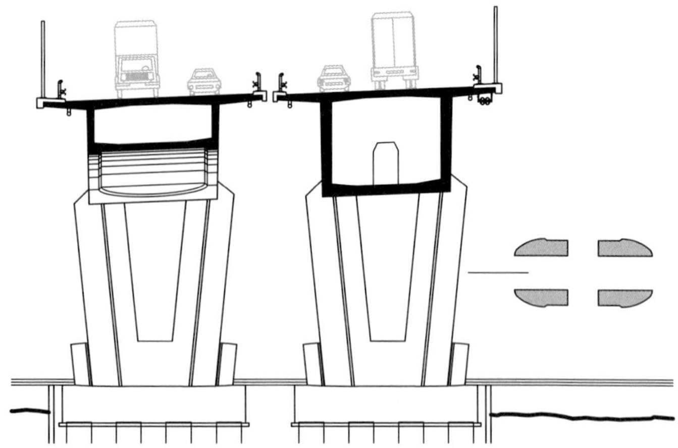

Figure 11.42 Bridge in Figure 11.40 – pier: (a) axonometric view; (b) elevation; (c) transverse elevation

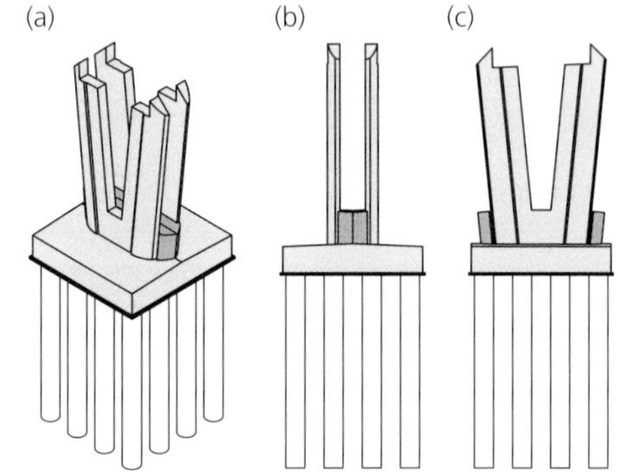

11.3. Deck above the piers

The decks of cantilever bridges are usually stiffened by support diaphragms that transfer the stresses from the deck into the substructure. In the transverse direction, the diaphragms function in a similar way to diaphragms in common bridges (see Chapter 8).

During cantilever construction, the bridge deck above the support is subjected to bending moments M_L and M_R acting on the support's left and right sides, respectively (Figure 11.44). The difference between these moments $\Delta M = M_L - M_P$ must be transferred into the substructure. Therefore, the erected deck is always supported by double lines of bearings, or is frame connected into the substructure (see Figure 11.7). The supporting system must guarantee the stability of the cantilevered structure, which is stressed by its self-weight G, the erection live load P and by the wind load W (see Figure 11.6). Also, seismic forces can originate during erection.

Figure 11.43 Construction of the Qurashiyah Viaduct, Riyadh, Saudi Arabia

Figure 11.44 Unbalanced moments of cantilever structures: (a) cross-section; (b) partial elevation; (c) elevation

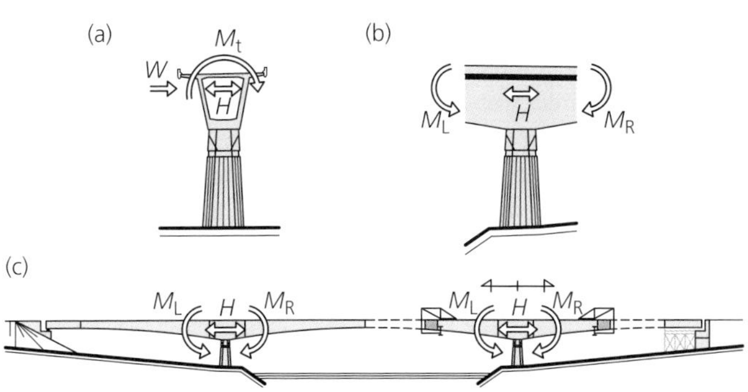

The unbalanced moment ΔM can be replaced by a pair of forces $T = C = \Delta M/h$. The force acting at the top slab must be transferred into the substructure, which is loaded by the reactions $A = -B = \Delta M/b$. The tension force T should be increased by a component of the normal force caused by the longitudinal wind and earthquake loads. The method of transferring the tensile force T depends on the design of the deck stiffening by support diaphragms.

Bridges in which the deck is supported in the final stage by a double line of bearings or through the frame connection to the piers, are stiffened by pier diaphragms arranged in one of two ways. They can be stiffened by inclined (sloping) diaphragms (Figure 11.45), or by vertical diaphragms that follow the transverse walls of the piers in the upward direction (Figures 11.46 and 11.47).

The supporting areas are subjected to complex stresses that can be determined through a three-dimensional (3D) analysis of the structures modelled by solid or shell elements (Figure 11.48). However, before this analysis is performed, it is useful to determine the stresses in a simple strut-and-tie analysis, which helps to understand the function of the individual structural elements.

11.3.1 Inclined (sloping) diaphragms

If the shell stiffness of the box girder webs is neglected, it can be assumed that the tension force T is transferred by the tension force T_p and the compression force C_p originating in the inclined diaphragms (Figure 11.49). The values of these forces are determined from the simple force diagram. It can be assumed that the tension force T is uniformly distributed across the width of the top slab situated between the girder webs, which is stressed by the stresses $t = T/w_s$.

$$T_p = -C_p = T/2 \sin \alpha \qquad (11.1)$$

Figure 11.45 Sloping diaphragms

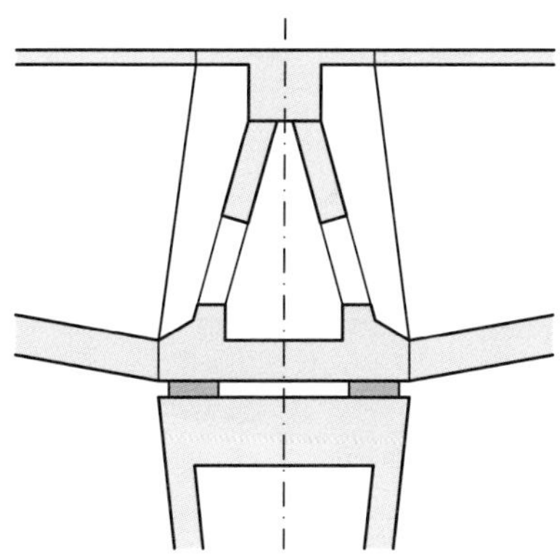

Figure 11.46 Double vertical diaphragms

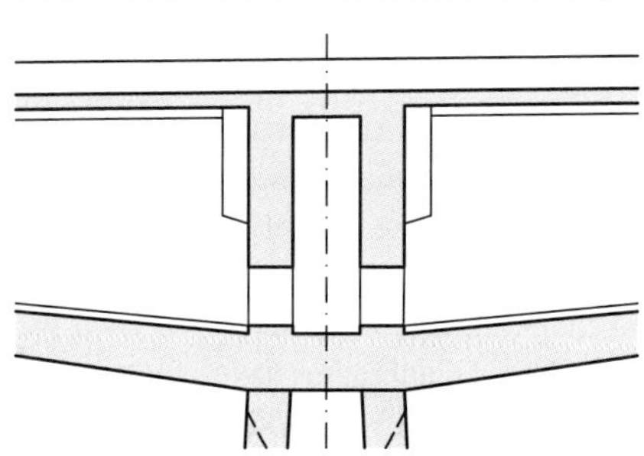

Figure 11.47 Bridge at Valy, Slovakia: pier diaphragm

Figure 11.48 Viaduct at Považká Bystrica, Slovakia: 3D analysis – model of the structure: (a) erected cantilever; (b) longitudinal section

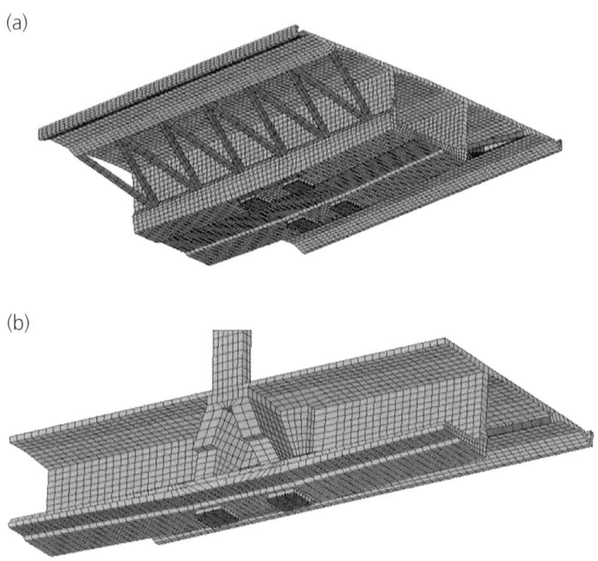

Figure 11.49 Static function of the sloping diaphragms

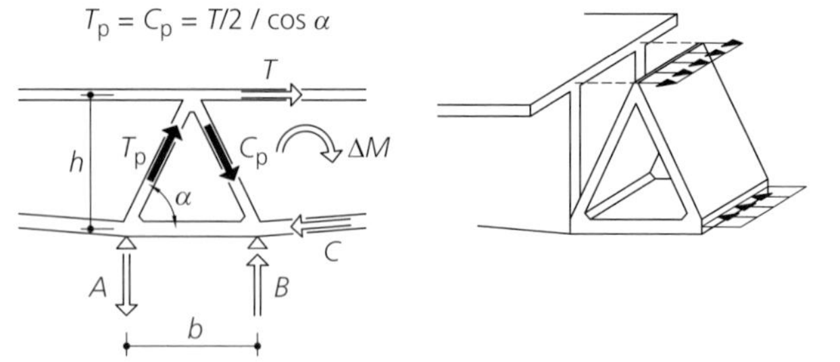

Figure 11.50 Sloping diaphragm: the transfer of stresses in the top deck slab

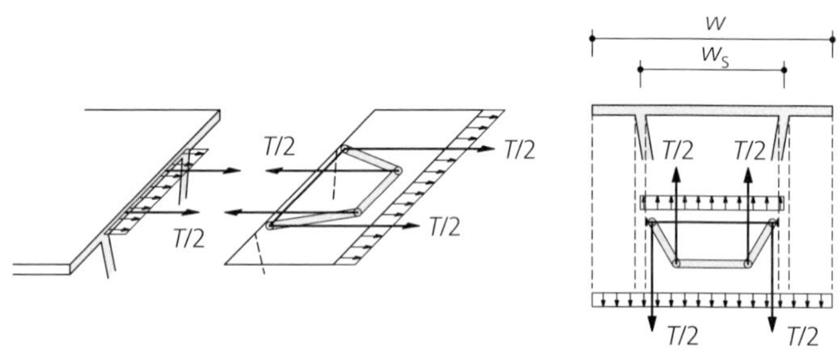

The method of the force transfer was analysed in detail by Professors Mathivat and Ghali (Mathivat, 1983), who studied differently supported 3D structures modelled as shell elements. Based on the analysis, they recommend a simplified procedure based on a strut-and-tie model. They recommend replacing the horizontal forces T and C with pairs of forces $T/2$ and $C/2$ acting in the upper and lower corners of the girder webs. The horizontal force $T/2$ is transmitted by a system of tension and compression members (Figure 11.51(b)). The tensile forces are resisted by additional vertical and horizontal web reinforcement; the compression force has to be resisted by the concrete compression capacity. In deep structures, the shear reinforcement can be determined from a typical strut-and-tie model.

As the tension force is distributed across the whole width of the top slab w, the tension force T must be transferred into the width w_s. This transfer requires additional transverse reinforcement, which is determined by a strut-and-tie analysis (Figure 11.50).

11.3.2 Double vertical diaphragms

While, from a static point of view, sloping diaphragms represent the optimum solution, they require a lot of labour, and therefore they are seldom used. They are mainly used in bridges assembled of precast segments, and the structures are usually stiffened by vertical diaphragms. As the vertical diaphragms have a high membrane stiffness but a low bending stiffness, the horizontal force T is transmitted into the substructure by the webs of the box girders (Figure 11.51).

It is necessary to realise that the normal forces T and C acting in the upper or lower slab must be transferred to the frame corners (Figure 11.52). The normal stresses are replaced by the force resultants, and transferred into the frame corners by a system of tension and compression members. The corresponding transverse reinforcement of the top and bottom slabs must resist the tension forces originating in the transverse ties.

11.3.3 Single vertical diaphragms

As simply supported structures (structures supported by one row of bearings) are supported by two rows of bearings (supporting blocks) during construction, the unbalanced moment during erection ΔM must be transferred into the erection supports. Therefore, the webs

Figure 11.51 Double vertical diaphragms: (a) static function; (b) force diagram

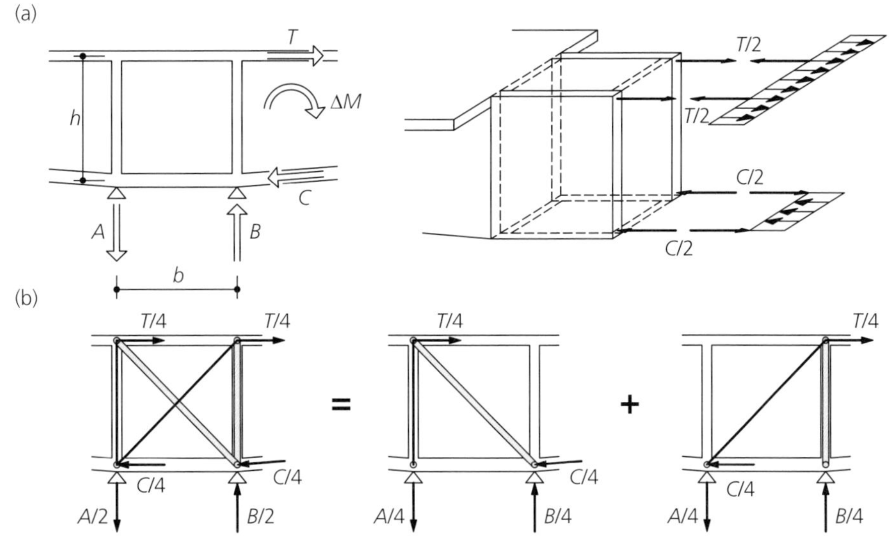

and the top and bottom slabs of the box girder must also be reinforced for the forces discussed in the paragraphs above (Figure 11.53). Temporary supports might require strengthening of the bottom slab and/or the design of span diaphragms.

11.4. Design of the prestressing and camber

During cantilever construction, the negative bending moments gradually increase (Figure 11.54). These cantilever moments are balanced by positive bending moments created by the cantilever tendons situated at the top slab. When the midspan closures are cast, the span (positive) tendons and continuity tendons are post-tensioned.

Figure 11.52 Double vertical diaphragms: transfer of stresses in the top and bottom slabs

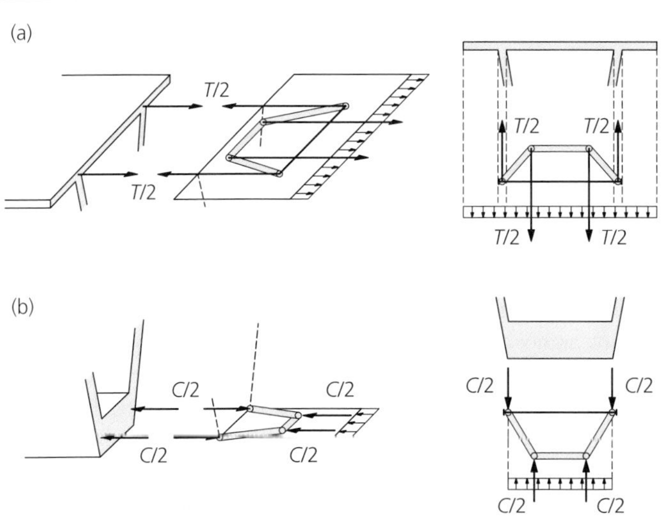

11.4.1 Layout of the prestressing cables

In the past, the cantilever tendons were bent and anchored at the girder webs, where they overlapped with draped continuity tendons anchored in the block-outs situated on the top slabs above the webs (Figure 11.55) (Mathivat, 1983; Podolny and Muller, 1982). These cables were supplemented by short-span tendons led at the bottom slab. Due to the inclination of the cantilever and the continuity tendons, the principal tension stresses in the webs are locally reduced. However, these tendons do not reduce the shear forces and corresponding shear stresses that originate not only in the webs but also in the top and bottom slabs of the box girders. Therefore, this arrangement, which complicates the construction and requires the provision of block-outs at the top surface of the girder, is seldom used nowadays.

As already discussed, the goal of the design of the prestressing is to create a state of stresses in which the concrete structure is stressed by a uniform compression for the dead load. From Figure 11.54(b), it is evident that the cantilever tendons can balance the bending moment caused by the self-weight (Figure 11.54(c), but they cannot balance the shear forces (Figure 11.54(d), which can also contribute to the long-term deformation of the deck.

Therefore, the arrangement of the prestressing has been modified (Figure 11.56). The cantilever tendons are led in and anchored at the top slab. The span tendons are situated in the bottom slabs, and are anchored at the bottom corners of the box sections. The cantilever and span tendons are supplemented by continuity tendons, which guarantee the integrity of the structure. The layout of these tendons corresponds to the course of the bending moments of the continuous girders. The tendons can be internal (bonded, led in the girder webs and anchored at the anchor blocks situated beyond the pier diaphragms) or can be formed of external unbonded cables deviated at the pier and span

Figure 11.53 Bridges supported by one row of bearings: (a) static function during erection; (b) force diaphragm

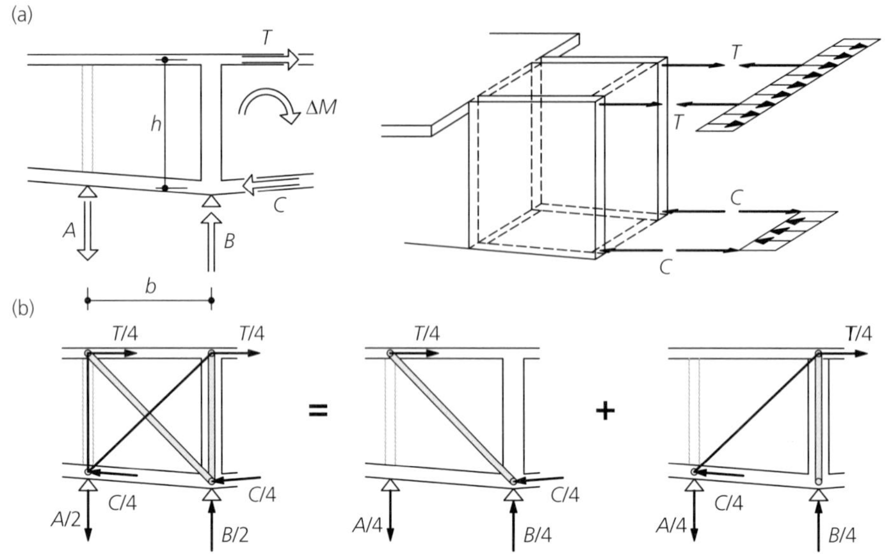

Figure 11.54 Balancing of cantilever moments by prestressing: (a) elevation; (b) layout of cantilever tendons; (c) bending moments; (d) shear forces

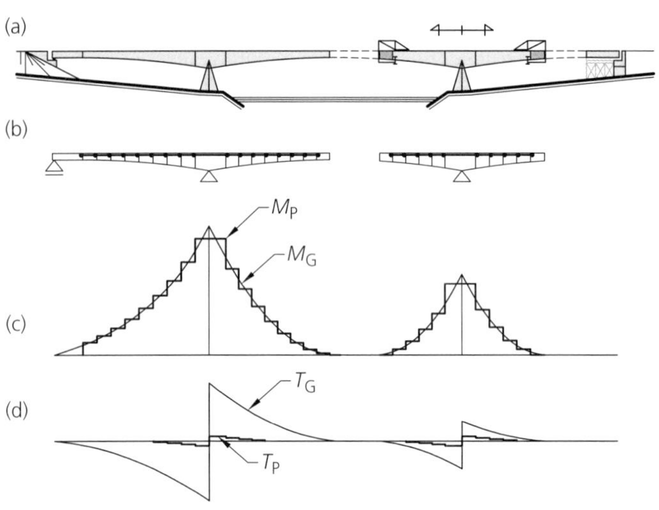

Figure 11.55 Former typical layout of tendons: (a) cantilever tendons; (b) span and continuity tendons; (c) overlaping of cantilever and continuity tendons

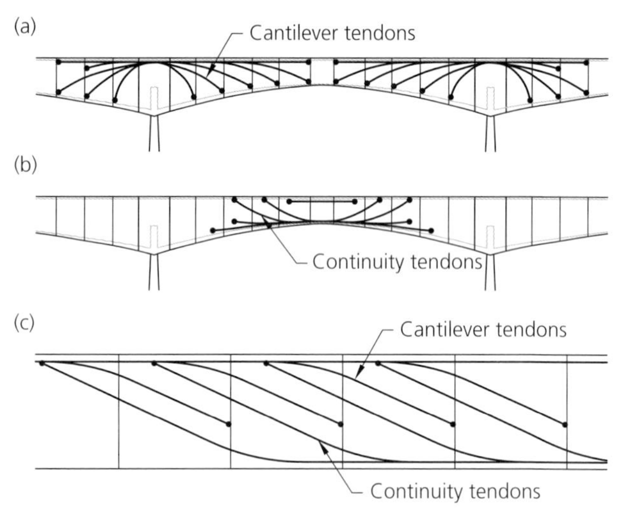

deviators. As the external cables are usually led in polyethylene ducts, the friction losses are relatively small. Therefore, the cables can be led across two or more spans, and be anchored at the blocks situated beyond the pier diaphragms (Figure 11.57). Both internal bonded tendons and external non-bonded cables overlap each other over the supports. If external continuity cables are used, there are no ducts in the webs, and therefore the webs can be very slender. Consequently, the deck can be lighter, and static effects due to the dead load are smaller. If necessary, the webs can be vertically prestressed by prestressing bars (Figure 11.58).

Figure 11.56 shows the detailed arrangement of the prestressing tendons, and Figure 11.58 shows the reinforcement of a typical cantilevered structure built on Freeway 1 in Taiwan and designed by the first author. The bridge has three spans (lengths 85.0 + 160.0 + 85.0 m), its width is 12.20 m, and its depth varies from 8.00 m at the piers to 3.50 m at the midspan. Figure 11.58 shows all the cables, including their anchors and vertical prestressing. As the prestressing strands may break or a tendon duct can be damaged during construction, international recommendations require that the possibility of additional prestressing by means of additional cantilever tendons

Figure 11.56 Typical layout of tendons: (a) scheme of the structure; (b) plan view of the cantilever tendons; (c) elevation with all tendons and prestressing bars

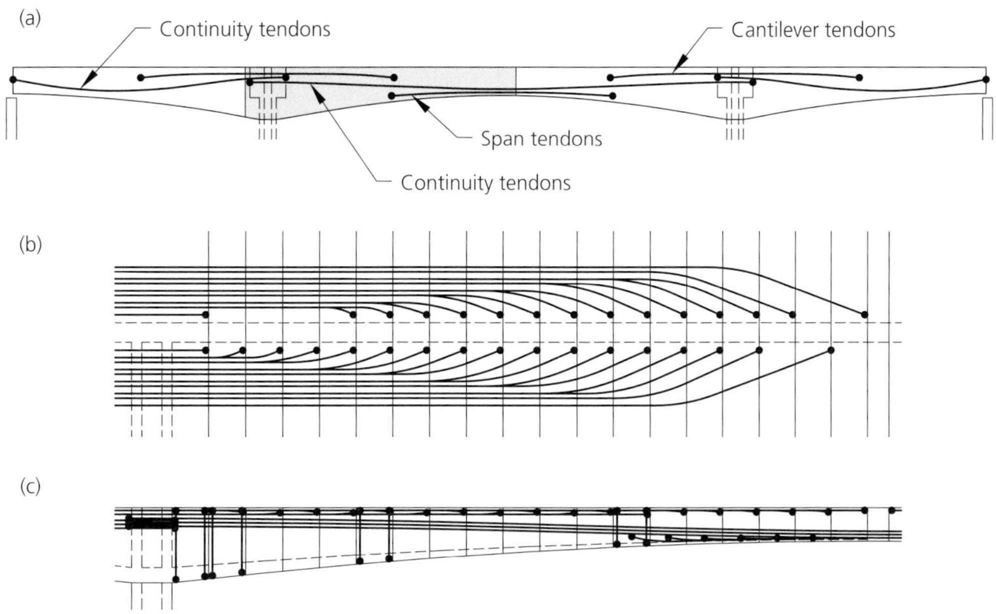

be considered. Therefore, it is necessary to create free cable ducts and spaces for anchoring additional tendons during construction. If these are not used, the tendon ducts are grouted. Similarly, additional ducts for positive tendons should also be designed.

International recommendations also require that the bridge structure can be reinforced in the future by the provision of additional external cables. The ducts for these cables must be prepared at the pier diaphragms and span deviators. The reinforcement of the structures must ensure the transfer of forces from these cable into the structure.

Figure 11.57 External cables

The described arrangement of the prestressing tendons has been used in all cantilever bridges on which the authors have had an opportunity to participate. When designing the cantilever structures, it is necessary to realise that, after the erection of the bridge deck, the moment due to the dead load is zero at the midspan sections. However, due to the different possible positions of the live load, the moments of both signs originate there. While in cast-in-place structures this bending moment can be resisted by reinforcing bars, in precast segmental structures, in which the joints are not reinforced by rebars, short prestressing tendons situated at the top slab should

Figure 11.58 Typical reinforcement of the box girder

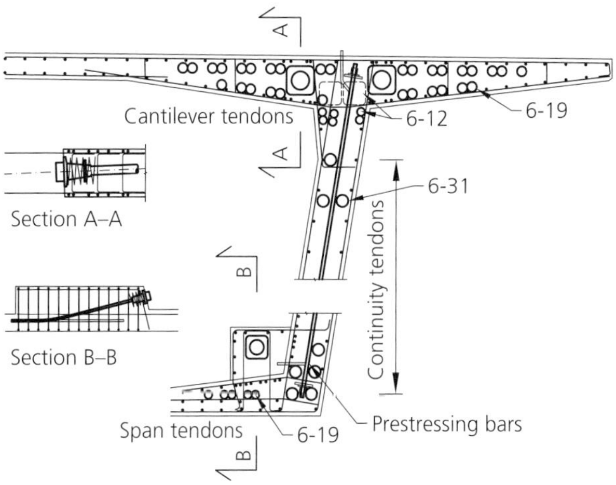

be also designed. These tendons are particularly necessary in bridges built in seismic areas.

Some contractors assume that avoiding the use of parabolic continuity tendons reduces the the amount of labour required for the construction of a bridge. However, they do not realise that these tendons significantly reduce the shear stresses, not only in the girder webs but also in the top and bottom slabs. As the reduction of shear stresses consequently reduces the amount of shear reinforcement required, and thus the amount of labour needed for their placement, parabolic continuity tendons should be always used.

As an example, Figure 11.59 shows the arrangement of the internal prestressing tendons and external continuity cables on the Roskilde Fjord viaduct in Denmark, the structural arrangement and erection of which are shown in Figures 11.19–11.22. The anchor blocks shown in the corners of the box section serve not only to anchor the span tendons, but also to anchor the prestressing bars used during the erection of a segment to creating uniform compression in the joints. The coupled prestressing bars were also used to controlling the compression stresses at the bottom fibres of the erected cantilevers.

Figure 11.59 Roskilde Fjord viaduct, Denmark – erection and final prestressing: (a) longitudinal section with erection prestressing bars; (b) longitudinal section with bonded tendons; (c) longitudinal section with external non-bonded tendons; (d) cross-section near the midspan; (e) cross-section near the piers

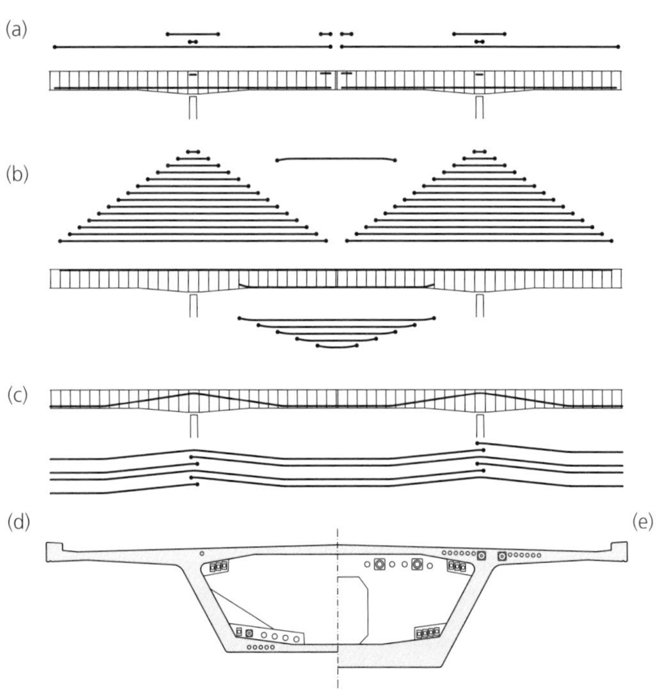

11.4.2 Redistribution of the bending moments

Both cast-in-place and precast cantilever bridge structures are progressively assembled from elements (segments) of different ages. The structures are progressively prestressed, and after casting the midspan joints (closures) their static system is changed. Therefore, there is a significant redistribution of internal forces in the structures. Before the midspan joints are cast, the bridges function as symmetrical cantilevers, while after these joints have been cast the bridges function as continuous structures. As the new midspan connections prevent free shrinkage and creep of the cantilevers, positive bending moments originate in the middle of the span. These moments consequently reduce the moments above the supports (see Chapter 7, Figures 7.11 and 7.12). The structure tends to function the same as a structure that has been cast in situ (Figure 11.60). At the same time, the stresses also redistribute from the younger segments to the older ones, and from the younger cantilever into the older one.

It is necessary to realise that not only the moments due to the self-weight but also the moments due to the prestressing redistribute over time. At first sight it may seem that if the self-weight is perfectly balanced by the prestressing, the bridge deck will be stressed by a zero bending moment and there will be no redistribution of stresses over time. However, due to the deck shortening caused by concrete creep and shrinkage, the value of the prestressing force and the corresponding moments are reducing, and therefore the internal forces redistribute over time.

It is evident that the above issues must be taken into consideration in the static analysis. It is necessary to carefully determine not only the stresses originating in the deck during its erection and service, but also the deformations of the deck. This means that the arrangement of the structure and the layout of the prestressing tendons have to guarantee the shape of the deck over time. As the stresses and the deflection of the deck depend on the load history, it is necessary to perform a detailed time-

Figure 11.60 Redistribution of the bending moments: (a) deflections of cantilevers of different ages; (b) connection of cantilevers which deflections were eliminated by their camber; (c) self weight bending moments at time t_0, t_∞; CG – continues girder

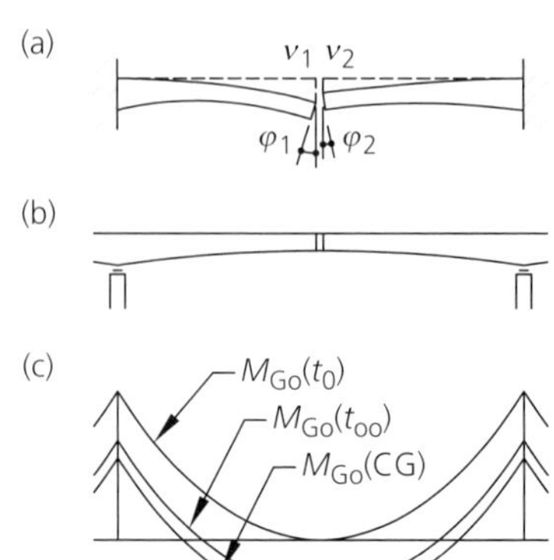

dependent analysis, taking into account the progress of the construction, loading and unloading of the structure by travellers, tendon tensioning, loading during erection and changing boundary conditions.

Modern software used for the analysis of structures is able to cover all the issues described. Therefore, the main problems in the analyses are

- preparing an appropriate analytical model that describes the bridge structure and boundary conditions
- checking if the theory used in the software being used is appropriate for the bridge structure being analysed
- understanding the methods of analyses

Figure 11.61 Bridge at Rieka, Slovakia – arrangement of the prestressing: (a) cross-section; (b) elevation

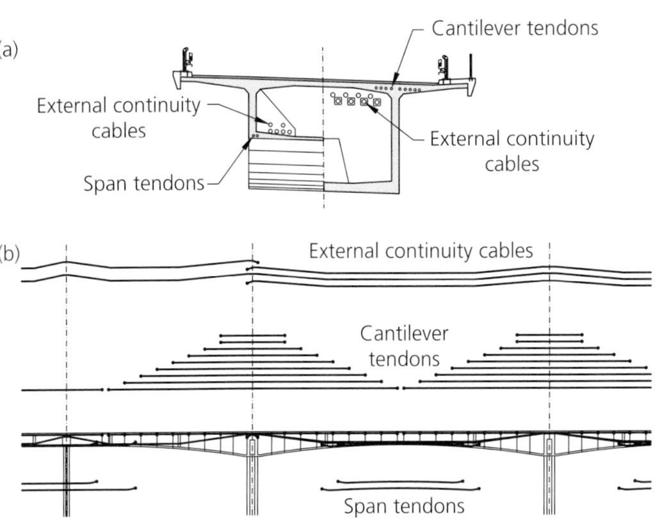

- preparing the input data that describes the progressive construction and the loading and unloading of the bridge deck by travellers
- properly expressing the changes in the static system
- correctly interpreting the results.

The analysis is different for structures that have modest or very long spans, and for structures that can be modelled as a 3D structure assembled of beam or shell elements.

11.4.3 Design of the camber

The bridges at Valy and Rieka, introduced in Chapter 4 and shown in Figures 4.96, 4.97, 11.38 and 11.39, are used here as an example of the analysis of a modest structure. During the progressive erection of the deck, the internal cantilever tendons were progressively installed and post-tensioned. Once the midspan closures had been cast, the internal span tendons and external continuity cables were placed and post-tensioned. The arrangement of the prestressing is shown in Figures 11.57 and 11.61. The calculation model comprises several typical construction steps, and is shown in Figure 11.62.

The goal of the design was not only to fulfil all the requirements of the Eurocode (CEN, 2005), but also to design a bridge structure that would have limited long-term deformations. At the end of construction, the designed prestressing created in the deck and the designed bending moments balanced the dead-load bending moments (Figure 11.63(a)). Due to the use of external cables, a substantial portion of the shear forces was also balanced.

As the long-term deformations are usually larger than the deformations determined using the fib Model Code 1990, which is used in the Eurocode (Pfuntner, 2015), the value of the designed camber at the midspans was taken as *L*/1000, where *L* is the length of the cantilever span.

Figure 11.62 Calculation model in several steps (a–f) for the bridge in Figure 11.61

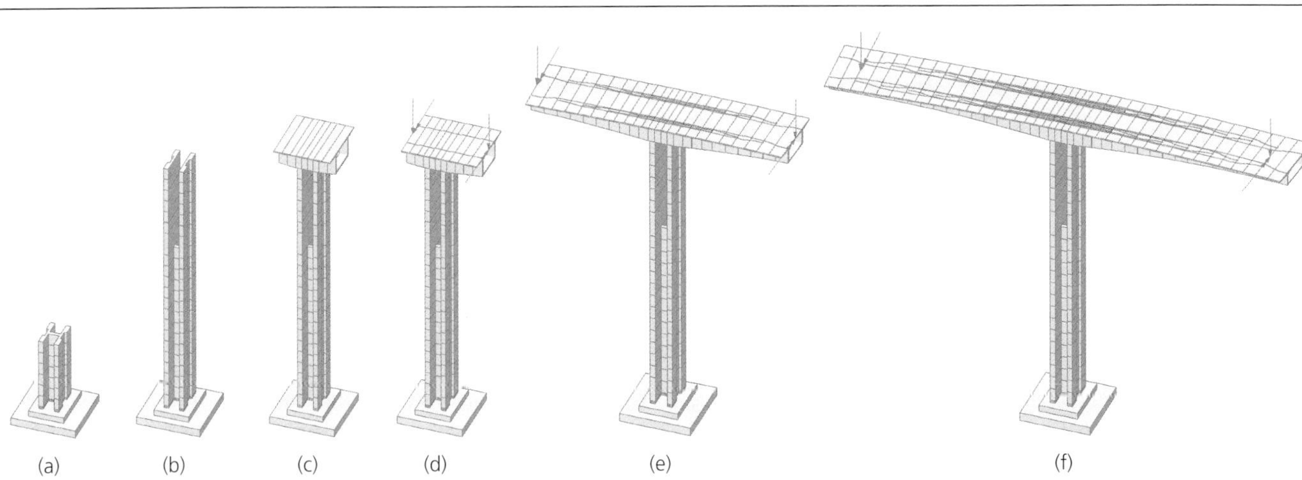

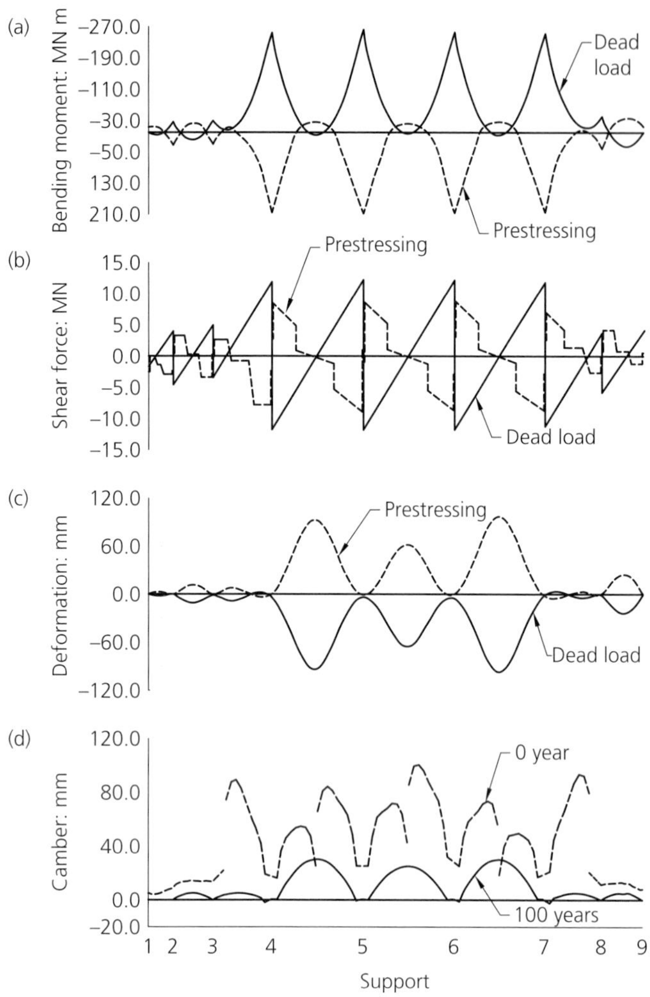

Figure 11.63 Bridge in Figure 11.61: (a) bending moments; (b) shear forces; (c) deformation; (d) camber

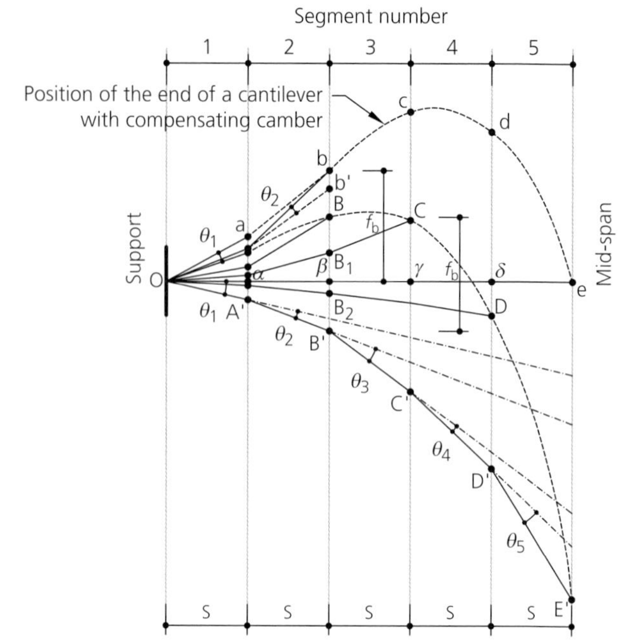

Figure 11.64 Cantilever deflection and compensating camber. (From Mathivat (1983))

The bridge structure has a tendency to function the same as a similar structure cast in situ, so the camber curve follows the shape of the affine curve to the deformation curve due to the dead load (Figure 11.63(c)). The construction camber was determined under the condition that at the end of its erection the cantilever will have no deflection (Figure 11.63(d)). The final camber, which is a result of the static action of the prestressing of the external cables, reduces over time. As shown by Figure 11.63(d), the camber is still positive, and therefore it is believed that the structure will keep the required shape.

The construction camber was determined according to the procedure described by Mathivat (1983). Figure 11.64 shows the deflections of the erected cantilever without any compensation, and the geometry of the erected segments that were adjusted to obtain the required shape of the structure.

11.4.4 The camber of long-span structures

While structures having modest spans and common widths can be analysed as 3D structures assembled of beam elements, bridges with large spans and/or large widths should be analysed as 3D structures assembled of shell or solid elements. The Kayak Bridge, which erection was recently completed on the D4 motorway near the city of Bratislava, Slovakia (see Figures 11.12–11.14), is used here as an example.

The bridge has three spans (lengths 127.5 + 210.0 + 127.5 m) and its deck is formed of a 35.0 m wide one-cell box girder with additionally cast overhangs supported by precast struts. The deck is longitudinally prestressed by bonded tendons situated at the top and bottom slabs. Transverse bonded tendons are installed and post-tensioned during the progressive casting of the overhangs.

The spine girder is being progressively cast in cantilevers starting at the intermediate piers (Figure 11.65(a)). The stability of the erected structure is guaranteed by temporary supports situated at the side spans. When symmetrical segments are cast, the cantilever tendons are post-tensioned. As the length of the side spans is longer than half the length of the main span, the last 25 m long parts of the side spans adjacent to the side piers were cast on the scaffolding and were supported by the temporary pier until the side span closure joint was cast. Once the joints were cast, the midspan closure was cast and the span tendons situated at the bottom slab were post-tensioned.

Figure 11.65 Kayak Bridge, Bratislava, Slovakia – analytical model: (a) progressive casting of the spine girder; (b) progressive casting of the overhangs

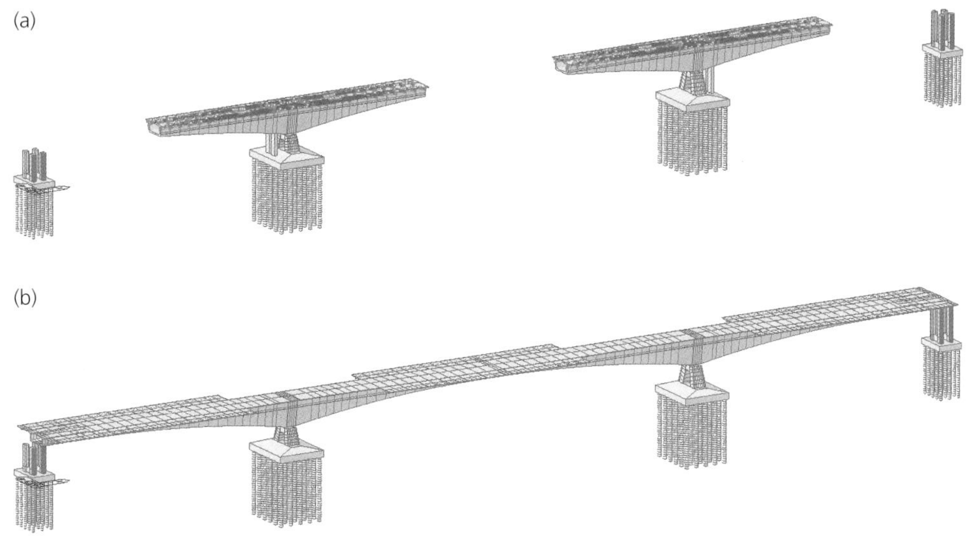

The overhangs were progressively cast in the direction from the side piers to the midspan (Figure 11.65(b)). When the overhangs were completed, the additional cantilever tendons situated at the top slab and span tendons situated at bottom slab were post-tensioned.

The bridge was carefully analysed for both the serviceability and ultimate limit states given in the Eurocode. A 3D model assembled of beam elements was used for the global analyses. The space function of the structure was checked on a 3D model assembled of shell elements (Figure 11.66). To check the stresses during construction, the calculation model (see Figure 11.65) in which the basic box girder and the overhangs were modelled by composite section beam elements was used.

The following creep and shrinkage functions were considered in order to predict the camber as accurately as possible

Figure 11.66 Bridge in Figure 11.65: 3D analytical model

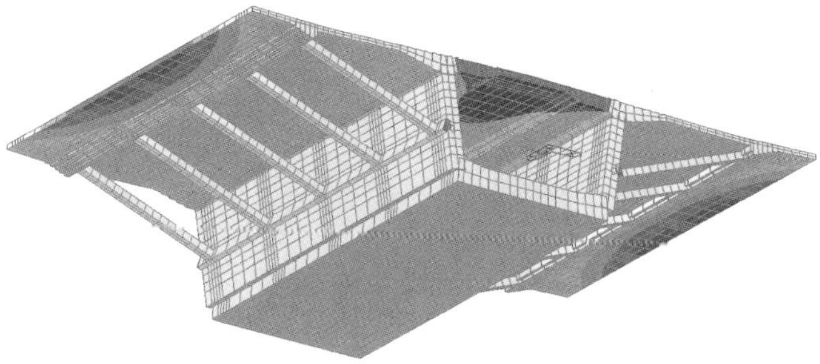

1. STN EN 1992, Appendix B (Eurocode (CEN, 2005))
2. CEB-FIP Model Code 1978
3. fib Model Code 2010.

The deflections calculated using different creep and shrinkage functions gave significantly different results. For example, the maximum deflection calculated using the CEB-FIP Model Code 1978 was 400 mm, and that calculated using the FIB Model Code (2010) was 288 mm, a difference of 112 mm (Figure 11.67).

As the beam elements are not able to express the shear deformation of the deep girder webs due to creep and the shear-lag effect of the overhangs, a combined model was also developed. The deck was modelled by a structure assembled of: a central core wall, which modelled the girder webs; beams, which modelled the function of

Figure 11.67 Bridge in Figure 11.65: deflections calculated using different creep and shrinkage functions

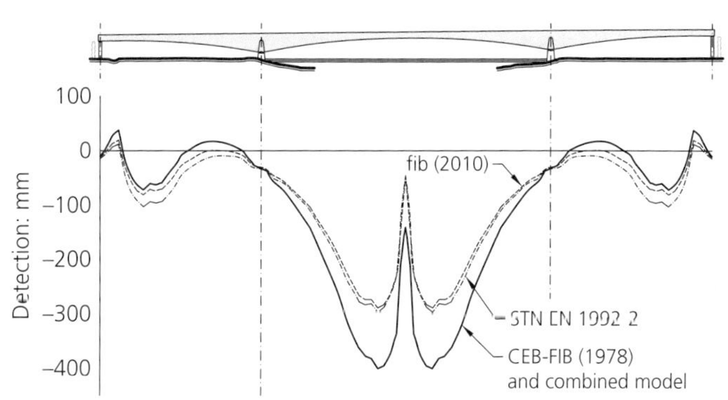

Figure 11.68 Bridge in Figure 11.65 – combined model: (a) cross-section; (b) axonometric view

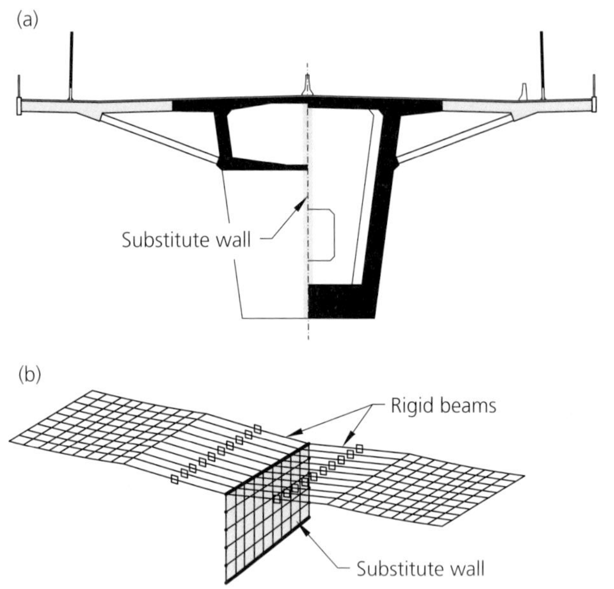

the top and bottom slabs; and outer slabs, which modelled the function of the additionally cast overhangs. The central wall and outer slabs were assembled of shell elements that are stressed by both membrane and bending stresses (Figure 11.68).

The simplified crack prediction for cracking of the concrete core walls and overhangs was also taken into account. The unfavourable deviations of actions from the mean values (r_{inf} = 0.9 for the prestressing force, γ_{lt} = 1.25 for creep and shrinkage, 1.025 for the self-weight, 1.4 for the pavement weight) were also considered.

The main span deflections obtained using the combined model, were similar to the results obtained when the beam model was used. Likewise, the creep and shrinkage determined using the fib Model Code 2010, including deviations of actions from the mean values, were similar to those determined using the CEB-FIP Model Code 1978 and the mean values.

From the above it is evident that, whether the fib Model Code 2010 or Eurocode is used, it is advisable to use a more sophisticated analytical model and to include possible adverse deviations of actions. The analysis done using the CEB-FIP Model Code 1978 gives a sufficiently conservative prediction of the deflection of cantilever bridges, even when a simplified beam model and mean values of actions are used.

REFERENCES

Brown DC (2007) Built to resist the big one. *ASPIRE* Winter: 12–17.

CEN (2015) EN 1992-2:2005. Eurocode 2 - Design of concrete structures. Part 2: Concrete bridges: Design and detailing rules. CEN, Brussels, Belgium.

Chalupec A, Ondroš M, Kopčák J, Suchár V and Novotný P (in press) Long span post-tensioned box girder bridge in Bratislava. *fib International Conference on Concrete Sustainability. Concrete – A Challenge for Sustainable and Resilient Built Environment, Prague, Czech Republic.*

fib (2018) Structural concrete in the Czech Republic 2014–2017. *5th International fib Congress, Melbourne, Australia.* National Report of the Czech Republic. fib, Lausanne, Switzerland.

Mathivat J (1983) *The Cantilever Construction of Prestressed Concrete Bridges.* John Wiley, New York, NY, USA.

Murugesh G and Cormier K (2007) When lighter is better. *ASPIRE* Summer: 20–24.

Novák R, Kocourek P, Pecha J and Stráský J (2019) *Roskilde Fjord Link*, High Bridge, Denmark. *Beton TKS* **4**.

Pauser A (1986) *Entwicklungsgeschichte des Massivbrückenbaues. Unter Berücksichtigung der Verhältnisse Österreichs.* Austrian Concrete Association, Vienna, Austria.

Pfuntner J (2015) What is the right approach for long term creep and shrinkage effects – CEB-FIP 78 vs. 90 Design Codes. In *27th Annual ASBI Convention. Dallas, TX, USA.* American Segmental Bridge Institute, Buda, TX, USA.

Podolny W and Muller J (1982) *Construction and Design of Prestressed Concrete Bridges.* John Wiley, New York, NY, USA.

Srinivasan S (2004) Innovative concepts and construction method in precast bridge design. In *Proceedings of the fib Symposium on Segmental Construction in Concrete, New Delhi, India.*

Stráský J, Navrátil J and Suský S (2001) Applications of time-dependent analysis of in the design of hybrid bridge structures. *PCI Journal* **46(4)**: 56–74.

Stráský J, Maťaščík M, Novák R and Táborská K (2010) Multi-span extradosed viaduct in Povazska Bystrica, Slovakia. *3rd fib International Congress 2010.* Washington, D.C.

ice Publishing

Stráský and Nečas
ISBN 978-0-7277-6385-3
https://doi.org/10.1680/dcpb.63853.227
ICE Publishing: All rights reserved

Index

Page numbers in *italics* refer to illustrations. Tables are indicated by an italic *t* following the page number.